The IMA Volumes
in Mathematics
and its Applications

Volume 48

Series Editors
Avner Friedman   Willard Miller, Jr.

Institute for Mathematics and
its Applications
IMA

The **Institute for Mathematics and its Applications** was established by a grant from the National Science Foundation to the University of Minnesota in 1982. The IMA seeks to encourage the development and study of fresh mathematical concepts and questions of concern to the other sciences by bringing together mathematicians and scientists from diverse fields in an atmosphere that will stimulate discussion and collaboration.

The IMA Volumes are intended to involve the broader scientific community in this process.

Avner Friedman, Director
Willard Miller, Jr., Associate Director

* * * * * * * * * *

IMA ANNUAL PROGRAMS

1982–1983	Statistical and Continuum Approaches to Phase Transition
1983–1984	Mathematical Models for the Economics of Decentralized Resource Allocation
1984–1985	Continuum Physics and Partial Differential Equations
1985–1986	Stochastic Differential Equations and Their Applications
1986–1987	Scientific Computation
1987–1988	Applied Combinatorics
1988–1989	Nonlinear Waves
1989–1990	Dynamical Systems and Their Applications
1990–1991	Phase Transitions and Free Boundaries
1991–1992	Applied Linear Algebra
1992–1993	Control Theory and its Applications
1993–1994	Emerging Applications of Probability
1994–1995	Waves and Scattering

IMA SUMMER PROGRAMS

1987	Robotics
1988	Signal Processing
1989	Robustness, Diagnostics, Computing and Graphics in Statistics
1990	Radar and Sonar (June 18 - June 29)
	New Directions in Time Series Analysis (July 2 - July 27)
1991	Semiconductors
1992	Environmental Studies: Mathematical, Computational, and Statistical Analysis
1993	Modeling, Mesh Generation, and Adaptive Numerical Methods for Partial Differential Equations

* * * * * * * * * *

SPRINGER LECTURE NOTES FROM THE IMA:

The Mathematics and Physics of Disordered Media

Editors: Barry Hughes and Barry Ninham
(Lecture Notes in Math., Volume 1035, 1983)

Orienting Polymers

Editor: J.L. Ericksen
(Lecture Notes in Math., Volume 1063, 1984)

New Perspectives in Thermodynamics

Editor: James Serrin
(Springer-Verlag, 1986)

Models of Economic Dynamics

Editor: Hugo Sonnenschein
(Lecture Notes in Econ., Volume 264, 1986)

Carl D. Meyer   Robert J. Plemmons
Editors

Linear Algebra,
Markov Chains,
and Queueing Models

With 55 Illustrations

Springer-Verlag
New York Berlin Heidelberg London Paris
Tokyo Hong Kong Barcelona Budapest

Carl D. Meyer
Mathematics Department
North Carolina State University
Raleigh, NC 27695-8205 USA

Robert J. Plemmons
Department of Mathematics
  and Computer Science
Wake Forest University
Winston-Salem, NC 27109 USA

Series Editors:
Avner Friedman
Willard Miller, Jr.
Institute for Mathematics and its Applications
University of Minnesota
Minneapolis, MN 55455
USA

Mathematics Subject Classifications (1991): 65F, 65U, 68M, 60J, 15A

Library of Congress Cataloging-in-Publication Data
Linear algebra, Markov chains, and queueing models / Carl D. Meyer,
    Robert J. Plemmons, editors ; with 55 illustrations.
        p.   cm. — (The IMA volumes in mathematics and its
    applications ; v. 48)
    Includes bibliographical references.
    ISBN 0-387-94085-5 (acid-free paper)
    1. Algebras, Linear—Congresses.   2. Markov processes—Congresses.
3. Queueing theory—Congresses.   I. Meyer, C. D. (Carl Dean)
II. Plemmons, Robert.   III. Series.
QA184.L545   1993
519.2′33—dc20   93-2100

QA
184
L545
1993

Printed on acid-free paper.

Production managed by Hal Henglein; manufacturing supervised by Genieve Shaw.
Camera-ready copy prepared from the IMA's AMSTeX files.
Printed and bound by Edwards Brothers, Inc., Ann Arbor, MI.
Printed in the United States of America.

9 8 7 6 5 4 3 2 1

ISBN 0-387-94085-5 Springer-Verlag New York Berlin Heidelberg
ISBN 3-540-94085-5 Springer-Verlag Berlin Heidelberg New York

The IMA Volumes
in Mathematics and its Applications

Current Volumes:

Volume 1: Homogenization and Effective Moduli of Materials and Media
   Editors: Jerry Ericksen, David Kinderlehrer, Robert Kohn,
   and J.-L. Lions

Volume 2: Oscillation Theory, Computation, and Methods of
      Compensated Compactness
   Editors: Constantine Dafermos, Jerry Ericksen,
   David Kinderlehrer, and Marshall Slemrod

Volume 3: Metastability and Incompletely Posed Problems
   Editors: Stuart Antman, Jerry Ericksen, David Kinderlehrer, and
   Ingo Muller

Volume 4: Dynamical Problems in Continuum Physics
   Editors: Jerry Bona, Constantine Dafermos, Jerry Ericksen, and
   David Kinderlehrer

Volume 5: Theory and Applications of Liquid Crystals
   Editors: Jerry Ericksen and David Kinderlehrer

Volume 6: Amorphous Polymers and Non-Newtonian Fluids
   Editors: Constantine Dafermos, Jerry Ericksen, and
   David Kinderlehrer

Volume 7: Random Media
   Editor: George Papanicolaou

Volume 8: Percolation Theory and Ergodic Theory of Infinite
      Particle Systems
   Editor: Harry Kesten

Volume 9: Hydrodynamic Behavior and Interacting Particle Systems
   Editor: George Papanicolaou

Volume 10: Stochastic Differential Systems, Stochastic Control Theory
      and Applications
   Editors: Wendell Fleming and Pierre-Louis Lions

Forthcoming Volumes:

FOREWORD

This IMA Volume in Mathematics and its Applications

LINEAR ALGEBRA, MARKOV CHAINS, AND QUEUEING MODELS

is based on the proceedings of a workshop which was an integral part of the 1991-92 IMA program on "Applied Linear Algebra".

We thank Carl Meyer and R.J. Plemmons for editing the proceedings.

We also take this opportunity to thank the National Science Foundation, whose financial support made the workshop possible.

Avner Friedman

Willard Miller, Jr.

PREFACE

This volume contains some of the lectures given at the workshop *Linear Algebra, Markov Chains, and Queueing Models* held January 13–17, 1992, as part of the Year of Applied Linear Algebra at the Institute for Mathematics and its Applications.

Markov chains and queueing models play an increasingly important role in the understanding of complex systems such as computer, communication, and transportation systems. Linear algebra is an indispensable tool in such research, and this volume collects a selection of important papers in this area. The articles contained herein are representative of the underlying purpose of the workshop, which was to bring together practitioners and researchers from the areas of linear algebra, numerical analysis, and queueing theory who share a common interest of analyzing and solving finite state Markov chains. The papers in this volume are grouped into three major categories—perturbation theory and error analysis, iterative methods, and applications regarding queueing models.

It is hoped that these contributions can provide the reader with an enlarged perspective of some of the major issues that are of current concern to both the pure and applied communities.

Carl D. Meyer

Robert J. Plemmons

CONTENTS

Queueing Theory and Applications

ERROR BOUNDS FOR THE COMPUTATION OF NULL VECTORS WITH APPLICATIONS TO MARKOV CHAINS

JESSE L. BARLOW*

Abstract. We consider the solution of the homogeneous system of linear equations $Ap = 0$ subject to $c^H p = 1$ where $A \in \mathbf{C}^{n \times n}$ is a singular matrix of rank $n-1$, $c, p \in \mathbf{C}^n$ and $c^H A = 0$. We assume that the vector c is known. An important applications context for this problem is that of finding the stationary distribution of a Markov chain. In that context, A is a real singular M-matrix of the form $A = I - Q^T$ where Q is row stochastic. In previous work by the author [SIAM J. Alg. Discrete Methods, 7(1986),pp. 414-424], it was shown that, for A an M-matrix, this problem could be solved by solving a nonsingular linear system B of degree $n - 1$ that was a principal submatrix of A. Moreover, this matrix B could always be chosen so that $\| B^{-1} \|_2 \approx \| A^\dagger \|_2$. Thus a stable algorithm to solve this problem could be developed for the Markov modeling problem using LU decomposition without pivoting and an update step that identified the correct submatrix.

In this paper, we improve that result in several ways. First, we show that many of the results can be generalized to the case where A is any complex matrix of degree n with rank $n - 1$, thus they can be generalized to the case where $A = \lambda I - Q^H$ and λ is a simple eigenvalue of Q^H. Secondly, we show that the update step is not necessary for the Markov modeling problem. For that problem, we can just use LU decomposition without pivoting. Finally, we show that a more elegant characterization and sharper error bounds are obtained for this algorithm in terms of the group inverse $A^\#$. In fact, we show that the error is bounded in terms of $\| A^\# \|_1 \| |A||p| \|_1$ which is smaller than the more traditional condition number $\| A^\# \|_1 \| A \|_1$. These bounds show that the procedure is unconditionally stable.

Key words. Group inverse, eigenvectors, condition numbers.

AMS(MOS) subject classifications. 65F05, 65F15,65F20,65G05.

1. Introduction. We consider the problem of solving the homogeneous system of linear equations

$$(1.1) \qquad\qquad Ap = 0$$

subject to the constraint

$$(1.2) \qquad\qquad c^H p = 1$$

where $A \in \mathbf{C}^{n \times n}$ is a singular matrix of rank $n - 1$, $c, p \in \mathbf{C}^n$ and

$$(1.3) \qquad\qquad c^H A = 0.$$

This problem arises naturally if $A = \lambda I - Q^H$ where $\lambda \in C$ is a simple eigenvalue of Q^H. Thus p is a right eigenvector of Q^H and c is a left eigenvector.

* Department of Computer Science, The Pennsylvania State University, University Park, PA 16802-6196, E-mail: barlow@cs.psu.edu, Supported by the National Science Foundation under grant CCR-9000526.

An important applications context for (1.1)-(1.2) is that of finding the stationary distribution of a Markov chain for which we have the definition.

DEFINITION 1.1. *The **Markov chain problem** is that of finding the vector p in (1.1)-(1.2) where $A = I - Q^T$, $c = (1, 1, \ldots, 1)^T$ and Q is row stochastic. For this problem A,c,p, and Q are assumed to be real.*

If the chain is ergodic, then Q is irreducible. Also, A will be a singular M-matrix, thus according to the Perron-Frobenius theorem [24, p.30] or [3, p.27] , p is unique and positive. For practical Markov chain problems, A will be large and sparse [18].

We consider a simple algorithm to solve (1.1)-(1.2) for p. Let A have the form

$$(1.4) \qquad A = P_j \begin{pmatrix} B & y \\ z^H & \alpha_{nn} \end{pmatrix} P_k^T$$

where P_j exchanges rows j and n, and P_k^T exchanges columns k and n. It is well known that there is always a choice of j and k such that B is nonsingular. Moreover, if A is an irreducible M-matrix or merely irreducible and diagonally dominant, any choice of j and k will make B nonsingular. Also, if $j = k$, then B is a nonsingular M-matrix.

If we know j and k *a priori*, we can solve for p using the following procedure discussed by Harrod and Plemmons[15,14] and Barlow [1].

ALGORITHM 1.1.
  1. *Find the permutation matrices P_j and P_k^T.*
  2. *Solve $B\hat{x} = -y$ by LU decomposition (or QR decomposition).*
  3. *Let $x = P_k(\hat{x}, 1)^T$.*
  4. *Set $p = x/c^H x$.*

In [1], it was shown that if A is an M-matrix, then there is always a choice of j and k such that

$$(1.5) \qquad \| A^\dagger \|_2 \leq \| B^{-1} \|_2 \leq (\sqrt{n} + 1)^2 \| A^\dagger \|_2$$

where $A^\dagger$ is the Moore-Penrose inverse of A [10, p.243]. Thus from the well-known fact that $\| B \|_2 \leq \| A \|_2$ we have that

$$(1.6) \qquad \kappa_2(B) \leq (\sqrt{n} + 1)^2 \kappa_2(A).$$

$$\kappa_2(B) = \| B \|_2 \| B^{-1} \|_2; \quad \kappa_2(A) = \| A \|_2 \| A^\dagger \|_2 .$$

This result explained experiments in Harrod and Plemmons[15].

In section two, we generalize this result to an arbitrary problem of the form (1.1)-(1.2) and to any generalized inverse A^- satisfying

$$(1.7) \qquad A = AA^- A.$$

A particularly important matrix from the class(1.7) is the group inverse, denoted $A^\#$. It is the unique matrix such that

$$(1.8) \qquad 1.\ AA^\# A = A;\ 2.\ A^\# AA^\# = A^\#;\ 3.\ AA^\# = A^\# A.$$

The group inverse exists if and only if $rank(A) = rank(A^2)$ and the latter condition holds since zero is a simple eigenvalue of A. As shown by Meyer[19], it yields a more elegant characterization of the problem (1.1)-(1.2) than does the Moore-Penrose inverse. From Meyer [19], we note that $A^\dagger = A^\#$ if and only if $c = p$.

In [1], we showed that for the Markov modeling problem, it is easy to choose j and k (with $j = k$) and Algorithm 1.1 would just have to be followed by an update step such as in [8] after choosing the value of k. We show here that the update step can be avoided. The results in section two are proven using different side conditions for (1.1). General side conditions for (1.1) are discussed by Meyer and Stewart[20].

In section three, we show that, with a slight modification, Algorithm 1.1 is stable for the Markov chain problem and for any problem of the form (1.1)-(1.2) as long as

    1. The left eigenvector c is known.

    2. The right eigenvector p is simple.

These results are characterized in terms of the group inverse and two measures of the condition in the 1-norm. The first is the "group" condition number

$$(1.9) \qquad \kappa_1^\#(A) = \| A \|_1 \| A^\# \|_1 .$$

and the second isthe "effective" condition number

$$(1.10) \qquad \kappa_1^E(A) = \| A^\# \|_1 \| |A||p| \|_1 .$$

Throughout the paper $|A|$ denotes the matrix $|A| = (|a_{ij}|)$ amd $|p|$ is the vector $|p| = (|p_1|, |p_2|, \ldots, |p_n|)^T$. Since $\| p \|_1 = 1$, for the Markov modeling problem, we have $\kappa_1^E(A) \le \kappa_1^\#(A)$. From our tests in section four, $\kappa_1^E(A)$ seems to be a more reliable estimate of the condition of the problem (1.1)-(1.2). Good estimates of $\kappa_1^\#(A)$ and $\kappa_1^E(A)$ can be obtained using the Hager-Higham[13,16] method. A condition number of this form for the problem (1.1)-(1.2) was introduced by Geurts [9].

The term "effective" condition number is similar to a term used by Chan and Fousler[4] and has similar implications here, but our definition is slightly different from theirs.

As shown by Harrod and Plemmons[15], we can use Gaussian elimination without pivoting[7] which was first advocated for this problem by Funderlic and Mankin[5]. Golub and Meyer[11] discuss the use of orthogonal factorization for dense matrices. It is easy to use standard sparse matrix software such as SPARSPAK-A or the Harwell code MA28 when A is large and sparse. Static storage allocation can be used throughout the computation. We omit all of the proofs in this paper. They will appear in an expanded version of this paper in another forum.

Section four gives the conclusions to our paper.

2. Analysis of Algorithm to Compute the Null Vector. In this section, we will show that Algorithm 1.1 obtains a solution to (1.1)-(1.2) that is as good as can be expected. We begin our analysis of the algorithm with a technical lemma that will shorten many of our arguments.

LEMMA 2.1. *Let $\tilde{x} \in \mathbf{C}^{n-1}$ be a vector that satisfies*

$$(2.1) \qquad B\tilde{x} = -y + r_B.$$

Let A have the form (1.4). Let $c = (c_1, \ldots, c_{n-1}, c_n)^T$ satisfy (1.3), assume $c_n \neq 0$, and define $d = -c_n^{-1}(c_1, \ldots, c_{n-1})^T$. Let $\| \cdot \|_s$ and $\| \cdot \|_q$ be Hölder norms. Then $\bar{x} = (\tilde{x}^T, 1)^T$ satisfies

$$(2.2) \qquad A\bar{x} = r_A$$

where

$$(2.3) \qquad \| r_A \|_s \leq (1 + \| d \|_q) \| r_B \|_s \quad s^{-1} + q^{-1} = 1.$$

We now give a lemma on the perturbation theory of the system (1.1).

LEMMA 2.2. *Assume the hypothesis and notation of Lemma 2.1 and let $\Delta x = (\tilde{x} - \hat{x}, 0)^T$ where $\hat{x}$ solves $B\hat{x} = -y$. Assume also $p_n \neq 0$ where $p = (p_1, \ldots, p_n)^T$ is the solution to (1.1)-(1.2). Then*

$$\| \Delta x \|_s = \| B^{-1} r_B \|_s \leq (1 + \frac{\| p \|_s}{|p_n|})(1 + \| d \|_q) \| A^- \|_s \| r_B \|_s \quad s^{-1} + q^{-1} = 1.$$
$$(2.4)$$

where A^- is any matrix satisfying (1.7).

There is an immediate corollary to this result that is relevant to the Markov chain problem.

COROLLARY 2.3. *Let P_j be the permutation matrix that exchanges rows j and n of A. Then there exist j and k such that if*

$$A = P_j \begin{pmatrix} B & y \\ z^H & \alpha_{nn} \end{pmatrix} P_k^T$$

then

$$(2.5) \qquad \| B^{-1} \|_1 \leq 2(1 + n) \| A^- \|_1$$

$$(2.6) \qquad \| B^{-1} \|_2 \leq (1 + \sqrt{n})(1 + \sqrt{n-1}) \| A^- \|_2$$

for any matrix A^- satisfying (1.7).

If we take $A^- = A^\dagger$, then (2.6) is a generalization of the bound (1.5) from [1].

For the Markov chain problem, $\| p \|_1 = 1$ and $c = e = (1, 1, \ldots, 1)^T$ so these conditions are always met. In general, this gives us

$$(2.7) \qquad \| B^{-1} \|_1 \leq 2(1 + \frac{1}{|p_n|}) \| A^- \|_1 .$$

Algorithm 1.1 actually solves (1.1)-(1.2) more accurately than Lemma 2.2 would suggest. The following analysis indicates this.

We can now prove the theorem that establishes the stability of Algorithm 1.1.

THEOREM 2.4. *Let $\tilde{x}$ be the solution to (2.1) and let $\bar{p}$ be the computed result of Algorithm 1.1 with $\bar{x} = (\tilde{x}^T, 1)^T$. Assume the hypothesis and notation of Lemma 2.1 and that r_B in (2.1) satisfies*

$$(2.8) \qquad \| r_B \|_1 \leq \psi(n) \, \| A \|_1 \| \, \bar{x} \, \|_1 \, u + O(u^2)$$

where $\psi(n)$ is a modestly growing function of n. Let p be the exact solution of (1.1)-(1.2) then

$$(2.9) \qquad \frac{\| \Delta p \|_1}{\| \bar{p} \|_1} \leq [(1 + \| d \|_\infty)\psi(n) + 5\sqrt{2}]\kappa_1^\#(A)$$

$$(2.10) \qquad + (n + 2\sqrt{2}) \, \| c \|_\infty \| p \|_1 \, u + O(u^2).$$

where $\kappa_1^\#(A) = \| A \|_1 \| A^\# \|_1$ is the "group" condition number. If $\| r_B \|_1$ satisfies

$$(2.11) \qquad \| r_B \|_1 \leq \phi(n) \, \| \, |A||\bar{x}| \, \|_1 \, u + O(u^2)$$

then

$$(2.12) \qquad \| \Delta p \|_1 \leq ([1 + \| d \|_\infty]\phi(n) + 5\sqrt{2})\kappa_1^E(A)u$$

$$(2.13) \qquad + (n + 2\sqrt{2}) \, \| c \|_\infty \| p \|_1 \, u + O(u^2)$$

where $\kappa_1^E(A)$ is as defined in (1.10).

Remark 1. *Meyer and Stewart[20] describe a strong relationship between $\| A^\# \|_2$ and the sep($\cdot$) function that is commonly used to bound the error in eigenvectors[22]. All of the bounds given in this paper can be easily reworked in terms of sep($\cdot$), but they are somewhat less elegant.*

Remark 2. *Funderlic and Meyer[6] use the condition number $\max_{(i,j)} |a_{ij}^\#|$. The result here can be made consistent with theirs from the observation that*

$$\| \Delta p \|_\infty \leq \| A^\# r \|_\infty + \epsilon \| p \|_\infty$$
$$\leq \| A^\# \|_{\infty,1} \| r \|_1 + |\epsilon| \| p \|_\infty$$

where $\| A^\# \|_{\infty,1} = \max_{\|x\|_1=1} \| A^\# x \|_\infty = \max_{1 \leq i,j \leq n} |a_{ij}^\#|$. Thus we can obtain bounds on $\| \Delta p \|_\infty$ instead of $\| \Delta p \|_1$ and define condition numbers $\kappa_{\infty,1}^\#(A) = \| A^\# \|_{\infty,1} \| A \|_1$ and $\kappa_{\infty,1}^E(A) = \| A \|_{\infty,1} \| |A||p| \|_1$. The comparisons between these two condition numbers will be identical to those between $\kappa_1^\#(A)$ and $\kappa_1^E(A)$. We choose the latter because known software [13,17] can be used to estimate them [2]. Also, this keeps as many of the bounds as possible in the same family of norms.

A similar line of analysis can obtain the componentwise bound

$$|(\Delta p)_i| \leq ([1 + \| d \|_\infty]\phi(n) + 5\sqrt{2}) \max_{1 \leq j \leq n} |a_{ij}^\#| \| |A||p| \|_1$$

$$+(n + 2\sqrt{2}) \| c \|_\infty |p_i| + O(u^2) \quad i = 1, 2, \ldots, n$$

under the assumption (2.11).

Since $c = (1, 1, \ldots, 1)^T$ for the Markov chain problem, the bound in Theorem 2.4 simplifies. The corollary below summarizes a simple version of this bound.

COROLLARY 2.5. *Assume the hypothesis of Theorem 2.4. If A arises out of the Markov chain problem in Definition 1.1 then*

$$\| \Delta p \|_1 \leq [2\psi(n) + 5\sqrt{2}]\kappa_1^\#(A)u + (n + 2\sqrt{2})u + O(u^2).$$

Under the assumption (2.11), we have that

$$\| \Delta p \|_1 \leq [2\phi(n) + 5\sqrt{2}]\kappa_1^E(A)u + (n + 2\sqrt{2})u + O(u^2).$$

The above results mean the for any problem of the form (1.1)-(1.2) where the right null vector p is simple and the left null vector c is known, we need only be concerned about the accuracy of solving the nonsingular linear system

$$(2.14) \qquad\qquad\qquad B\hat{x} = -y.$$

That problem is considered below.

The recommendations in the remainder of this section concern only the Markov modeling problem where A is real and has the form $A = I - Q^T$ where Q is row stochastic and is thus diagonally dominant. In that case

$$B = (b_{ij})$$

where $b_{jj} > 0$ and $b_{ij} \leq 0, i \neq j$, thus B is a nonsingular M-matrix. Varga and Cai[25] show that we can perform Gaussian elimination without pivoting and the diagonal entries b_{jj} should remain positive while the off-diagonal entries $b_{ij}, i \neq j$ remain non-positive, throughout the computation. Thus, if, for some k, b_{kk} becomes negative or zero during the course of the algorithm, then B must be singular to machine precision.

Since B is diagonally dominant, combining results of Wilkinson[26] and Reid[21], we have that Gaussian elimination without pivoting yields an LU decomposition such that

$$L(U \tilde{y}) = (B y) + E, \quad E = (e_{ij})$$

where

$$|e_{ij}| \leq 6.02|a_{jj}|m_{ij}u \quad i = 1, 2, \ldots, n-1, j = 1, 2, \ldots, n$$

Here m_{ij} is the number of operations performed on the entry b_{ij}. Thus Gaussian elimination on B satisfies a bound of the form (2.8). Neglecting the backsubstitution errors which satisfy similar bounds[27], we have

$$(B\,y)p = r_B = -Ep.$$

We can now say that

(2.15) $\|\, r_B\, \|_1 = \|\, Ep\, \|_1 \leq 6.02\phi_0(n)\, \|\, diag(a_{11}, \ldots, a_{nn})p\, \|_1\, u$

where

(2.16) $$\phi_0(n) = \max_{(i,j)} m_{ij} n_j$$

(2.17) $n_j = $ number of nonzeroes in column j of L and U

We can now give a theorem that shows that Algorithm 1.1 obtains answers that are as good as can be expected for the Markov modeling problem.

THEOREM 2.6. *Let $A = I - Q^T$ where Q is an irreducible row stochastic matrix. Let $Ap = 0$ be solved by Algorithm 1.1 where the system $Bx = -y$ is solved by Gaussian elimination without pivoting. Neglecting the error from back substitution, if no diagonal entry of B becomes negative or zero, then*

(2.18) $\|\, \Delta p\, \|_1 \leq \quad [\phi_0(n) + 5\sqrt{2}]\kappa_1^E(A)u$

(2.19) $\qquad\qquad\qquad +(n + 2\sqrt{2})u + O(u^2)$

where

$$\phi_0(n) = 6.02 max_{(i,j)} m_{ij} n_j$$
$$m_{ij} = \text{number of operations performed on entry } a_{ij}$$
$$n_j = \text{number of nonzeroes in column } j \text{ of } L \text{ and } U$$

We consider the implementation of Algorithm 1.1 for the Markov modeling problem. In the general case, we recommend the use of the orthogonal factorization approach of Golub and Meyer[11].

The main remaining difficulty in solving (2.14) is if B turns out to be singular to machine precision. For the solution of (2.14), we have that B is a nonsingular M-matrix. Thus $B = (b_{ij})$ where $b_{jj} > 0$ and $b_{ij}, i \neq j$ and moreover, in exact arithmetic, these entries will not change sign thorough the course of elimination[25]. However, if B is ill-conditioned, a diagonal entry could become negative or zero, because of rounding errors. If that occurs, if, say $b_{kk} = a_{kk} \leq 0$, then we can use the fact that $a_{kk} = -\sum_{i=k+1}^{n} a_{ik}$ when it is time to eliminate column k. This technique was introduced by Grassman, Taksar, and Heyman[12]. A special case of that method for nearly uncoupled problems was analyzed by Stewart and Zhang[23]. We then proceed with the elimination. Thus breakdown in Gaussian elimination can be completely avoided and the error bounds given here are unconditional.

3. Tests Using the Algorithm. The test program was written in FORTRAN 77 and the test runs were done on a SUN 4 in the author's office at the Pennsylvania State University. The estimated machine precision was 5.960×10^{-8}. We did one large random test set plus two particular ill-conditioned problems.

EXAMPLE 3.1. *We tested Algorithm 4.1 with 500 randomly generated dense 50×50 matrices. These matrices were of the form*

$$a_{jj} = 0.8e^{(j-1)\omega} \quad \omega = ln(10^{-7})/49 \quad j = 1, 2, \ldots, n$$

$$a_{ij} = -\alpha_{ij} \cdot a_{jj}/\beta_j$$

where $\beta_j = \sum_{i \neq j} \alpha_{ij}$ and α_{ij} are random numbers between 0 and 1. This method gives a random matrix A such that A is a singular irreducible M-matrix. This always insured that $\| A \|_1 = 1.6$.

Using the Hager-Higham[13,16] method as discussed in [2], our estimates of $\| A^\# \|_1$ ranged from 3.309×10^6 to 1.136×10^7. In the traditional sense, these matrices are very ill-conditioned. According to the $\kappa_1^\#(A)$ bound, one should expect $\| \Delta p \|_1$ to be about 10^{-1} or 10^{-2} However, the $\kappa_1^E(A)$ estimates are much smaller. They ranged from 7.431 to 25.30, thus Algorithm 4.1 should obtain accurate answers which it did.

TABLE 3.1 (RESIDUALS AND FORWARD ERROR FOR RANDOM MATRICES).

Quantity	Maximum	Average
$\| r_B \|_1$	1.663E-13	1.307E-13
$\| r_A \|_1$	1.884E-13	1.450E-13
$\| \Delta p \|_1$	8.313E-7	2.859E-7

TABLE 3.2 (RATIOS OF RESIDUALS AND ESTIMATES TO ACTUAL ERRORS).

Quantity	Maximum	Average	Minimum
$\frac{\|r_A\|_1}{\|r_B\|_1}$	1.392	1.112	1.001
$\frac{\|\Delta p\|_1}{est.A}$	1.415	5.343E-1	5.524E-2
$\frac{\|\Delta p\|_1}{est.B}$	1.970	6.239E-1	5.585E-2
$\frac{\|\Delta p\|_1}{est.Z}$	1.880	6.094E-1	6.354E-2

Here

(3.1)
$$est.A = \| r_A \|_1 * cond.est.$$

(3.2)
$$est.B = \| r_B \|_1 * cond.est.$$

(3.3)
$$est.Z = \| |A||p| \|_1 * cond.est. * u$$

where cond.est. is the Hager[13] estimate of $\| A^\# \|_1$ *and u is the estimated machine precision. Thus we note that both est. A and est. B are good estimates of* $\| \Delta p \|_1$. *None of the 500 linear systems were determined to have a rank deficiency of greater than one.*

The remaining two examples are taken from Harrod and Plemmons[15]. For both examples, est. A, est. B, and est. Z are as defined in (3.1),(3.2), and (3.3).

EXAMPLE 3.2.

$$A = \begin{pmatrix} 1.0E-06 & -0.4 & -5.0E-07 & -5.0E-07 & -2.0E-07 \\ -1.0E-07 & 0.7 & 0 & 0 & -3.0E-07 \\ -2.0E-07 & 0 & 1.0E-06 & 0 & -1.0E-07 \\ -3.0E-07 & 0 & 0 & 1.0E-06 & -4.0E-07 \\ -4.0E-07 & -0.3 & -5.0E-07 & -5.0E-07 & 1.0E-06 \end{pmatrix}$$

We obtained the following computational results.

$$\| A \|_1 = 1.4; \quad \| A^\dagger \|_2^{-1} = 1.007E-06$$

The value for $\| A^\dagger \|_2^{-1}$ *is from [15].*

The Hager estimate of $\| A^\# \|_1 = 4.444E+05$. Thus, from the $\kappa_1^\#(A)$ estimate one should expect $\| \Delta p \|_1$ to be about 10^{-2} or 10^{-3}. However, the Hager estimate of $\kappa_1^E(A)$ is only 1.011, thus we should expect and do obtain good answers.

$$\| \Delta p \|_1 = 2.061E-08;$$
$$\| r_B \|_1 = 4.441E-14; \quad \| r_A \|_1 = 6.561E-14$$
$$\| r_A \|_1 / \| r_B \|_1 = 1.489$$
$$\| \Delta p \|_1 /est.A = 0.7068; \quad \| \Delta p \|_1 /est.B = 1.052$$
$$\| \Delta p \|_1 /est.Z = 0.4952$$

The solution p is

$$p = (0.315217, 1.95652E-07, 9.81884E-02, 0.235145, 0.351449)^T.$$

EXAMPLE 3.3. *Let*

$$\bar{Q}(\epsilon) = \begin{pmatrix} Q_{11} & 0 \\ 0 & Q_{22} \end{pmatrix} + \epsilon(e_1 e_6^T + e_6 e_1^T)$$

where

$$Q_{11} = \begin{pmatrix} 0.1 & 0.3 & 0.1 & 0.2 & 0.3 \\ 0.2 & 0.1 & 0.1 & 0.2 & 0.4 \\ 0.1 & 0.2 & 0.2 & 0.4 & 0.1 \\ 0.4 & 0.2 & 0.1 & 0.2 & 0.1 \\ 0.6 & 0.3 & 0.0 & 0.0 & 0.1 \end{pmatrix}$$

$$Q_{22} = \begin{pmatrix} 0.1 & 0.2 & 0.2 & 0.4 & 0.1 \\ 0.2 & 0.2 & 0.1 & 0.3 & 0.2 \\ 0.1 & 0.5 & 0.0 & 0.2 & 0.2 \\ 0.5 & 0.2 & 0.1 & 0.0 & 0.2 \\ 0.1 & 0.2 & 0.2 & 0.3 & 0.2 \end{pmatrix} .$$

Then $A = I - Q^T$ where $Q = D\bar{Q}(\epsilon)$ and $D = diag(d_1, d_2, \ldots, d_{10})$ is chosen so that Q satisfies $Qc = c$, $c = (1, 1, \ldots, 1)^T$. This is a very ill-conditioned problem of a Markov chain with two loosely coupled sets of states. We chose $\epsilon = 1.0E - 06$.

The Hager estimate of $\| A^{\#} \|_1 = 2.278E + 06$. Thus according to the $\kappa_1^{\#}(A)$ estimate, one should expect $\| \Delta p \|_1$ to be about 10^{-1} or 10^{-2}. In this case, that is accurate, but that is because the $\kappa_1^E(A)$ is estimated to be $4.032E + 06$, so it is not much more optimistic.

$$\| \Delta p \|_1 = 8.423E - 01;$$
$$\| r_B \|_1 = 4.504E - 08; \quad \| r_A \|_1 = 6.891E - 08$$
$$\| r_A \|_1 / \| r_B \|_1 = 1.530$$
$$\| \Delta p \|_1 / est.A = 0.5365$$
$$\| \Delta p \|_1 / est.B = 0.8210$$
$$\| \Delta p \|_1 / est.Z = 0.3505$$

The solution p is

$$p = (\quad 0.124286, 9.87915E - 02, 3.71796E - 02, 7.43592E - 02, 9.772923E$$
$$-02, 0.124286, 0.135159, 7.20948E - 02, 0.135012, 0.101102)^T$$

The above tests confirm that this algorithm will always yield a small residual for the problem (1.1)-(1.2). Note that although $\kappa_1^{\#}(A)$ is large for all of the above problems, only the last example obtains poor results. That is because it is the only one for which $\kappa_1^E(A)$ is large.

We note that the last example is a loosely coupled Markov chain. Stewart and Zhang[23] describe a variant of Gaussian elimination that obtains better error bounds for this particular class of problems. That procedure has a more restrictive pivoting strategy.

4. Conclusion. The partition algorithm described by Harrod and Plemmons[15] is stable provided that we use the diagonal adjustment technique of Grassman, Taksar, and Heyman[12]. Thus for the large sparse problem that normally arise in queueing models, we can fully exploit the nonzero structure.

Error bounds in terms of $\| A^{\#} \|_1 \| |A||p| \|_1$ seem to give very accurate estimates of the error in computing the null vector p.

Acknowledgements. The author would like to thank Bob Plemmons for his careful reading of the manuscript and helpful comments. He also thanks the referee who pointed out reference [6] and made other helpful comments that resulted in making the manuscript more readable. Lastly, he acknowledges the helpful comments of Guodang Zhang, Bob Funderlic, Daniel Heyman, and Francoise Chatelin.

REFERENCES

[1] J.L. Barlow. On the smallest positive singular value of a singular M-matrix with applications to ergodic Markov chains. *SIAM J. Alg. Dis. Methods*, 7:414–424, 1986.

[2] J.L. Barlow. Error bounds and condition estimates for the computation of null vectors with applications to Markov chains. Technical Report CS-91-20, The Pennsylvania State University, Department of Computer Science, University Park, PA, 1991.

[3] A. Berman and R.J. Plemmons. *Nonnegative Matrices in the Mathematical Sciences.* Academic Press, New York, 1979.

[4] T.F. Chan and D.E. Fousler. Effectively well-conditioned linear systems. *SIAM J. Sci. Stat. Computing*, 9:963–969, 1988.

[5] R.E. Funderlic and J.B. Mankin. Solution of homogeneous systems of linear equations arising in compartmental models. *SIAM J. Sci. Stat. Computing*, 2:375–383, 1981.

[6] R.E. Funderlic and C.D. Meyer, Jr. Sensitivity of the stationary distribution for an ergodic Markov chain. *Linear Alg. Appl.*, 76:1–17, 1986.

[7] R.E. Funderlic and R.J Plemmons. LU decomposition of M-matrices by elimination without pivoting. *Linear Alg. Appl.*, 41:99–110, 1981.

[8] R.E. Funderlic and R.J Plemmons. Updating LU factorizations for computing stationary distributions. *SIAM J. Alg. Disc. Meth.*, 7:30–42, 1986.

[9] A.J. Geurts. A contribution to the theory of condition. *Numerische Mathematik*, 39:85–96, 1982.

[10] G.H. Golub and C.F. Van Loan. *Matrix Computations: Second Edition.* The Johns Hopkins Press, Baltimore, 1989.

[11] G.H. Golub and C.D. Meyer, Jr. Using the QR factorization and group inverse to compute, differentiate and estimate the sensitivity of stationary probabilities of Markov chains. *SIAM J. Alg. Dis. Methods*, 7:273–281, 1986.

[12] W.K. Grassman, M.I. Taksar, and D.P. Heyman. Regenerative analysis and steady state distribution for Markov chains. *Operations Research*, 33:1107–1116, 1985.

[13] W.W. Hager. Condition estimates. *SIAM J. Sci. Stat. Computing*, 5:311–316, 1984.

[14] W.J. Harrod. *Rank modification methods for certain systems of linear equations.* PhD thesis, Department of Mathematics, University of Tennessee, 1984.

[15] W.J. Harrod and R.J. Plemmons. Comparison of some direct methods for computing stationary distributions of Markov chains. *SIAM J. Sci. Stat. Computing*, 5:453–469, 1984.

[16] N.J. Higham. A survey of condition number estimation for triangular matrices. *SIAM Review*, 29:575–598, 1987.

[17] N.J. Higham. Fortran codes for estimating the one-norm of a real or complex matrix, with applications to condition estimation. *ACM Trans. Math. Software*, 14:381–396, 1988.

[18] L.C. Kaufman. Matrix methods for queueing problems. *SIAM J. Sci. Stat. Computing*, 4:525–552, 1983.

[19] C.D. Meyer, Jr. The role of the group inverse in the theory of finite Markov chains. *SIAM Review*, 17:443–464, 1975.

[20] C.D. Meyer, Jr. and G.W. Stewart. Derivatives and perturbations of eigenvectors. *SIAM J. Numer. Anal.*, 25:679–691, 1988.

[21] J.K. Reid. A note on the stability of Gaussian elimination. *J. Inst. Math. Appl.*, 8:374–375, 1971.

[22] G.W. Stewart. Error bounds for approximate invariant subspaces for close linear operators. *SIAM J. Num. Anal.*, 8:796–808, 1971.

[23] G.W. Stewart and G. Zhang. On a direct method for the solution of nearly uncoupled Markov chains. *Numerische Mathematik*, 59:1–11, 1991.

[24] R.S. Varga. *Matrix Iterative Analysis*. Prentice-Hall, Englewood Cliffs, N.J., 1962.

[25] R.S. Varga and D.-Y. Cai. On the LU factorization of M-matrices. *Numerische Mathematik*, 38:179–192, 1981.

[26] J.H. Wilkinson. Error analysis of direct methods of matrix inversion. *J. Assoc. Comput. Mach.*, 8:281–330, 1961.

[27] J.H. Wilkinson. *The Algebraic Eigenvalue Problem*. Oxford University Press, London, 1965.

THE INFLUENCE OF NONNORMALITY ON MATRIX COMPUTATIONS

FRANÇOISE CHATELIN*

Abstract. We consider the simple problem of solving the linear system $Ax = b$. It is well known that we might have some computational difficulties if A is "nearly singular". We wish to make this statement more precise by introducing the *condition number of a singularity*. Then we show how, for matrix computations, the departure from normality is an essential parameter which affects this condition number. In this paper we propose to quantify the influence of a singularity on a neighboring computation by means of a condition number for the singularity. In matrix computations, when the singularity is either a singular system, or a multiple defective eigenvalue, we present several numerical experiments which examplify various ways in which the departure from normality may affect computations in finite precision arithmetic.

1. Condition Number of a Singularity.

The presentation to follow is restricted to **normwise absolute** condition numbers. It can be extended without difficulty to componentwise and relative condition numbers (see [8]).

1.1. Linear systems. Consider the system $Ax = b$. Its normwise distance to singularity is $\min(\|\Delta A\|; A + \Delta A$ singular), where $\|\cdot\|$ is an arbitrary subordinate norm. This is a *backward* distance since it is expressed in terms of the data.

There is a difficulty to define the *forward* distance to singularity, since at the singularity, the solution either does not exist, or is not unique. The matrix A would be singular if and only if one of its eigenvalues μ were equal to zero. One way to resolve the above difficulty is then to represent the forward distance by means of the spectral distance $\min(|\mu|, \mu \in sp(A)) = \text{dist}(sp(A), 0)$. $B = A + \Delta A$ is exactly singular, that is B admits the eigenvalue $\lambda = 0$.

If λ is a **simple** eigenvalue of B, then $|\Delta \lambda| = |\mu - 0|$ which represents the variation of λ under the variation ΔA of A such that $B = A + \Delta A$, is proportional to $\|\Delta A\|$, for $\|\Delta A\|$ small enough.

Then the quotient $\frac{|\Delta \lambda|}{\|\Delta A\|}$ measures the ratio of the forward to the backward distance to the singularity. The quantity $C(\lambda) = \lim_{\|\Delta A\| \to 0} \frac{|\Delta \lambda|}{\|\Delta A\|}$ is the condition number of λ as a closeby singularity of A. It defines the region $\{z \in \mathbb{C}; |z - \lambda| \le C(\lambda)\|\Delta A\|\}$ of $\mathbb{C}$ which is affected by the singularity at the level $\|\Delta A\|$ of perturbation amplitude on the datum A. The larger $C(\lambda)$, the stronger the influence of the singularity. We remark that $C(\lambda)$ is also, by definition, the condition number of λ as a simple eigenvalue of $A + \Delta A$. As such, we shall see in the next section that it is strongly affected by the departure from normality of A.

---

* University of Paris IX - Dauphine and LCR, Thomson - CSF 91 404 Orsay - Cedex, France. chatelin@thomson-lcr.fr

Remarks. 1. If $\lambda = 0$ is a multiple eigenvalue with index ℓ, $1 < \ell \leq m$, than $|\Delta\lambda|$ is proportional to $\|\Delta A\|^{1/\ell}$. (see [2], p. 106).

2. Other choices are possible to represent the forward distance to singularity.

The singularity we have just presented corresponds to the stiff transition from a smooth well-posed problem to an ill-posed one (x is C^1 to x is not C^0). Let us look at a more gentle transition where the problem remains well-posed, but the regularity of the solution decreases from order 1 to order $1/m$.

1.2. A multiple root. Let x be a simple solution for a nonlinear equation; for example a simple root for the polynomial $p(x) = 0$. The backward distance to singularity is now

$$\min(\|\Delta p\|; p + \Delta p \quad \text{has a multiple root})$$

where $\|\cdot\|$ is a norm on the set of polynomials.

In contrast with section 1.1, the solution (i.e. a multiple root denoted ξ) **exists** at the singularity and we have no problem to define the forward distance to singularity as $\|x - \xi\|$. If m is the multiplicity of ξ, then $\|\Delta x\| = \|x - \xi\|$ is proportional to $\|\Delta p\|^{1/m}$, and we get back the condition number of ξ defined as the condition number of a root of multiplicity m (see [12], p.40).

2. Spectral Condition Numbers. Let λ be a simple eigenvalue of A with right and left eigenvectors denoted by x and x_* respectively. Then

$$C(\lambda) = \frac{\|x_*\|_* \|x\|}{|x_*^* x|}$$

If we choose the euclidian norm $\|\cdot\|_2$, than $C(\lambda) = 1$ for any normal matrix ($x = x_*$). In general, with $\|x\|_2 = x_*^* x = 1$, $\|x_*\|_2 = \frac{1}{\cos\theta} > 1$, where θ is the acute angle between the directions spanned by x and x_*. When A is not normal, these 2 vectors need not be identical, and may even be almost orthogonal, depending on the departure from normality of A.

The departure from normality can be quantified by

$$\nu(A) = \|AA^* - A^*A\|_F, \quad \text{with } \frac{\nu(A)}{\|A\|_F^2} \leq 2.$$

But $\frac{\nu(A)}{\|A\|_F}$ and $\frac{\nu(A)}{\|A^2\|_F}$ can be arbitrarily large [6].

$\nu(A)$ is related to the Schur form in the following way. Let $A = QSQ^*$ Q unitary, with $S = D + N$ where N is strictly upper triangular, then $\|N\|_F^2 = \|A\|_F^2 - \sum_{i=1}^n |\lambda_i|^2$, $\frac{\|N\|_F}{\|A\|_F} \leq 1$, and $\frac{1}{6}\left(\frac{\nu(A)}{\|A\|_F}\right)^2 \leq \|N\|_F^2 \leq \sqrt{\frac{n^3-n}{12}}\nu(A)$, [9]. If A is diagonalizable, $A = XDX^{-1}$, then

$$\text{cond}_2(X) \geq \left[1 + \frac{1}{2}\left(\frac{\nu(A)}{\|A^2\|_F}\right)^2\right]^{1/4}.$$

If λ is simple, with $\delta = dist(sp(A) - \{\lambda\}, \lambda)$, then $C(\lambda) \leq \left[1 + \frac{1}{n-1}\left(\frac{1}{\delta}\|N\|_F\right)^2\right]^{\frac{n-1}{2}}$, [10]. This implies that if λ is ill-conditioned, then $\frac{1}{\delta}\|N\|_F$ must be large: λ becomes multiple and/or the departure from normality is large. It follows that for a diagonalizable matrix, an ill-conditioned eigenbasis may not be an indicator of a large departure from normality: it may merely indicate that the matrix is close to defectiveness [2]. On the other hand, we shall see that a diagonizable matrix, with well separated eigenvalues, may behave computationally like one Jordan block.

3. The Influence of Nonnormality. We consider the matrix S of size 20, under real Schur form, and depending on the parameter ν which controls the departure from normality ($\nu(S) > \nu^2$):

$$
S = \begin{pmatrix}
x_1 & y_1 & & & & \\
-y_1 & x_1 & \nu & & & \\
& & x_2 & y_2 & & \\
& & -y_2 & x_2 & \nu & \\
& & & & \ddots & \nu \\
& & & & & x_p & y_p \\
& & & & & -y_p & x_p
\end{pmatrix}, \quad n = 2p = 20
$$

For $k = 1, \ldots, 10$, $x_k = -\frac{(2k-1)^2}{1000}$ and $y_k = \frac{2k-1}{100}$ are chosen such that the exact spectrum is on the parabole $x = -10y^2$. The exact spectral radius is $\rho(S) = 0.40795$.

Finally $M = QSQ$, where Q is the symmetric orthonormal matrix defined by $q_{ij} = \sqrt{\frac{2}{n+1}} \sin \frac{ij\pi}{n+1}$. And we intend to solve $Ax = b$ with $A = I - M$, first by an iterative method, then by a direct one.

3.1. Successive iteration. We consider the iteration

$$x_0, x_i = M x_{i+1} + b, \quad i \geq 1$$

to approximate the solution $x = \lim_{i \to \infty} x_i$. Convergence history for various values of ν is shown on figures 1 to 4, where the relative error is $\frac{\|x - \tilde{x}_i\|_\infty}{\|x\|_\infty}$, the relative residual is $\frac{\|A\tilde{x}_i - b\|_\infty}{\|A\|_\infty \|\tilde{x}_i\|_\infty + \|b\|_\infty}$, and $\tilde{x}_i$ denotes the computed ith iterate x_i.

The results examplify a double phenomeneon as ν increases:

1. the condition number of A increases, so that the error increases,
2. the quality of the convergence deteriorates until there is complete divergence on the error, while the residual remains between 10^{-2} and 10^{-3}.

The first aspect is well known, but the second is less appreciated: even though the exact spectral radius is significantly less than 1, the iteration becomes weakly backward unstable. This is only marginally explained by the fact that the **computed** spectral radius becomes larger than 1.

3.2. A direct method. When we solve $Ax = b$ by Gaussian Elimination with partial pivoting, we have no surprise. The error increases with ν, but the algorithm remains backward stable to machine precision.

Unlike successive iteration, such a direct method is not affected by the departure from normality of M. However such a clear cut difference between direct and iterative methods may not remain true for large systems. One can find evidence of the contrary in experiments with Arnoldi in [6].

3.3. The completed spectrum of M. Figure 5 displays the exact (+) and compluted (0) spectra of four matrices M corresponding to $\nu = 1, 10, 10^2$ and 10^3. The $*$ represents the exact and computed arithmetic means of the eigenvalues, which are identical within machine precision. For $\nu = 10^2$ and 10^3, apart from 2 of them, 18 of the so computed eigenvalues lie on a circle centered at the arithmetic mean. This suggests that the matrix M behaves approximately like one Jordan block of size 18, completed by 2 diagonal elements.

In order to test this hypothesis, we compute the spectra of a sample of matrices $M' = M + \Delta M$, randomly perturbed from M, in the following componentwise way: m_{ij} becomes $m'_{ij} = m_{ij}(1 + \alpha t)$, where α is a random variable taking the values ± 1 with probability $1/2$, and $t = 2^{-k}$, the integer k varying from 40 to 50. Therefore $|\Delta M| = t|M|$ where t ranges from $2^{-40} \sim 10^{-12}$ to $2^{-50} \sim 10^{-15}$. For each t, the sample size is 30, so that the total number of matrices is $30 \times 11 = 330$. The superposition of the corresponding 330 spectra is plotted in Figure 6. The transformation of the perturbed spectra as ν varies is dramatic. The 36 spikes around the mean $\hat{\lambda}$ for $\nu = 10^2$ and 10^3 confirm the hypothesis that M becomes computationally close to a Jordan form as ν increases. The computed eigenvalues $\tilde{\lambda}'$ are solutions of $(\tilde{\lambda}' - \hat{\lambda})^{18} = \epsilon = 0(t)$, ϵ being positive or negative with equal probability. The computed eigenvalues appear at the vertices of two families of regular polygons with 18 sides, which are symmetric with respect to the vertical axis; hence the 36 spikes. Note that, although the spectra for $\nu = 10^2$ and 10^3 are qualitatively similar, the one for $\nu = 10^3$ is more ill-conditioned than the one for $\nu = 10^2$. The fact that the matrix approximately behaves like a Jordan block of size 18 (rather than 20 for example) is **experimental** only. It remains true under normwise perturbations [8].

The perturbed spectra are part of the componentwise ϵ-pseudospectrum, with $\epsilon = 2^{-40}$. But the information on the underlying Jordan structure, in finite arithmetic, of M (which is diagonalisable in exact arithmetic) is too detailed to be retrievable by looking at the *global* pseudospectrum [11]. The interested reader can find other suggestive examples in [3].

3.4. Normwise versus componentwise pseudo spectra. The normwise absolute ϵ-pseudospectrum defined by Trefethen [11] is $\{z \in \mathbb{C}; z$ is an eigenvalue of $A + \Delta A, \|\Delta A\| \leq \epsilon\}$.

One could define similarly componentwise and/or relative ϵ-pseudospectra, which contain the associated perturbed spectra.

For the above family of matrices M, the normwise and componentwise perturbed spectra are similar, but this need not be the case in general, as we show now on a calculation taken from [1].

We consider the LU factorization of the family of matrices of varying size $n = 1, 2, \ldots$:

$$A(\beta, \gamma) = \begin{pmatrix} \gamma & \beta & & & 0 \\ 1 & \ddots & \ddots & & \\ & \ddots & \ddots & & \beta \\ 0 & & 1 & & \gamma \end{pmatrix}_{n \times n} = A(\beta, 0) + \gamma I$$

where $\beta \geq 0$ and γ are real parameters, and the left and right factors are chosen under the form

$$L = \begin{pmatrix} \alpha_1 & & & \\ 1 & \ddots & & \\ & \ddots & \ddots & \\ 0 & & 1 & \alpha_n \end{pmatrix}_{n \times n} \quad and \quad U = \begin{pmatrix} 1 & x_1 & & & 0 \\ & \ddots & \ddots & & \\ & & \ddots & & x_{n-1} \\ & & & & 1 \end{pmatrix}_{n \times n}.$$

The α_i and x_i can be computed by the following nonlinear recurrence:

$$\begin{cases} \alpha_1 = \gamma, \quad x_1 = \frac{\beta}{\gamma}, \\ \alpha_i = \gamma - \frac{\beta}{\alpha_{i-1}}, \quad x_i = \frac{\beta}{\alpha_i}, \quad i \geq 2. \end{cases}$$

For $|\gamma| \geq 2\sqrt{\beta}$, there exists a unique limit for α_i as $i \to \infty$. But if $|\gamma| < 2\sqrt{\beta}$, α_i may have a periodic behaviour for certain values of γ. Indeed the values $\gamma_{r_p} = 2\sqrt{\beta} \cos \frac{r\pi}{p}$ for $r = 1, \cdots, p-1$, lead to oscillations of period p [1].

Let $B_n = A(\beta, 0) = A(B, \gamma) - \gamma I$. Its spectrum Σ_n consists of the values of γ for which $A(\beta, \gamma)$ is singular, allowing a periodic behaviour for the recurrence. We have $\Sigma_n = \{2\sqrt{\beta} \cos \frac{r\pi}{n+1}, r = 1, \ldots, n\}$. In the limit when $n \to \infty$, we get $\Sigma = \overline{\cup_n \Sigma_n} = [-2\sqrt{\beta}, 2\sqrt{\beta}]$.

Now the limit, as $n \to \infty$ and $\epsilon \to 0$, of the ϵ-pseudospectrum is the domain limited by the ellipse $\mathcal{E}$ of major horizontal axis $\beta + 1$, and minor vertical axis $\beta - 1$, centered at 0. This is the *spectrum of the family of matrices B_n*, as defined by Bakhvalov [1]:

$$\{\lambda \in \mathbb{C}; \forall \epsilon > 0, \exists B_n, x_n \neq 0 \text{ such that } \|B_n x_n - \lambda x_n\| \leq \epsilon \|x_n\|\}.$$

Plotted on figure 8 are the normwise perturbed spectra of B_n for $n = 30, 50, 120, 170$ and $\beta = 10$. We see that the shape of the exact spectrum

Σ_n (indicated by stars) is lost, and that, as n increases, the perturbed spectra tend to occupy the interior of the ellipse $\mathcal{E}$, also drawn.

Such a behaviour leads to conjecture that the computation of α_i, which is, in exact arithmetic, convergent for $|\gamma| \geq 2\sqrt{\beta}$, might show instabilities for real γ in the intervals $[2\sqrt{\beta}, \beta + 1]$ and $[-\beta - 1, -2\sqrt{\beta}]$, if the finite precision computation was ruled by the ϵ-pseudospectrum.

Figure 7 rules out this hypothesis. For $\beta = 1, 10^2, 10^5$ and 10^8, and for 500 values of γ around $2\sqrt{\beta}$, we compute the first 500 iterates for α_i, then we plot the 50 following iterates, that is $i = 501$ to 550. The difference between the apparently chaotic behaviour of the computed iterates for $\gamma < 2\sqrt{\beta}$ and the exact one for $\gamma \geq 2\sqrt{\beta}$ is striking.

In this example, the limit **normwise** pseudospectrum is not the right key. The proper key is the limit **componentwise** pseudospectrum which is equal to Σ. Indeed because B_n is tridiagonal, its spectrum Σ_n is not affected by componentwise perturbations.

The true difficulty of computing the spectrum in the absence of normality has been known to mathematicians for a long time. Its practical implications should not be overlooked, since such examples appear in essential industrial applications, as well as in physics (fluid dynamics and plasma physics, for example).

4. Acknowledgement. The author is indebted to Valérie Fraysse for performing the numerical experiments as part of her Ph.D. thesis [8], at CERFACS, Centre Européen de Recherche et de Formation Avancée au Calcul Scientifique, in Toulouse, France. The author wishes to thank the IMA for giving her the opportunity of a fruitful research visit.

REFERENCES

[1] L. Brugnano, D. Trigiante. Sulle soluzioni di equazioni alle differenze dell' algebra lineare numerica *Atti della Acad. Sc. Bologna*, Ser V., 93–108, 1990.

[2] F. Chatelin. *Valeurs propres de matrices*, Masson, Paris, 1988.

[3] F. Chatelin. *Résolution approchée d'équations sur ordinateur*, Lect. Notes, Université Paris VI, 1989.

[4] F. Chatelin. *Matrix eigenvalues*, Wiley, Chichester, (to appear) 1993.

[5] F. Chatelin and V. Fraysse. Elements of a condition theory for the computational analysis of algorithms in *Iterative methods in linear algebra*, R. Beauwens and P. de Groen eds., 15-25, North-Holland - IMACS, 1992.

[6] F. Chatelin and S. Godet-Thobie. *Stability analysis in Aeronautical Industries* in *High Performance Computing II* M. Durand and F.El Dabaghi eds., pp. 415-422, North-Holland, Amsterdam, 1991.

[7] J.W. Demmel. On condition numbers and the distance to the nearest ill-posed problem, *Numer. Math.*, 51, 251–289, 1987.

[8] V. Fraysse. Reliability of computer solutions, Ph.D. Thesis, INP Toulouse, 1992.

[9] G.H. Golub and C.F. Van Loan. *Matrix computations*, 2nd ed. John Hopkins U. Press, 1989.

[10] R.A. Smith. The condition numbers of the matrix eigenvalue problem. *Numer. Math.*, 10, 232–240, 1967

[11] L.N. Trefethen. Pseudospectra of matrices in *14th Dundee Biennal Conf. on Num. Anal.*, D.F. Griffiths and G.A. Watson, eds. 1991.

[12] J. Wilkinson. *Rounding errors in algebraic processes*, HMSO, London, 1963.

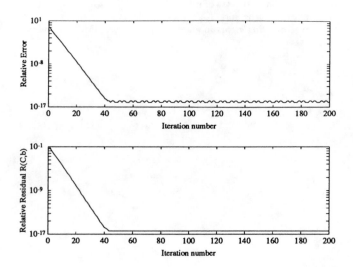

Successive iterations : $\nu = 1$

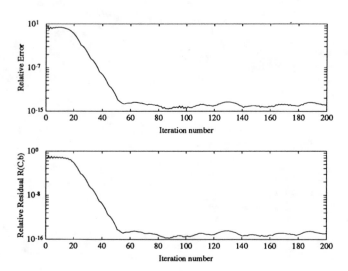

Successive iterations : $\nu = 10$

Figure 1

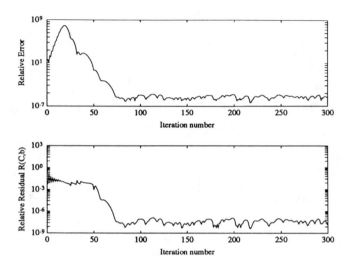

Successive iterations : $\nu = 10^2$

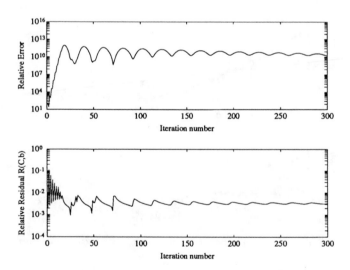

Successive iterations : $\nu = 2.83 \; 10^2$

Figure 2

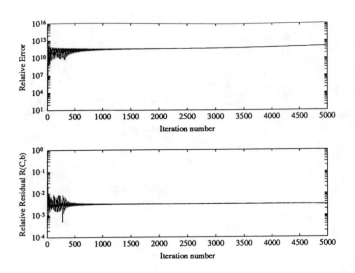

Successive iterations : $\nu = 2.92 \; 10^2$

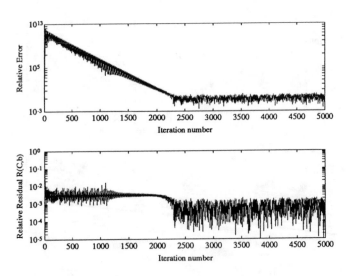

Successive iterations : $\nu = 2.96 \; 10^2$

Figure 3

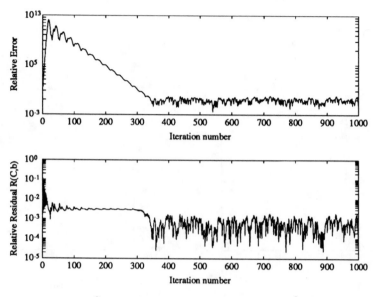

Successive iterations : $\nu = 3 \ 10^2$

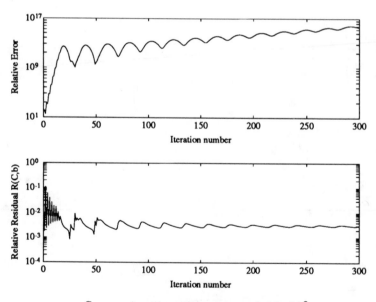

Successive iterations : $\nu = 3.16 \ 10^2$

Figure 4

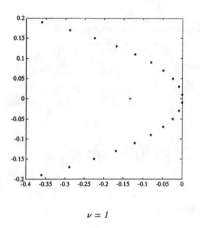

$\nu = 1$

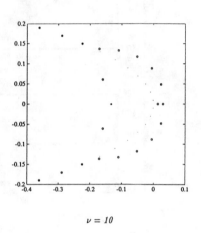

$\nu = 10$

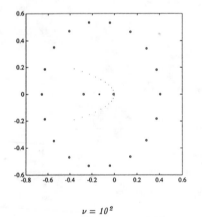

$\nu = 10^2$

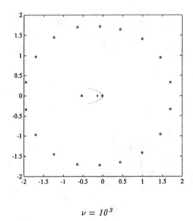

$\nu = 10^3$

Exact and computed spectra

Figure 5

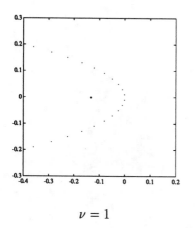

$$\nu = 1$$

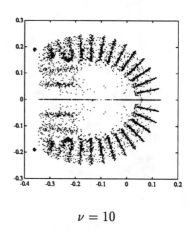

$$\nu = 10$$

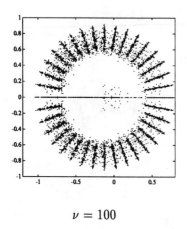

$$\nu = 100$$

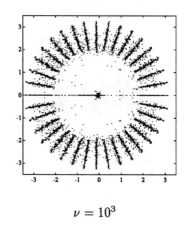

$$\nu = 10^3$$

Perturbed spectra

Figure 6

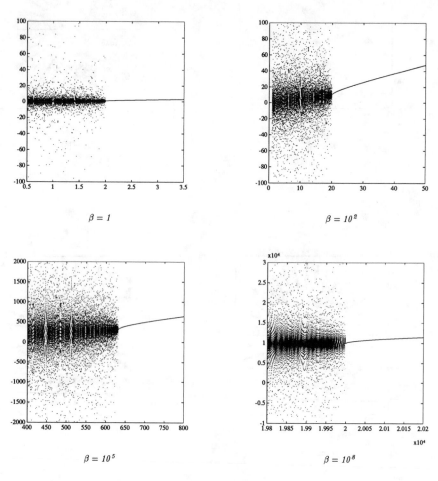

$\beta = 1$ $\beta = 10^2$

$\beta = 10^5$ $\beta = 10^8$

Iterations for computing α_i

Figure 7

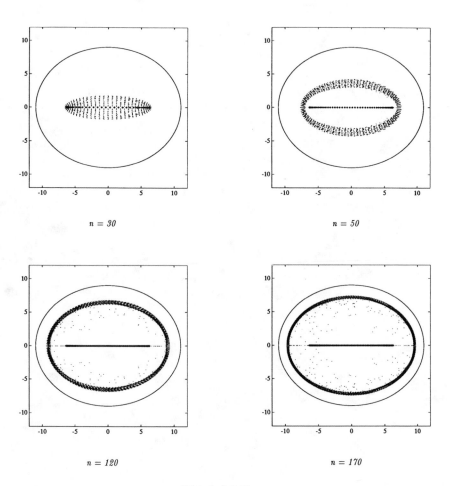

$n = 30$

$n = 50$

$n = 120$

$n = 170$

Perturbed spectra

Figure 8

COMPONENTWISE ERROR ANALYSIS FOR STATIONARY ITERATIVE METHODS

NICHOLAS J. HIGHAM* AND PHILIP A. KNIGHT†

Abstract. How small can a stationary iterative method for solving a linear system $Ax = b$ make the error and the residual in the presence of rounding errors? We give a componentwise error analysis that provides an answer to this question and we examine the implications for numerical stability. The Jacobi, Gauss-Seidel and successive over-relaxation methods are all found to be forward stable in a componentwise sense and backward stable in a normwise sense, provided certain conditions are satisfied that involve the matrix, its splitting, and the computed iterates. We show that the stronger property of componentwise backward stability can be achieved using one step of iterative refinement in fixed precision, under suitable assumptions.

Key words. stationary iteration, Jacobi method, Gauss-Seidel method, successive over-relaxation, error analysis, numerical stability.

AMS(MOS) subject classifications. primary 65F10, 65G05

1. Introduction. The effect of rounding errors on LU and QR factorization methods for solving linear systems is well understood. Various backward and forward error bounds are known that are informative and easy to interpret. Many of the bounds can be approximate equalities, to within factors depending on the problem dimension, and so they give a useful guide to the accuracy that can be attained in practice. (Note that it is still not well understood why the growth factor for LU factorization with partial pivoting is usually small, but progress towards explaining this phenomenon has been made recently [22].)

In contrast, there is little published error analysis for iterative methods. This is surprising, since iterative methods for computer solution of $Ax = b$ have at least as long a history as direct methods [26]. One reason for the paucity of error analysis may be that in many applications accuracy requirements are modest and are satisfied without difficulty. Nevertheless, we believe the following question is an important one, and in this work we attempt to answer it for a particular class of methods.

> How accurate a solution can we obtain using an iterative
> method in floating point arithmetic?

To be more precise, how small can we guarantee that the backward or forward error will be over all iterations $k = 1, 2, \ldots$? Without an answer to this question we cannot be sure that a convergence test of the form $\|b - A\widehat{x}_k\| \leq \epsilon$ (say) will ever be satisfied, for any given value of $\epsilon < \|b - Ax_0\|$!

---

* Nuffield Science Research Fellow. Department of Mathematics, University of Manchester, Manchester, M13 9PL, England (na.nhigham@na-net.ornl.gov).

† Supported by a SERC Research Studentship. Department of Mathematics, University of Manchester, Manchester, M13 9PL, England (mbbgppk@cms.mcc.ac.uk).

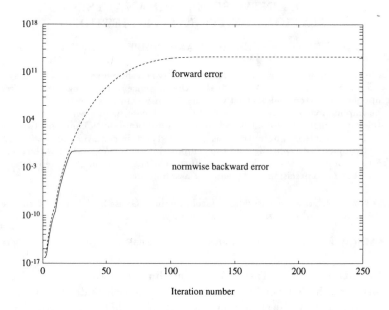

FIG. 1.1. *SOR iteration*

As an indication of the potentially devastating effects of rounding errors we present an example constructed and discussed by Hammarling and Wilkinson [11]. Here, A is the 100×100 lower bidiagonal matrix with $a_{ii} \equiv 1.5$ and $a_{i,i-1} \equiv 1$, and $b_i \equiv 2.5$. The successive over-relaxation (SOR) method is applied in Matlab with parameter $\omega = 1.5$, starting with the rounded version of the exact solution x, given by $x_i = 1 - (-2/3)^i$. The forward errors $\|\widehat{x}_k - x\|_\infty / \|x\|_\infty$ and the normwise backward errors $\eta_\infty(\widehat{x}_k)$ (defined in (1.1)) are plotted in Figure 1.1. The SOR method converges in exact arithmetic, since the iteration matrix has spectral radius $1/2$, but in the presence of rounding errors it diverges. The iterate $\widehat{x}_{238}$ has a largest element of order 10^{13}, $\widehat{x}_{k+2} \equiv \widehat{x}_k$ for $k \geq 238$, and for $k > 100$, $\widehat{x}_k(60{:}100) \approx (-1)^k \widehat{x}_{100}(60{:}100)$. The divergence is not a result of ill-conditioning of A, since $\kappa_\infty(A) \approx 5$. The reason for the initial rapid growth of the errors in this example is that the iteration matrix is far from normal; this allows the norms of the powers to become very large before they ultimately decay by a factor $\approx 1/2$ with each successive power. The effect of rounding errors in this example is to cause the forward error curve in Figure 1.1 to level off near $k = 100$, instead of decaying to zero as it would in exact arithmetic. More insight into the initial behaviour of the errors can be obtained using the notion of pseudo-eigenvalues [21].

To establish what we should try to prove, we review some normwise and componentwise backward error results and perturbation theory. If y is an approximate solution to $Ax = b$ then the *normwise (relative) backward error* is

$$\eta(y) = \min\{\epsilon : (A+\Delta A)y = b+\Delta b, \ \|\Delta A\| \leq \epsilon\|A\|, \ \|\Delta b\| \leq \epsilon\|b\|\},$$
(1.1)

where $\|\cdot\|$ denotes any vector norm and the corresponding subordinate matrix norm. Rigal and Gaches [20] show that $\eta(y)$ is given by the explicit formula

$$(1.2) \qquad\qquad \eta(y) = \frac{\|r\|}{\|A\|\|y\| + \|b\|}.$$

The forward error of y can be bounded using the standard perturbation result

$$(1.3) \qquad\qquad \frac{\|x - y\|}{\|x\|} \leq \kappa(A)\eta(y) + O(\eta(y)^2),$$

where $\kappa(A) = \|A\|\|A^{-1}\|$ is the matrix condition number.

The *componentwise (relative) backward error* is defined by

$$\omega(y) = \min\{\epsilon : (A + \Delta A)y = b + \Delta b, \ |\Delta A| \leq \epsilon|A|, \ |\Delta b| \leq \epsilon|b|\},$$

where the absolute values and inequalities are interpreted componentwise. This is a more stringent measure of backward error than the normwise measure, since each perturbation Δa_{ij} and Δb_i is measured relative to the entry it perturbs. The constraints ensure that $a_{ij} = 0 \Rightarrow \Delta a_{ij} = 0$, and similarly for b. This latter property is particularly attractive in the context of iterative solvers, where A is usually sparse and it may not be meaningful to perturb the zero entries of A [1]. Oettli and Prager [19] derive the convenient formula

$$\omega(y) = \max_i \frac{|b - Ay|_i}{(|A||y| + |b|)_i},$$

where $\xi/0$ is interpreted as zero if $\xi = 0$ and infinity otherwise.

A perturbation bound involving componentwise quantities is, for any monotonic norm[1],

$$
\begin{aligned}
\frac{\|y - x\|}{\|x\|} &\leq \frac{\| |A^{-1}|(|A||x| + |b|) \|}{\|x\|} \omega(y) + O(\omega(y)^2) \\
&\leq 2\mathrm{cond}(A, x)\omega(y) + O(\omega(y)^2),
\end{aligned}
$$
(1.4)

where $\mathrm{cond}(A, x) = \| |A^{-1}||A||x| \|/\|x\|$. The condition number $\mathrm{cond}(A, x)$ is independent of row scalings $A \rightarrow \mathrm{diag}(d_i)A$, and it satisfies $\mathrm{cond}_\infty(A, x) \leq \kappa_\infty(A)$, so the bound (1.4) is potentially much smaller than (1.3).

---

[1] A monotonic norm is one for which $|x| \leq |y| \Rightarrow \|x\| \leq \|y\|$ [16, p. 285].

Ideally a numerical method for solving $Ax = b$ will produce a computed solution $\widehat{x}$ that satisfies $\omega(\widehat{x}) = O(u)$, where u is the unit roundoff. Such a method is said to be *componentwise backward stable*. A method that is not componentwise backward stable may still satisfy a bound of the form

$$(1.5) \qquad \frac{\|\widehat{x} - x\|}{\|x\|} \leq c_n \operatorname{cond}(A, x)u + O(u^2),$$

where c_n is a constant; this is the same type of forward error bound as holds for a method that is componentwise backward stable. We will call a method for which (1.5) holds *componentwise forward stable*. Similarly, if $\eta(\widehat{x}) = O(u)$ we call the method *normwise backward stable*, and if (1.5) holds with $\operatorname{cond}(A, x)$ replaced by $\kappa(A)$ we call the method *normwise forward stable*.

In this paper we analyse stationary iterative methods: those that iterate according to $Mx_{k+1} = Nx_k + b$, using the splitting $A = M - N$. Even though this is a relatively straightforward class of iterative methods (as regards determining the rate of convergence in exact arithmetic, say), our error analysis is not as concise as that which can be done for LU factorization (see, e.g., [8, Sec. 3.3], [12]), and our conclusions are less clear-cut. It seems inherently more difficult to obtain useful rounding error bounds for iterative methods than it is for direct methods. An important feature of our analysis is that we use sharp inequalities wherever possible, so as to obtain the best possible bounds. Our results should be of use in guiding the choice of stopping criterion for an iterative solver; cf. the discussion in [2].

In section 2 we derive the basic recurrences for the error and residual. Forward error bounds are developed in section 3, and these are specialised to the Jacobi, Gauss-Seidel and SOR methods in section 4. We find that all three methods are componentwise forward stable if a certain product $c(A)\theta_x$ is not too large (and ω is not too close to zero for the SOR iteration). Here, $c(A) \geq 1$ (defined in (3.11)) depends on the matrix and the splitting, and $\theta_x \geq 1$ (defined in (3.8)) describes how "well-behaved" the iterates are. For each method, $c(A) = 1$ if A is an M-matrix (assuming $0 \leq \omega \leq 1$ for the SOR iteration), and $c(A)$ can be expected to be of modest size in many applications. This forward stability result is quite strong, for apart from triangular matrices and certain classes of tridiagonal matrices [14], the only class of matrices we know for which LU factorization is guaranteed to be componentwise forward stable is the class of totally nonnegative matrices [5].

In section 5 we derive bounds for the residual. We show that any stationary iterative method is normwise backward stable under conditions which include the requirement that the spectral radius of NM^{-1} is not too close to 1. Unfortunately, it does not seem possible to prove that a small componentwise backward error will be obtained under any reasonable assumptions. However, we explain why one step of iterative refinement in fixed precision does lead to a small componentwise backward error provided certain assumptions are satisfied.

We briefly survey existing error analyses for iterative methods. For symmetric positive definite systems, Golub [7] derives both statistical and non-statistical bounds for the forward error and residual of the Richardson method. Benschop and Ratz [3] give a statistical analysis of the effect of rounding errors on stationary iteration, under the assumption that the rounding errors are independent random variables with zero mean. Lynn [18] presents a statistical analysis for the SOR method with a symmetric positive definite matrix.

Hammarling and Wilkinson [11] give a normwise error analysis for the SOR method. With the aid of numerical examples, they emphasise that while it is the spectral radius of the iteration matrix $M^{-1}N$ that determines the asymptotic rate of convergence, it is the norms of the powers of this matrix that govern the behaviour of the iteration in the early stages. This point is also elucidated by Trefethen [21], using the tool of pseudospectra.

Dennis and Walker [6] obtain bounds for $\|x - \hat{x}_{k+1}\|/\|x - \hat{x}_k\|$ for stationary iteration as a special case of error analysis of quasi-Newton methods for nonlinear systems. The bounds in [6] do not readily yield information about normwise or componentwise forward stability.

Bollen [4] analyses the class of "descent methods" for solving $Ax = b$, where A is required to be symmetric positive definite; these are obtained by iteratively using exact line searches to minimize the quadratic function $F(x) = (A^{-1}b - x)^T A(A^{-1}b - x)$. The choice of search direction $p_k = b - Ax_k \equiv r_k$ yields the steepest descent method, while $p_k = e_j$ (unit vector), where $|r_k|_j = \|r_k\|_\infty$, gives the Gauss-Southwell method. Bollen shows that both these methods are normwise backward stable as long as a condition of the form $c_n \kappa(A)u < 1$ holds. If the p_k are cyclically chosen to be the unit vectors $e_1, e_2, \ldots, e_n$ then the Gauss-Seidel method results, but unfortunately no results specific to this method are given in [4].

Woźniakowski [23] shows that the Chebyshev semi-iterative method is normwise forward stable but not normwise backward stable. In [25] Woźniakowski analyses a class of conjugate gradient algorithms (which does not include the usual conjugate gradient method). He obtains a forward error bound proportional to $\kappa(A)^{3/2}$ and a residual bound proportional to $\kappa(A)$, from which neither backward nor forward normwise stability can be deduced. We note that as part of the analysis in [25] Woźniakowski obtains a residual bound for the steepest descent method that is proportional to $\kappa(A)$, and is therefore much weaker than the bound obtained by Bollen [4].

Greenbaum [9] presents a detailed error analysis of the conjugate gradient method, but her concern is with the rate of convergence rather than the attainable accuracy. An excellent survey of work concerned with the effects of rounding error on the conjugate gradient method (and the Lanczos method) is given in the introduction of [10].

The work most closely related to ours is [24], wherein Woźniakowski gives a normwise error analysis of stationary iterative methods. Some of the assumptions in [24] are difficult to justify, as we explain in section 3.

Finally, we mention that extension of the analysis reported here to singular systems is being investigated and will be reported elsewhere.

2. Basic Equations. We are concerned with the iteration

$$M x_{k+1} = N x_k + b,$$

where $A = M - N \in \mathbb{R}^{n \times n}$ is nonsingular, and M is nonsingular. We assume throughout that the spectral radius $\rho(M^{-1}N) < 1$, so that in exact arithmetic the iteration converges for any starting vector. We assume that x_{k+1} is computed by forming $N x_k + b$ and then solving a linear system with M. The computed vectors $\widehat{x}_k$ therefore satisfy

$$(2.1) \qquad (M + \Delta M_{k+1})\widehat{x}_{k+1} = N\widehat{x}_k + b + f_k,$$

where f_k is the error in forming $fl(N\widehat{x}_k + b)$ and ΔM_{k+1} represents the error in solving the linear system. We will use the following model of floating point arithmetic, where u is the unit roundoff:

$$(2.2) \qquad \begin{matrix} fl(x \pm y) & = & x(1+\alpha) \pm y(1+\beta), \\ fl(x \operatorname{op} y) & = & (x \operatorname{op} y)(1+\delta), \end{matrix} \quad \begin{matrix} |\alpha|, |\beta| \le u, \\ |\delta| \le u, \quad \operatorname{op} = *, /. \end{matrix}$$

This model is valid for machines that do not use a guard digit in addition/subtraction. Under this model we have the standard result that

$$(2.3) \qquad |f_k| \le c_n u(|N||\widehat{x}_k| + |b|),$$

where c_n is a constant of order n. If N has at most m nonzeros per row then c_n can be taken to be of order m, but sparsity does not otherwise affect the analysis below.

We will assume that

$$(2.4) \qquad |\Delta M_{k+1}| \le c'_n u |M|.$$

This is valid if M is triangular (see, e.g., [13]), which is the case for the Jacobi, Gauss-Seidel and SOR iterations. It also holds if M is (a) totally nonnegative [5] or (b) tridiagonal and either symmetric positive definite, an M-matrix, or diagonally dominant [14], assuming in both cases (a) and (b) that Gaussian elimination without pivoting is used.

We now derive recurrences for the residuals $r_k = b - A\widehat{x}_k$ and errors $e_k = x - \widehat{x}_k$. Using (2.1),

$$(2.5) \qquad \begin{matrix} r_{k+1} & = & b - (M - N)\widehat{x}_{k+1} \\ & = & \Delta M_{k+1}\widehat{x}_{k+1} - N(\widehat{x}_k - \widehat{x}_{k+1}) - f_k. \end{matrix}$$

From (2.1) again,

$$r_k = b - (M - N)\widehat{x}_k$$
$$= -M\widehat{x}_k + (M + \Delta M_{k+1})\widehat{x}_{k+1} - f_k,$$

so that

$$\widehat{x}_k - \widehat{x}_{k+1} = M^{-1}(\Delta M_{k+1}\widehat{x}_{k+1} - f_k - r_k).$$

Substituting into (2.5) gives

(2.6) $$r_{k+1} = NM^{-1}r_k + (I - NM^{-1})(\Delta M_{k+1}\widehat{x}_{k+1} - f_k),$$

which is our recurrence for the residuals. Since $e_k = A^{-1}r_k$, we have from (2.6)

$$e_{k+1} = A^{-1}NM^{-1}Ae_k + A^{-1}(I - NM^{-1})(\Delta M_{k+1}\widehat{x}_{k+1} - f_k).$$

Simple manipulation shows that $A^{-1}NM^{-1}A = M^{-1}N$ and $A^{-1}(I - NM^{-1}) = M^{-1}$, so the recurrence for the errors is

(2.7) $$e_{k+1} = M^{-1}Ne_k + M^{-1}(\Delta M_{k+1}\widehat{x}_{k+1} - f_k).$$

The following lemmas are needed in the next section.

LEMMA 2.1. *(a) If $|B| \leq E \in \mathbb{R}^{n \times n}$ then $|Bx| \leq E|x|$, with equality for some B with $|B| = E$.*

(b) If $B_j \in \mathbb{R}^{n \times n}$ and $|x_j| \leq h_j$, $j = 0, \ldots, m$, then $\sum_{j=0}^{m} B_j x_j \leq \sum_{j=0}^{m} |B_j| h_j$, and there is equality in the ith component for some $x_0, \ldots, x_m$ with $|x_j| = h_j$. In particular,

$$\| \sum_{j=0}^{m} B_j x_j \|_\infty \leq \| \sum_{j=0}^{m} |B_j| h_j \|_\infty,$$

and equality is attainable.

Proof. (a) The inequality is straightforward. Equality is obtained when $b_{ij} = \text{sign}(x_j)e_{ij}$. (b) The vector inequality is straightforward. Equality is obtained in the ith component for $x_j(k) = \text{sign}(b_{ik}^{(j)})h_j(k)$, where $B_j = (b_{ik}^{(j)})$. The norm results follows easily. ∎

LEMMA 2.2. *If $B \in \mathbb{R}^{n \times n}$ and $\rho(B) < 1$, then the series $\sum_{k=0}^{\infty} |B^k|$ and $\sum_{k=0}^{\infty} \|B^k\|$ are both convergent, where $\| \cdot \|$ is any consistent norm.*

Proof. Since $\rho(B) < 1$, a standard result [16, Lemma 5.6.10] guarantees the existence of a norm $\| \cdot \|_\rho$ for which $\|B\|_\rho < 1$. The series $\sum_{k=0}^{\infty} \|B^k\|_\rho \leq \sum_{k=0}^{\infty} \|B\|_\rho^k = (1 - \|B\|_\rho)^{-1}$ is clearly convergent, and so, by the equivalence of norms, $\sum_{k=0}^{\infty} \|B^k\|$ is convergent for any norm.

Since $(|B^k|)_{ij} \leq \|B^k\|_\infty$, the convergence of $\sum_{k=0}^{\infty} \|B^k\|_\infty$ ensures that of $\sum_{k=0}^{\infty} |B^k|$. (The convergence of $\sum_{k=0}^{\infty} |B^k|$ can also be proved directly using the Jordan canonical form.) ∎

3. Forward Error Analysis. The basic equation from which we work is (2.7), which we write as

$$(3.1) \qquad e_{k+1} = Ge_k + M^{-1}\xi_k,$$

where $G = M^{-1}N$ and

$$(3.2) \qquad \xi_k = \Delta M_{k+1}\widehat{x}_{k+1} - f_k.$$

A componentwise bound for ξ_k is, from (2.3) and (2.4),

$$(3.3) \qquad |\xi_k| \le d_n u(|M||\widehat{x}_{k+1}| + |N||\widehat{x}_k| + |b|) \equiv \mu_k,$$

where $d_n = \max(c_n, c_n')$, and this bound is sharp modulo the multiplicative constant d_n, as we now explain. Note that each element of M, N and b takes part in at least one floating point operation. Using Lemma 2.1 (a) it is easy to see that if we associate with each element of M, N and b a single rounding error of maximal modulus and the appropriate sign, equality is attained in (2.3), (2.4) and (3.3), modulo the multiplicative constants c_n, c_n' and d_n. In other words, there exists a set of rounding errors for which there is equality in (3.3) if we ignore the constant d_n.

Now we return to the recurrence (3.1), which has the solution

$$e_{m+1} = G^{m+1}e_0 + \sum_{k=0}^{m} G^k M^{-1}\xi_{m-k}.$$

The first term, $G^{m+1}e_0$, is the error of the iteration in exact arithmetic. This term tends to zero as $m \to \infty$, since $\rho(G) < 1$, so Lemma 2.1 (b) shows that each component of the following inequality is sharp for large m:

$$(3.4) \qquad |e_{m+1}| \le |G^{m+1}e_0| + \sum_{k=0}^{m} |G^k M^{-1}|\mu_{m-k},$$

where μ_k is the sharp bound for ξ_k defined in (3.3). As $m \to \infty$ the accuracy that can be guaranteed by the analysis is determined by the last term in (3.4), and it is this term on which the rest of the analysis focuses.

At this point we can proceed by using further componentwise inequalities or by using norms. First, we consider the norm approach. In place of (3.4) we use the normwise bound

$$(3.5) \qquad \|e_{m+1}\|_\infty \le \|G^{m+1}e_0\|_\infty + \|\sum_{k=0}^{m} |G^k M^{-1}|\mu_{m-k}\|_\infty,$$

which is sharp for large m, in view of Lemma 2.1 (b). Defining

$$\gamma_x = \sup_k \frac{\|\widehat{x}_k\|_\infty}{\|x\|_\infty},$$

we obtain from (3.5) and (3.3)

$$\|e_{m+1}\|_\infty \ \leq \ \|G^{m+1}e_0\|_\infty + \max_{0 \leq k \leq m} \|\mu_k\|_\infty \sum_{k=0}^{m} \|G^k M^{-1}\|_\infty$$

$$\leq \ \|G^{m+1}e_0\|_\infty + d_n u(1 + \gamma_x)(\|M\|_\infty$$

$$(3.6) \qquad\qquad + \|N\|_\infty)\|x\|_\infty \sum_{k=0}^{\infty} \|G^k M^{-1}\|_\infty,$$

where the existence of the sum is assured by Lemma 2.2, since $\rho(G) < 1$. This is similar to a result in the analysis of Woźniakowski [24] (see (3.11) and (3.12) therein). Woźniakowski's result has $\|x\|$ in place of $\sup_k \|\widehat{x}_k\|$ (i.e., $\gamma_x \equiv 1$), which we believe is erroneous, despite an extra "$+O(u^2)$" term. (Like (3.6), the bounds for the SOR iteration in [11] also contain a factor $\sup_k \|\widehat{x}_k\|$.) Also, Woźniakowski's bound contains $\|G^k\|$ rather than $\|G^k M^{-1}\|$; this is because he assumes $\widehat{x}_{k+1}$ is computed as

$$\widehat{x}_{k+1} \ = \ fl(H\widehat{x}_k + h) = (H + \Delta H)\widehat{x}_k + h + \Delta h, \quad \|\Delta H\| \leq c_n u\|H\|,$$

$$(3.7) \qquad\qquad \|\Delta h\| \leq c'_n u\|h\|.$$

Comparison with (2.1)–(2.4) shows that (3.7) is not valid if the iteration is implemented in the natural way described in section 2. Woźniakowski assumes that the vectors ξ_k (defined slightly differently in his analysis) are arbitrary subject to a bound on $\|\xi_k\|$, and by taking the ξ_k to be eigenvectors of G he shows that his bound analogous to (3.6) is sharp when G is symmetric. However, under our assumptions on how x_{k+1} is computed, ξ_k satisfies (3.3), and the structure imposed by this inequality may preclude ξ_k from being an eigenvector of G. (Moreover, G is almost always unsymmetric for the SOR method). The bound for $\|e_{m+1}\|_\infty$ in (3.5) cannot be weakened without losing the sharpness.

If $\|G\|_\infty = \|M^{-1}N\|_\infty = q < 1$ then (3.6) yields

$$\|e_{m+1}\|_\infty \leq \|G^{m+1}e_0\|_\infty + d_n u(1 + \gamma_x)(\|M\|_\infty + \|N\|_\infty)\|x\|_\infty \frac{\|M^{-1}\|_\infty}{1 - q}.$$

Thus if q is not too close to 1 ($q \leq 0.9$, say), and γ_x and $\|M^{-1}\|_\infty$ are not too large, this bound guarantees a small forward error.

Of more interest for us is the following componentwise development of (3.4). Defining

$$(3.8) \qquad\qquad \theta_x = \sup_k \max_{1 \leq i \leq n} \left(\frac{|\widehat{x}_k|_i}{|x_i|} \right),$$

so that $|\widehat{x}_k| \leq \theta_x |x|$ for all k, we have from (3.3),

$$(3.9) \qquad\qquad |\mu_k| \leq d_n u(1 + \theta_x)(|M| + |N|)|x|.$$

Hence (3.4) yields

$$(3.10) \quad |e_{m+1}| \leq |G^{m+1}e_0| + d_n u(1 + \theta_x)\left(\sum_{k=0}^{\infty} |G^k M^{-1}|\right)(|M| + |N|)|x|.$$

Again, the existence of the sum is assured by Lemma 2.2, since $\rho(G) = \rho(M^{-1}N) < 1$. Since $A = M - N = M(I - M^{-1}N)$ we have

$$A^{-1} = \left(\sum_{k=0}^{\infty}(M^{-1}N)^k\right)M^{-1}.$$

The sum in (3.10) is clearly an upper bound for $|A^{-1}|$. Defining $c(A) \geq 1$ by

$$c(A) = \min\left\{\epsilon : \sum_{k=0}^{\infty}|(M^{-1}N)^k M^{-1}| \leq \epsilon \left|\sum_{k=0}^{\infty}(M^{-1}N)^k M^{-1}\right| = \epsilon|A^{-1}|\right\},$$

(3.11)

we have our final bound

$$(3.12) \quad |e_{m+1}| \leq |G^{m+1}e_0| + d_n u(1 + \theta_x)c(A)|A^{-1}|(|M| + |N|)|x|.$$

Unlike (3.4), this bound is not sharp in general, but it is optimal in the sense that it is the smallest bound that can be used to assess the componentwise forward stability.

An interesting feature of stationary iteration methods is that if the elements of M and N are linear combinations of those of A, then any scaling of the form $Ax = b \rightarrow D_1 A D_2 \cdot D_2^{-1}x = D_1^{-1}b$ (D_i diagonal) leaves the eigenvalues of $M^{-1}N$ unchanged; hence the asymptotic convergence rate is independent of row and column scaling. This scale independence applies to the Jacobi and SOR iterations, but not, for example, to the stationary Richardson iteration, for which $M = I$. One of the benefits of doing a componentwise analysis is that under the above assumptions on M and N the bound (3.12) largely shares the scale independence. In (3.12) the scalar $c(A)$ is independent of the row and column scaling of A, and the term $|A^{-1}|(|M| + |N|)|x|$ scales in the same way as x. Furthermore, θ_x can be expected to depend only mildly on the row and column scaling, because the bounds in (2.3) and (2.4) for the rounding error terms have the correct scaling properties.

What can be said about $c(A)$? In general, it can be arbitrarily large. Indeed, $c(A)$ is infinite for the Jacobi and Gauss-Seidel iterations for any $n \geq 3$ if A is the symmetric positive definite matrix with $a_{ij} = \min(i, j)$, because A^{-1} is tridiagonal and $(M^{-1}N)^k M^{-1}$ is not.

If M^{-1} and $M^{-1}N$ both have nonnegative elements then $c(A) = 1$; as we will see in the next section, this condition holds in some important instances.

Some further insight into $c(A)$ can be obtained by examining the cases where $M^{-1}N$ has rank 1 or is diagonal. The rank 1 case is motivated by the fact that if $B \in \mathbb{R}^{n \times n}$ has a unique eigenvalue λ of largest modulus then $B^k \approx \lambda^k xy^T$, where $Bx = \lambda x$ and $y^T B = \lambda y^T$ with $y^T x = 1$. If we set $(M^{-1}N)^k M^{-1} \equiv \lambda^k xy^T M^{-1} = \lambda^k xz^T$, where $|\lambda| < 1$, then

$$c(A) = \min\{\epsilon : \sum_{k=0}^{\infty} |\lambda^k xz^T| \le \epsilon |\sum_{k=0}^{\infty} \lambda^k xz^T|\},$$

or $\sum_{k=0}^{\infty} |\lambda^k| = c(A)|\sum_{k=0}^{\infty} \lambda^k|$, that is, $(1-|\lambda|)^{-1} = c(A)(1-\lambda)^{-1}$. Hence $c(A) = (1-\lambda)/(1-|\lambda|)$, and so $c(A)$ is of modest size unless $M^{-1}N$ has an eigenvalue close to -1. If $M^{-1}N \in \mathbb{C}^{n \times n}$ is diagonal with eigenvalues λ_i then it is easy to show that $c(A) = \max_i |1 - \lambda_i|/(1-|\lambda_i|)$, so $c(A)$ can be large only if $\rho(M^{-1}N)$ is close to 1. Although $M^{-1}N$ cannot be diagonal for the Jacobi or Gauss-Seidel methods, this formula can be taken as being indicative of the size of $c(A)$ when $M^{-1}N$ is diagonalizable with a well-conditioned matrix of eigenvectors. These considerations suggest the heuristic inequality, for general A,

$$(3.13) \qquad c(A) \ge \max_i \frac{|1 - \lambda_i|}{1 - |\lambda_i|}, \qquad \lambda_i = \lambda_i(M^{-1}N).$$

In practical problems where stationary iteration is used we would expect $c(A)$ to be of modest size ($O(n)$, say), for two reasons. First, to achieve a reasonable convergence rate $\rho(M^{-1}N)$ has to be safely less than 1, which implies that the heuristic lower bound (3.13) for $c(A)$ is not too large. Second, even if A is sparse, A^{-1} will usually be full, and so there are unlikely to be zeros on the right-hand side of (3.11). (Such zeros are dangerous because they can make $c(A)$ infinite.)

Note that in (3.12) the only terms that depend on the history of the iteration are $|G^{m+1}e_0|$ and θ_x. In using this bound we can redefine x_0 to be any iterate $\hat{x}_k$, thereby possibly reducing θ_x. This is a circular argument if used to obtain a priori bounds, but it does suggest that the potentially large θ_x term will generally be innocuous. Note that if $x_i = 0$ for some i then θ_x is infinite unless $(\hat{x}_k)_i = 0$ for all k. This difficulty with zero components of x can usually be overcome by redefining

$$\theta_x = \sup_k \max_{1 \le i \le n} \frac{((|M| + |N|)|\hat{x}_k|)_i}{((|M| + |N|)|x|)_i},$$

for which the above bounds remain valid if θ_x is replaced by $2\theta_x$.

Finally, we note that (3.12) implies

$$(3.14) \quad \|e_{m+1}\|_\infty \le \|G^{m+1}e_0\|_\infty + d_n u(1+\theta_x)c(A)\| |A^{-1}|(|M| + |N|)|x| \|_\infty.$$

If $\theta_x c(A) = O(1)$ and $|M| + |N| \le \alpha |A|$, with $\alpha = O(1)$, this bound is of the form (1.5) as $m \to \infty$ and we have componentwise forward stability.

TABLE 4.1
Jacobi method, $a = 1/2 - 8^{-j}$

	$\rho(M^{-1}N)$	Iters.	$\text{cond}_\infty(A, x)$	$\min_k \phi_\infty(\widehat{x}_k)$	$\min_k \eta_\infty(\widehat{x}_k)$
$j = 1$	0.75	90	3.40	2.22e-16	1.27e-16
$j = 2$	0.97	352	4.76	1.78e-15	9.02e-16
$j = 3$	0.996	1974	4.97	1.42e-14	7.12e-15
$j = 4$	1.00	11226	5.00	1.14e-13	5.69e-14
$j = 5$	1.00	55412	5.00	9.10e-13	4.55e-13

4. Forward Error Bounds for Specific Methods.
In this section we specialize the forward error bound (3.14) of the previous section to the Jacobi, Gauss-Seidel and SOR iterations.

4.1. Jacobi's Method. For the Jacobi iteration $M = D = \text{diag}(A)$ and $N = \text{diag}(A) - A$. Hence $|M| + |N| = |M - N| = |A|$, and so (3.14) yields

$$(4.1) \quad \|e_{m+1}\|_\infty \leq \|G^{m+1}e_0\|_\infty + d_n u(1 + \theta_x)c(A)\| |A^{-1}||A||x| \|_\infty.$$

If A is an M-matrix then $M^{-1} \geq 0$ and $M^{-1}N \geq 0$, so $c(A) = 1$. Hence in this case we have componentwise forward stability as $m \to \infty$ if θ_x is suitably bounded.

Woźniakowski [24, Example 4.1] cites the symmetric positive definite matrix

$$A = \begin{bmatrix} 1 & a & a \\ a & 1 & a \\ a & a & 1 \end{bmatrix}, \quad 0 < a < \frac{1}{2}, \quad \kappa_2(A) = \frac{1 + 2a}{1 - a}, \quad \rho(M^{-1}N) = 2a,$$

as an example where the Jacobi method can be unstable, in the sense that there exist rounding errors such that no iterate has a relative error bounded by $c_n \kappa_\infty(A)u$. It is interesting to compare our bound (4.1) with the observed errors for this example. Straightforward manipulation shows that if $a = 1/2 - \epsilon$ ($\epsilon > 0$), then $c(A) \approx (3\epsilon)^{-1}$, so $c(A) \to \infty$ as $\epsilon \to 0$. (The heuristic lower bound (3.13) is $\approx 3(2\epsilon)^{-1}$ in this case.) Therefore (4.1) suggests that the Jacobi iteration can be unstable for this matrix. To confirm the instability we applied the Jacobi method to the problem with $x = (1, 1, \ldots, 1)^T$ and $a = 1/2 - 8^{-j}$, $j = 1:5$. We used Matlab, for which the unit roundoff $u \approx 1.1 \times 10^{-16}$, and we took a random x_0 with $\|x - x_0\|_2 = 10^{-10}$. The iteration was terminated when there was no decrease in the norm of the residual for 50 consecutive iterations. Table 4.1 reports the smallest value of $\phi_\infty(\widehat{x}_k) = \|x - \widehat{x}_k\|_\infty / \|x\|_\infty$ over all iterations, for each j; the number of iterations is shown in the column "Iters."

The ratio $\min_k \phi_\infty(\widehat{x}_k)_{j+1} / \min_k \phi_\infty(\widehat{x}_k)_j$ takes the values 8.02, 7.98, 8.02, 7.98 for $j = 1:4$, showing excellent agreement with the behaviour

TABLE 4.2
Jacobi method, $a = -(1/2 - 8^{-j})$

	$\rho(M^{-1}N)$	Iters.	$\text{cond}_\infty(A, x)$	$\min_k \phi_\infty(\widehat{x}_k)$	$\min_k \eta_\infty(\widehat{x}_k)$
$j = 1$	0.75	39	7.00	4.44e-16	5.55e-17
$j = 2$	0.97	273	6.30e1	4.88e-15	7.63e-17
$j = 3$	0.996	1662	5.11e2	4.22e-14	8.24e-17
$j = 4$	1.00	9051	4.09e3	3.41e-13	8.32e-17
$j = 5$	1.00	38294	3.28e4	2.73e-12	8.33e-17

predicted by (4.1), since $c(A) \approx 8^j/3$. Moreover, $\theta_x \approx 1$ in these tests and setting $d_n \approx 1$ the bound (4.1) is at most a factor 13.3 larger than the observed error, for each j.

If $-1/2 < a < 0$ then A is an M-matrix and $c(A) = 1$. The bound (4.1) shows that if we set $a = -(1/2 - 8^{-j})$ and repeat the above experiment then the Jacobi method will perform in a componentwise forward stable manner (clearly, $\theta_x \approx 1$ is to be expected). We carried out the modified experiment, obtaining the results shown in Table 4.2. All the $\min_k \phi_\infty(\widehat{x}_k)_j$ values are less than $\text{cond}_\infty(A, x)u$, so the Jacobi iteration is indeed componentwise forward stable in this case. Note that since $\rho(M^{-1}N)$ and $\|M^{-1}N\|_2$ take the same values for a and $-a$, the usual rate of convergence measures cannot distinguish between these two examples.

4.2. Successive Over-Relaxation. The SOR method can be written in the form $Mx_{k+1} = Nx_k + b$, where

$$M = \frac{1}{\omega}(D + \omega L), \quad N = \frac{1}{\omega}((1 - \omega)D - \omega U),$$

and where $A = D + L + U$, with L and U strictly lower triangular and upper triangular, respectively. The matrix $|M| + |N|$ agrees with $|A|$ everywhere except, possibly, on the diagonal, and the best possible componentwise inequality between these two matrices is

$$(4.2) \qquad |M| + |N| \leq \frac{1 + |1 - \omega|}{\omega}|A| \equiv f(\omega)|A|.$$

Note that $f(\omega) = 1$ for $1 \leq \omega \leq 2$, and $f(\omega) \to \infty$ as $\omega \to 0$. From (3.14) we have

$$\|e_{m+1}\|_\infty \leq \|G^{m+1}e_0\|_\infty + d_n u(1 + \theta_x)c(A)f(\omega)\| |A^{-1}||A||x| \|_\infty.$$

If A is an M-matrix and $0 \leq \omega \leq 1$ then $M^{-1} \geq 0$ and $M^{-1}N \geq 0$, so $c(A) = 1$. The Gauss-Seidel method corresponds to $\omega = 1$, and it is interesting to note that for this method the above results have exactly the same form as those for the Jacobi method (though $c(A)$ and θ_x are, of course, different for the two methods).

5. Backward Error Analysis. In this section we obtain bounds for the residual vector $r_k = b - A\widehat{x}_k$. We write (2.6) as

$$r_{k+1} = Hr_k + (I - H)\xi_k,$$

where $H = NM^{-1}$ and ξ_k is defined and bounded in (3.2) and (3.3). This recurrence has the solution

(5.1) $$r_{m+1} = H^{m+1}r_0 + \sum_{k=0}^{m} H^k(I - H)\xi_{m-k}.$$

Using the same reasoning as in the derivation of (3.5), we obtain the sharp bound

$$\|r_{m+1}\|_\infty \leq \|H^{m+1}r_0\|_\infty + \|\sum_{k=0}^{m} |H^k(I - H)|\,\mu_{m-k}\,\|_\infty,$$

where μ_{m-k} is defined in (3.3). Taking norms in the second term gives, similarly to (3.6),

(5.2) $\|r_{m+1}\|_\infty \leq \|H^{m+1}r_0\|_\infty + d_n u(1+\gamma_x)(\|M\|_\infty + \|N\|_\infty)\|\overline{H}\|_\infty\|x\|_\infty,$

where

$$\overline{H} = \sum_{k=0}^{\infty} |H^k(I - H)|.$$

The following bound shows that $\|\overline{H}\|_\infty$ is small if $\|H\|_\infty = q < 1$, with q not too close to 1:

$$\|\overline{H}\|_\infty \leq \|I - H\|_\infty \sum_{k=0}^{\infty} \|H\|_\infty^k = \frac{\|I - H\|_\infty}{1 - q}.$$

A potentially much smaller bound can be obtained under the assumption that H is diagonalisable. If $H = XDX^{-1}$, with $D = \mathrm{diag}(\lambda_i)$, then

$$\begin{aligned}
\overline{H} &= \sum_{k=0}^{\infty} |X(D^k - D^{k+1})X^{-1}| \\
&\leq |X|\Big(\sum_{k=0}^{\infty} \mathrm{diag}(|1 - \lambda_i||\lambda_i^k|)\Big)|X^{-1}| \\
&= |X|\mathrm{diag}\Big(\frac{|1 - \lambda_i|}{1 - |\lambda_i|}\Big)|X^{-1}|.
\end{aligned}$$

Hence

(5.3) $$\|\overline{H}\|_\infty \leq \kappa_\infty(X)\max_i \frac{|1 - \lambda_i|}{1 - |\lambda_i|}.$$

Note that $\lambda_i = \lambda_i(H) = \lambda_i(NM^{-1}) = \lambda_i(M^{-1}N)$, so we see the reappearance of the term in the heuristic bound (3.13). The bound (5.3) is of modest size if the eigenproblem for H is well-conditioned ($\kappa_\infty(X)$ is small) and $\rho(H)$ is not too close to 1. Note that real eigenvalues of H near $+1$ do not affect the bound for $\|\overline{H}\|_\infty$, even though they may cause slow convergence.

To summarise, (5.2) and (1.2) show that for large m the normwise backward error $\eta_\infty(\widehat{x}_m)$ is certainly no larger than

$$d_n u(1 + \gamma_x)\left(\frac{\|M\|_\infty + \|N\|_\infty}{\|A\|_\infty}\right)\|\overline{H}\|_\infty.$$

Note that $\|M\|_\infty + \|N\|_\infty \le 2\|A\|_\infty$ for the Jacobi and Gauss-Seidel methods, and also for the SOR method if $\omega \ge 1$.

To investigate the componentwise backward error we use the bound, from (5.1),

$$|r_{m+1}| \le |H^{m+1}r_0| + \sum_{k=0}^{m} |H^k(I - H)|\mu_{m-k}.$$

Lemma 2.1 (b) shows that there can be equality in any component of this inequality, modulo the term $|H^{m+1}r_0|$. With θ_x defined as in (3.8) we have, using (3.9),

$$(5.4) \qquad |r_{m+1}| \le |H^{m+1}r_0| + d_n u(1 + \theta_x)\overline{H}(|M| + |N|)|x|.$$

To bound the componentwise backward error we need an inequality of the form

$$|\overline{H}| \le \overline{c}(A)\left|\sum_{k=0}^{\infty} H^k(I - H)\right| = \overline{c}(A)I.$$

Unfortunately, no such bound holds, since $\overline{H}$ has nonzero off-diagonal elements unless $H = NM^{-1}$ is diagonal. It therefore does not seem possible to obtain a useful bound for the componentwise backward error. However, the bound (5.4) does have one interesting implication, as we now explain.

Consider any linear equation solver that provides a computed solution $\widehat{x}$ to $Ax = b$ that satisfies

$$|b - A\widehat{x}| \le uE|\widehat{x}|,$$

where E is a nonnegative matrix depending on any or all of A, b, n and u (with $E = O(1)$ as $u \to 0$). Suppose one step of iterative refinement is done in fixed precision; thus, $r = b - A\widehat{x}$ is formed (in the working precision), $Ad = r$ is solved, and the update $y = \widehat{x} + d$ is computed. Theorem 2.1 of [15] shows that $\widehat{y}$ satisfies

$$(5.5) \qquad |b - A\widehat{y}| \le c_n(|A||\widehat{y}| + |b|) + O(u^2).$$

Theorem 2.2 of [15] shows further that if

$$E = \overline{E}|A|, \text{ and } 2|||A||A^{-1}|||_\infty \sigma(A, \widehat{y})(||\overline{E}||_\infty + n + 2)^2/(n+1) < u^{-1},$$

where $\sigma(A, x) = \max_i(|A||x|)_i / \min_i(|A||x|)_i$, then

$$(5.6) \qquad\qquad |b - A\widehat{y}| \le 2(n+2)u|A||\widehat{y}|.$$

The bound (5.5) shows that iterative refinement relegates E to the second order term, yielding asymptotic componentwise stability; but, since it is only an asymptotic result no firm conclusion can be drawn about the stability. Under further assumptions (that A is not too ill-conditioned, and that the components of $|A||\widehat{y}|$ do not vary too much in magnitude) (5.6) shows that the componentwise backward error is indeed small.

If m is so large that $H^{m+1}r_0$ can ignored in (5.4), then we can apply these results on fixed precision iterative refinement with $E = d_n(1 + \theta_x)\overline{H}(|M| + |N|)$ and $\overline{E} = \gamma d_n(1 + \theta_x)|\overline{H}|$, where $\gamma = 1$ for the Jacobi and Gauss-Seidel methods, and $\gamma = f(\omega)$ in (4.2) for the SOR method. We conclude that even though the basic iterative methods are not guaranteed to produce a small componentwise backward error, one step of iterative refinement in fixed precision is enough to achieve this desirable property, provided the iteration and the problem are both sufficiently "well-behaved" for the given data A and b. When A and b are sparse $\sigma(A, x)$ can be very large; see [1] for a discussion of the performance of iterative refinement in this case. Of course, in the context of iterative solvers, iterative refinement may be unattractive because the solution of $Ad = r$ in the refinement step will in general be as expensive as the computation of the original solution $\widehat{x}$. We note that iterative refinement is identified as a means of improving the *normwise* backward error for iterative methods in [17].

To conclude, we return to our numerical examples. For the SOR example in section 1, $c(A) = O(10^{45})$ and $||\overline{H}||_\infty = O(10^{30})$, so our error bounds for this problem are all extremely large. In this problem $\max_i |1 - \lambda_i|/(1 - |\lambda_i|) = 3$, where $\lambda_i = \lambda_i(M^{-1}N)$, so (3.13) is very weak; (5.3) is not applicable since $M^{-1}N$ is defective.

For the first numerical example in section 4.1, Table 4.1 reports the minimum normwise backward errors $\eta_\infty(\widehat{x}_k)$. For this problem it is straightforward to show that $||\overline{H}||_\infty = (1 - \epsilon)/\epsilon = 8^j(1 - 8^{-j})$. The ratios of backward errors for successive value of j are 7.10, 7.89, 7.99, 8.00, so we see excellent agreement with the behaviour predicted by the bounds. Table 4.2 reports the normwise backward errors for the second numerical example in section 4.1. The backward errors are all less than u, which again is close to what the bounds predict, since it can be shown that $||\overline{H}||_\infty \le 23/3$ for $-1/2 \le a \le 0$. In both the examples of section 4.1 the componentwise backward error $\omega(\widehat{x}_k) \approx \eta_\infty(\widehat{x}_k)$, and in our practical experience this behaviour is typical for the Jacobi and SOR iterations.

Acknowledgements. We thank Des Higham for his valuable comments on the manuscript, Sven Hammarling for providing us with a copy of [11], and Gene Golub for pointing out [18].

REFERENCES

[1] M. Arioli, J.W. Demmel and I.S. Duff, Solving sparse linear systems with sparse backward error, SIAM J. Matrix Anal. Appl., 10 (1989), pp. 165–190.

[2] M. Arioli, I.S. Duff and D. Ruiz, Stopping criteria for iterative solvers, SIAM J. Matrix Anal. Appl., 13 (1992), pp. 138–144.

[3] N.F. Benschop and H.C. Ratz, A mean square estimate of the generated roundoff error in constant matrix iterative processes, J. Assoc. Comput. Mach., 18 (1971), pp. 48–62.

[4] J.A.M. Bollen, Numerical stability of descent methods for solving linear equations, Numer. Math., 43 (1984), pp. 361–377.

[5] C. de Boor and A. Pinkus, Backward error analysis for totally positive linear systems, Numer. Math., 27 (1977), pp. 485–490.

[6] J.E. Dennis, Jr., and H.F. Walker, Inaccuracy in quasi-Newton methods: Local improvement theorems, Math. Prog. Study, 22 (1984), pp. 70–85.

[7] G.H. Golub, Bounds for the round-off errors in the Richardson second order method, BIT, 2 (1962), pp. 212–223.

[8] G.H. Golub and C.F. Van Loan, Matrix Computations, Second Edition, Johns Hopkins University Press, Baltimore, Maryland, 1989.

[9] A. Greenbaum, Behavior of slightly perturbed Lanczos and conjugate-gradient recurrences, Linear Algebra and Appl., 113 (1989), pp. 7–63.

[10] A. Greenbaum and Z. Strakos, Predicting the behavior of finite precision Lanczos and conjugate gradient computations, SIAM J. Matrix Anal. Appl., 13 (1992), pp. 121–137.

[11] S.J. Hammarling and J.H. Wilkinson, The practical behaviour of linear iterative methods with particular reference to S.O.R., Report NAC 69, National Physical Laboratory, England, 1976.

[12] N.J. Higham, How accurate is Gaussian elimination?, in Numerical Analysis 1989, Proceedings of the 13th Dundee Conference, D.F. Griffiths and G.A. Watson, eds., Pitman Research Notes in Mathematics 228, Longman Scientific and Technical, 1990, pp. 137–154.

[13] N.J. Higham, The accuracy of solutions to triangular systems, SIAM J. Numer. Anal., 26 (1989), pp. 1252–1265.

[14] N.J. Higham, Bounding the error in Gaussian elimination for tridiagonal systems, SIAM J. Matrix Anal. Appl., 11 (1990), pp. 521–530.

[15] N.J. Higham, Iterative refinement enhances the stability of QR factorization methods for solving linear equations, BIT, 31 (1991), pp. 447–468.

[16] R.A. Horn and C.R. Johnson, Matrix Analysis, Cambridge University Press, 1985.

[17] M. Jankowski and H. Woźniakowski, Iterative refinement implies numerical stability, BIT, 17 (1977), pp. 303–311.

[18] M.S. Lynn, On the round-off error in the method of successive over-relaxation, Math. Comp., 18 (1964), pp. 36–49.

[19] W. Oettli and W. Prager, Compatibility of approximate solution of linear equations with given error bounds for coefficients and right-hand sides, Numer. Math., 6 (1964), pp. 405–409.

[20] J.L. Rigal and J. Gaches, On the compatibility of a given solution with the data of a linear system, J. Assoc. Comput. Mach., 14 (1967), pp. 543–548.

[21] L.N. Trefethen, Pseudospectra of matrices, Technical Report 91/10, Oxford University Computing Laboratory, Oxford, 1991; to appear in the Proceedings of the 14th Dundee conference, D.F. Griffiths and G.A. Watson, eds.

[22] L.N. Trefethen and R.S. Schreiber, Average-case stability of Gaussian elimination, SIAM J. Matrix Anal. Appl., 11 (1990), pp. 335–360.

[23] H. Woźniakowski, Numerical stability of the Chebyshev method for the solution of large linear systems, Numer. Math., 28 (1977), pp. 191–209.

[24] H. Woźniakowski, Roundoff-error analysis of iterations for large linear systems, Numer. Math., 30 (1978), pp. 301–314.

[25] H. Woźniakowski, Roundoff-error analysis of a new class of conjugate-gradient algorithms, Linear Algebra and Appl., 29 (1980), pp. 507–529.

[26] D.M. Young, A historical overview of iterative methods, Computer Physics Communications, 53 (1989), pp. 1–17.

THE CHARACTER OF A FINITE MARKOV CHAIN

CARL D. MEYER *

Abstract. The purpose of this article is to present the concept of the *character* of a finite irreducible Markov chain. It is demonstrated how the sensitivity of the stationary probabilities to perturbations in the transition probabilities can be gauged by the use of the character.

1. Introduction. It is well-known that if a finite, irreducible, homogeneous Markov chain has a subdominant eigenvalue which is close to 1, then the chain is ill-conditioned in the sense that the stationary probabilities can be sensitive to small perturbations in the transition probabilities. However, the converse of this statement has been an open question. The purpose of this article is to help resolve this issue in terms of a spectral measure referred to as the *character* of the chain. Before defining the character, it is instructive to review the situation concerning the sensitivity of the stationary distribution.

If $\mathbf{P}_{n \times n}$ is the transition probability matrix for such a chain, and if $\boldsymbol{\pi}^T = (\pi_1, \pi_2, \cdots, \pi_n)$ is the stationary distribution vector satisfying $\boldsymbol{\pi}^T \mathbf{P} = \boldsymbol{\pi}^T$ and $\sum_{i=1}^{n} \pi_i = 1$, the goal is to describe the effect on $\boldsymbol{\pi}^T$ when $\mathbf{P}$ is perturbed by a matrix $\mathbf{E}$ such that $\tilde{\mathbf{P}} = \mathbf{P} + \mathbf{E}$ is the transition probability matrix of another irreducible Markov chain. The problem can be considered as a perturbed eigenvector problem, or it can be analyzed as a perturbed linear system $\boldsymbol{\pi}^T \mathbf{A} = \mathbf{0}$, $\boldsymbol{\pi}^T \mathbf{e} = 1$, where $\mathbf{A} = \mathbf{I} - \mathbf{P}$, and $\mathbf{e}$ is a column of 1's.

If $\sigma(\mathbf{P}) = \{1, \lambda_2, \lambda_3, \cdots, \lambda_n\}$ denotes the spectrum of $\mathbf{P}$, then the traditional perturbation theory for eigenvectors says that if a subdominant eigenvalue λ_i is close to 1, then we expect $\boldsymbol{\pi}^T$ to be sensitive. But for general eigenvector problems, the converse is not true—i.e., well-separated eigenvalues do not guarantee a well-conditioned eigenvector. For example, consider

$$T = \begin{pmatrix} 1 & -1 & -1 & \cdots & -1 \\ 0 & 2 & -2 & \cdots & -2 \\ 0 & 0 & 3 & \cdots & -3 \\ \vdots & \vdots & \vdots & \ddots & \vdots \\ 0 & 0 & 0 & \cdots & n \end{pmatrix} \quad \text{and the eigenvector} \quad \mathbf{v} = \begin{pmatrix} 1 \\ 0 \\ 0 \\ \vdots \\ 0 \end{pmatrix}.$$

We will show that the Markov chain problem is special in that there is a stronger relationship between the separation of the subdominant eigenvalues from 1 and the condition of $\boldsymbol{\pi}^T$.

* Mathematics Department, North Carolina State University, Raleigh, NC 27695-8205. This work was supported in part by the National Science Foundation under grants DMS-9020915 and DDM-8906248.

From the point of view of a perturbed linear system, the sensitivity of π^T can be described in terms of singular values along traditional lines. But doing so reveals very little concerning the structure of sensitive chains, and, in practice, the singular values of the problem are generally not available nor are they easy to estimate. A better approach was devised by Schweitzer [16] in terms of Kemeny and Snell's [11] "fundamental matrix" $Z = (A + e\pi^T)^{-1}$, and Meyer [13] cast these results in terms of the group inverse of A. If

$$P = Q^{-1} \begin{pmatrix} 1 & 0 \\ 0 & C \end{pmatrix} Q \quad \text{where} \quad 1 \notin \sigma(C),$$

then

(1.1) $$A = I - P = Q^{-1} \begin{pmatrix} 0 & 0 \\ 0 & I-C \end{pmatrix} Q,$$

and the group inverse of A is given by

(1.2) $$A^{\#} = Q^{-1} \begin{pmatrix} 0 & 0 \\ 0 & (I-C)^{-1} \end{pmatrix} Q.$$

Kemeny and Snell's fundamental matrix and the group inverse are related by the equation

$$Z = (A + e\pi^T)^{-1} = A^{\#} + e\pi^T.$$

For a complete development of other properties of $A^{\#}$, see [3] or [13]. In nearly all applications involving Z, the term $e\pi^T$ is redundant—i.e., all relevant information is contained in $A^{\#}$. In particular, if $\tilde{\pi}^T = (\tilde{\pi}_1, \tilde{\pi}_2, \cdots, \tilde{\pi}_n)$ is the stationary distribution for $\tilde{P} = P + E$, then

(1.3) $$\tilde{\pi}^T = \pi^T \left(I + EA^{\#}\right)^{-1}$$

so that

(1.4) $$\left\| \pi^T - \tilde{\pi}^T \right\| \leq \|E\| \, \|A^{\#}\|$$

in which $\|\star\|$ can be either the 1-, 2-, or ∞-norm. If the jth column and the (i,j)-entry of $A^{\#}$ are denoted by $A_{*j}^{\#}$ and $a_{ij}^{\#}$, respectively, then

(1.5) $$|\pi_j - \tilde{\pi}_j| \leq \|E\| \, \|A_{*j}^{\#}\|,$$

and

(1.6) $$\max_j |\pi_j - \tilde{\pi}_j| \leq \|E\|_\infty \max_{i,j} |a_{ij}^{\#}|.$$

There are chains for which equality in (1.6) is realized—e.g., consider

$$P = \begin{pmatrix} 1/2 & 1/2 \\ 1/2 & 1/2 \end{pmatrix} \quad \text{with} \quad E = \begin{pmatrix} \epsilon & -\epsilon \\ \epsilon & -\epsilon \end{pmatrix}.$$

Moreover, if the transition probabilities are analytic functions of a parameter t so that $\mathbf{P} = \mathbf{P}(t)$, then

(1.7)
$$\frac{d\boldsymbol{\pi}^T}{dt} = \boldsymbol{\pi}^T \frac{d\mathbf{P}}{dt} \mathbf{A}^{\#} \quad \text{and} \quad \frac{d\pi_j}{dt} = \boldsymbol{\pi}^T \frac{d\mathbf{P}}{dt} \mathbf{A}_{*j}^{\#}.$$

The formulas (1.3) and (1.4) are derived in [14], and (1.5) appears in [6]. The inequality (1.6) was given in [5], and the formulas (1.7) are developed in [6] and [12]. These facts make it clear that the entries in $\mathbf{A}^{\#}$ determine the extent to which $\boldsymbol{\pi}^T$ is sensitive to small changes in $\mathbf{P}$, so, on the basis of (1.6), it is natural to adopt the following definition (first given in [5]).

DEFINITION 1.1. *The condition of a Markov chain with a transition matrix* $\mathbf{P}$ *is measured by the size of its condition number which is defined to be*

$$\kappa = \max_{i,j} \left| a_{ij}^{\#} \right|$$

where $a_{ij}^{\#}$ *is the* (i,j)*-entry in the group inverse* $\mathbf{A}^{\#}$ *of* $\mathbf{A} = \mathbf{I} - \mathbf{P}$. *It follows from (1.2) that* κ *is invariant under permutations of the states of the chain.*

The question of interest can now be phrased as follows. *"How is* κ *related to the subdominant eigenvalues of* $\mathbf{A}$?" In one direction, the answer is quite easy. The structure of the group inverse makes it clear that if λ_i is a subdominant eigenvalue of $\mathbf{P}$, then

$$(1 - \lambda_i) \in \sigma(\mathbf{A}) \quad \Longrightarrow \quad \frac{1}{1 - \lambda_i} \in \sigma(\mathbf{A}^{\#})$$

$$\Longrightarrow \quad \rho(\mathbf{A}^{\#}) = \frac{1}{\min_{\lambda_i \neq 1} |1 - \lambda_i|} \leq \|\mathbf{A}^{\#}\|$$

$$\Longrightarrow \quad \frac{1}{n \, \min_{\lambda_i \neq 1} |1 - \lambda_i|} \leq \kappa.$$

In other words, as long as n is not too big, a small $|1 - \lambda_i|$ implies a large κ. But this simply corroborates the result derived from the eigenvector problem saying that poorly separated eigenvalues produce sensitive eigenvectors. The converse is what we are really interested in—i.e., *to what extent does a large* $\mathbf{A}^{\#}$ *force eigenvalues of* $\mathbf{A}$ *to be small?* The notion that $\max_{i,j} |a_{ij}^{\#}|$ somehow controls the size of the nonzero eigenvalues of $\mathbf{A}$ is contrary to what holds for general matrices. A large inverse does not generally guarantee the existence of small eigenvalues or a small determinant. A standard example is the triangular matrix

$$\mathbf{T}_{n \times n} \; = \; \begin{pmatrix} 1 & -2 & 0 & \cdots & 0 & 0 \\ 0 & 1 & -2 & \cdots & 0 & 0 \\ 0 & 0 & 1 & \ddots & 0 & 0 \\ \vdots & \vdots & \vdots & \ddots & \ddots & \vdots \\ 0 & 0 & 0 & \cdots & 1 & -2 \\ 0 & 0 & 0 & \cdots & 0 & 1 \end{pmatrix}$$

(1.8)

$$\text{with} \quad \mathbf{T}^{-1} \; = \; \begin{pmatrix} 1 & 2 & 4 & \cdots & 2^{n-2} & 2^{n-1} \\ 0 & 1 & 2 & \cdots & 2^{n-3} & 2^{n-2} \\ 0 & 0 & 1 & \ddots & 2^{n-4} & 2^{n-3} \\ \vdots & \vdots & \vdots & \ddots & \ddots & \vdots \\ 0 & 0 & 0 & \cdots & 1 & 2 \\ 0 & 0 & 0 & \cdots & 0 & 1 \end{pmatrix}$$

for which $\max_{i,j} [\mathbf{T}^{-1}]_{ij}$ is immense for even moderate values of n, but the eigenvalues of $\mathbf{T}$ provide no clue whatsoever that this occurs. The fact that the eigenvalues are repeated or that $\mathbf{T}$ is nonsingular is irrelevant—consider a small perturbation of $\mathbf{T}$ or the matrices

$$\tilde{\mathbf{T}} = \begin{pmatrix} 0 & 0 \\ 0 & \mathbf{T} \end{pmatrix} \quad \text{and} \quad \tilde{\mathbf{T}}^{\#} = \begin{pmatrix} 0 & 0 \\ 0 & \mathbf{T}^{-1} \end{pmatrix}.$$

Our goal is to prove that, unlike the situation illustrated above, irreducible stochastic matrices $\mathbf{P}$ possess enough special structure to guarantee that growth of the entries in $\mathbf{A}^{\#}$ is related to the magnitude of the nonzero eigenvalues of $\mathbf{A} = \mathbf{I} - \mathbf{P}$ by means of the character function which is defined below.

2. The Character of a Markov Chain

DEFINITION 2.1. *For an irreducible Markov chain whose eigenvalues are* $\{1, \lambda_2, \lambda_3, \cdots, \lambda_n\}$, *the* **character** *of the chain is defined to be the (necessarily real) number*

$$\chi = (1 - \lambda_2)(1 - \lambda_3) \cdots (1 - \lambda_n).$$

Before making the connection between the character of a chain and the sensitivity of its stationary probabilities, it is necessary to enumerate some basic properties of the character function. Some of these properties are self-evident, but others require some proof. Detailed arguments are not presented in this article, but they are given in [15].

LEMMA 2.2. *The character* χ *and its inverse* χ^{-1} *are the respective determinants of the nonsingular parts of* $\mathbf{A}$ *and* $\mathbf{A}^{\#}$ *in the sense that*

$$\chi = \det \left(\mathbf{A} \big/_{\mathcal{R}(\mathbf{A})} \right) \quad \text{and} \quad \chi^{-1} = \det \left(\mathbf{A}^{\#} \big/_{\mathcal{R}(\mathbf{A})} \right).$$

It is also true that $\chi^{-1} = det(\mathbf{Z})$ where $\mathbf{Z}$ is Kemeny and Snell's "fundamental matrix."

LEMMA 2.3. If $\mathbf{P}_{n \times n}$ is an irreducible stochastic matrix and if $\mathbf{A}_i$ denotes the principal submatrix of $\mathbf{A} = \mathbf{I} - \mathbf{P}$ obtained by deleting the ith row and column from $\mathbf{A}$, then

$$\chi = \sum_{i=1}^{n} det\mathbf{A}_i.$$

LEMMA 2.4. If $\mathbf{A}_i$ denotes the principal submatrix of $\mathbf{A} = \mathbf{I} - \mathbf{P}$ obtained by deleting the ith row and column from $\mathbf{A}$, then

$$\chi = \frac{det(\mathbf{A})_i}{\pi_i}$$

where π_i is the ith stationary probability.

LEMMA 2.5. If R_k denotes the mean return time for the kth state, then

$$0 < \max_{k} det(\mathbf{A}_k) < \chi \leq \min_{k} R_k \leq n.$$

COROLLARY 2.6. For every n-state irreducible chain, $0 < \chi < n$.

DEFINITION 2.7. Hereafter, a chain is said to be of weak character when χ is close to 0, and the chain is said to have a strong character when χ is significantly larger than 0.

If $\mathbf{A} = \mathbf{I} - \mathbf{P}$ where $\mathbf{P}_{n \times n}$ is an irreducible stochastic matrix, then $\mathbf{A}$ as well as each principal submatrix of $\mathbf{A}$ has strictly positive diagonal entries, and the off-diagonal entries are non-positive. Furthermore, it is well-known [2] that $\mathbf{A}$ is a singular M-matrix of rank $n - 1$. Consequently, the $k \times k$ $(k < n)$ principal submatrices of $\mathbf{A}$ are known to possess the following properties—see [2] or [10].

LEMMA 2.8. If $\mathbf{B}_{k \times k}$ $(k < n)$ is a principal submatrix of $\mathbf{A} = \mathbf{I} - \mathbf{P}$, then each of the following statements is true.

(a) $\mathbf{B}$ is a nonsingular M-matrix.

(b) $\mathbf{B}^{-1} \geq 0$.

(c) $det(\mathbf{B}) > 0$.

(d) $\mathbf{B}$ is diagonally dominant.

(e) $det(\mathbf{B}) \leq b_{11}b_{22} \cdots b_{kk} \leq 1$.   (A Hadamard Inequality)

The next lemma is not a well-known property, but it plays a pivotal role in subsequent developments. The result can be derived from the fact that the (i,j)-entry in $\mathbf{B}^{-1}$ represents the mean number of times the chain is in state j before first hitting any of the states complementary to those associated with $\mathbf{B}$ given that the process starts in state i (see [15] for a detailed proof).

LEMMA 2.9. If $\mathbf{A} = \mathbf{I} - \mathbf{P}$ where $\mathbf{P}_{n \times n}$ is an irreducible stochastic matrix, and if $\mathbf{B}_{k \times k}$ $(k < n)$ is a principal submatrix of $\mathbf{A}$, then for $j =$

$1, 2, \cdots, k$, *it must be the case that*

$$[\mathbf{B}^{-1}]_{jj} \geq [\mathbf{B}^{-1}]_{ij} \quad \text{for each} \quad i \neq j.$$

That is, the largest entry in each column of $\mathbf{B}^{-1}$ *is the diagonal entry.*

By using the Hadamard inequality of Lemma (2.8) the following related inequality can be derived.

LEMMA 2.10. *Let* $\mathbf{A} = \mathbf{I} - \mathbf{P}$ *where* $\mathbf{P}_{n \times n}$ *is an irreducible stochastic matrix, and let* $\mathbf{B}_{k \times k}$ *($k < n$) be a principal submatrix of* $\mathbf{A}$. *If* $\delta_r(\mathbf{B})$ *denotes the deleted product*

$$\delta_r(\mathbf{B}) = \prod_{k \neq r} b_{kk},$$

then

$$0 < det(\mathbf{B}) \leq \frac{\max_i \delta_i(\mathbf{B})}{\max_{i,j} [\mathbf{B}^{-1}]_{ij}} \leq \frac{1}{\max_{i,j} [\mathbf{B}^{-1}]_{ij}} \leq 1.$$

As a consequence,

$$\max_{i,j} \beta_{ij} \leq \frac{\max_i \delta_i(\mathbf{B})}{det(\mathbf{B})}.$$

3. Condition Number and Character. We now return to the original objective—to bound the condition number $\kappa = \max_{i,j} |a_{ij}^{\#}|$ with some function of the character χ. In order to do so, the following representation of the group inverse of $\mathbf{A}$ is needed.

LEMMA 3.1. *For an irreducible stochastic matrix* $\mathbf{P}_{n \times n}$, *let* $\mathbf{A}_j$ *be the principal submatrix of* $\mathbf{A} = \mathbf{I} - \mathbf{P}$ *obtained by deleting the* j^{th} *row and column from* $\mathbf{A}$, *and let* $\mathbf{Q}$ *be the permutation matrix such that*

$$\mathbf{Q}^T \mathbf{A} \mathbf{Q} = \begin{pmatrix} \mathbf{A}_j & \mathbf{c}_j \\ \mathbf{d}_j^T & a_{jj} \end{pmatrix}.$$

If the stationary distribution for $\mathbf{Q}^T \mathbf{P} \mathbf{Q}$ *is written as* $\psi^T = \pi^T \mathbf{Q} = (\overline{\pi}^T, \pi_j)$, *then the group inverse of* $\mathbf{A}$ *is given by*

$$\mathbf{A}^{\#} = \mathbf{Q} \begin{pmatrix} (\mathbf{I} - \mathbf{e}\overline{\pi}^T)\mathbf{A}_j^{-1}(\mathbf{I} - \mathbf{e}\overline{\pi}^T) & -\pi_j(\mathbf{I} - \mathbf{e}\overline{\pi}^T)\mathbf{A}_j^{-1}\mathbf{e} \\ -\overline{\pi}^T \mathbf{A}_j^{-1}(\mathbf{I} - \mathbf{e}\overline{\pi}^T) & \pi_j \overline{\pi}^T \mathbf{A}_j^{-1}\mathbf{e} \end{pmatrix} \mathbf{Q}^T$$

where $\mathbf{e}$ *is a column of 1's whose size is determined by the context in which it appears.*

There are two cases to consider when deriving an upper bound for κ— the first case is when $\max_{i,j} |a_{ij}^{\#}|$ occurs on the diagonal of $\mathbf{A}^{\#}$ and the other is when $\max_{i,j} |a_{ij}^{\#}|$ is generated by an off-diagonal element. In what follows, δ will denote the quantity

$$\delta = \max_{i,j} \prod_{k \neq i,j} a_{kk} < 1.$$

Case 1: If $\max_{i,j} |a_{ij}^{\#}|$ occurs when $i = j$, then the results of §2 can be applied to yield the following inequalities.

$$
\begin{aligned}
|a_{jj}^{\#}| &= \pi_j |\bar{\pi}^T \mathbf{A}_j^{-1} \mathbf{e}| \leq \pi_j \|\bar{\pi}\|_1 \|\mathbf{A}_j^{-1}\mathbf{e}\|_\infty \\
&< \pi_j (n-1) \max_{r,s} [\mathbf{A}_j^{-1}]_{rs} \\
&\leq \frac{\pi_j (n-1) \max_i \delta_i(\mathbf{A}_j)}{det(\mathbf{A})_j} \leq \frac{\pi_j (n-1)\delta}{det(\mathbf{A})_j} \\
&= \frac{\delta(n-1)}{\chi} \leq \frac{(n-1)}{\chi}
\end{aligned}
$$

Case 2: If $\max_{i,j} |a_{ij}^{\#}|$ occurs when $i \neq j$, then the following inequalities hold.

$$
\begin{aligned}
|a_{ij}^{\#}| &= \pi_j |\mathbf{e}_k^T (\mathbf{I} - \mathbf{e}\bar{\pi}^T)\mathbf{A}_j^{-1}\mathbf{e}| \leq \pi_j \|\mathbf{e}_k - \bar{\pi}\|_1 \|\mathbf{A}_j^{-1}\mathbf{e}\|_\infty \\
&< 2\pi_j (n-1) \max_{r,s} [\mathbf{A}_j^{-1}]_{rs} \\
&\leq \frac{2\pi_j (n-1) \max_i \delta_i(\mathbf{A}_j)}{det(\mathbf{A})_j} \leq \frac{2\pi_j (n-1)\delta}{det(\mathbf{A})_j} \\
&= \frac{2\delta(n-1)}{\chi} \leq \frac{2(n-1)}{\chi}
\end{aligned}
$$

Combining the results of these two cases produces the following theorem.

THEOREM 3.2. *For an irreducible chain whose eigenvalues are* $1, \lambda_2, \cdots,$ λ_n, *the condition number* $\kappa = \max_{i,j} |a_{ij}^{\#}|$ *is bounded by*

$$\frac{1}{n \min_{\lambda_i \neq 1} |1 - \lambda_i|} \leq \kappa < \frac{2\delta(n-1)}{\chi} \leq \frac{2(n-1)}{\chi}$$

where

$$\chi = \prod_{\lambda_i \neq 1} (1 - \lambda_i) \quad \text{and} \quad \delta = \max_{i,j} \prod_{k \neq i,j} a_{kk}.$$

In other words,

- *If an irreducible chain is well-conditioned, then all subdominant eigenvalues must be well-separated from 1.*

- *If all subdominant eigenvalues are well-separated from 1 in the sense that the chain has a strong character, then the chain must be well-conditioned.*

Remarks. Although the upper bound explicitly involves n, it is generally not the case that the term $2\delta(n-1)/\chi$ grows in proportion to n. Except in the special case when the diagonal entries of $\mathbf{P}$ are 0, the term δ somewhat mitigates the presence of n because as n becomes large, δ becomes small. Computational experience suggests that $2\delta(n-1)/\chi$ is usually a rather conservative estimate of κ, and the term δ/χ by itself, although not always an upper bound for κ, is often of the same order of magnitude as κ. However, there exist pathological cases for which even δ/χ severely overestimates κ. This seems to occur for chains which are not too badly conditioned, and no single eigenvalue is extremely close to 1, but enough eigenvalues are within range of 1 to force χ^{-1} to be too large. This suggests that for the purposes of bounding κ above, perhaps not all of the subdominant eigenvalues need to be taken into account.

4. Using an LU Factorization. Except for chains which are too large to fit into a computer's main memory, the stationary distribution $\boldsymbol{\pi}^T$ is generally computed by direct methods—i.e., either an LU or QR factorization of $\mathbf{A} = \mathbf{I} - \mathbf{P}$ (or $\mathbf{A}^T$) is computed—see [2], [5], [6], [7], [8]. Even for very large chains which are nearly uncoupled, direct methods are usually involved—they can be the basis of the main algorithm [17], or they can be used to solve the aggregated and coupling chains in iterative aggregation/disaggregation algorithms—see [4] or [9].

On the basis of computational experience, Golub and Meyer [6] conjecture that the condition of the chain is directly related to the magnitude of the nonzero pivots which emerge when a triangular factorization of $\mathbf{A}_{n \times n}$ is used to compute the stationary probabilities. If this is indeed true, then it should be possible to estimate κ with little or no extra cost beyond that incurred in computing $\boldsymbol{\pi}^T$, and this would represent a significant savings over the $O(n^2)$ operations demanded by traditional condition estimators. Of course, this is contrary to the situation which exists for general nonsingular matrices because the absence of small pivots (or the existence of a large determinant) is not a guarantee of a well-conditioned matrix—consider the matrix in (1.9). A mathematical formulation and proof (or even an intuitive explanation) of Golub and Meyer's observation has heretofore not been given, but the results of §2 and §3 lead to the following theorem which provides a more precise statement and a rigorous proof of the Golub-Meyer conjecture.

THEOREM 4.1. *If the LU factorization of* $\mathbf{A}$ *(or* $\mathbf{A}^T$*) is computed to be*

$$\mathbf{A} \ (or \ \mathbf{A}^T) = \mathbf{LU} = \begin{pmatrix} \mathbf{L}_n & \mathbf{0} \\ \mathbf{r}^T & 1 \end{pmatrix} \begin{pmatrix} \mathbf{U}_n & \mathbf{c} \\ \mathbf{0} & 0 \end{pmatrix},$$

then the stationary distribution is given by

$$\boldsymbol{\pi}^T = \frac{1}{1 + \|\mathbf{x}\|_1} (\mathbf{x}^T, 1) \quad where \quad \begin{cases} \mathbf{x}^T \mathbf{L}_n = -\mathbf{r}^T & \textit{if } \mathbf{A} \textit{ is used,} \\ \mathbf{U}_n \mathbf{x} = -\mathbf{c} & \textit{if } \mathbf{A}^T \textit{ is used,} \end{cases}$$

and the character of the chain is an immediate by-product because

$$\chi = \frac{det \mathbf{L}_n \mathbf{U}_n}{\pi_n} = (1 + \|\mathbf{x}\|_1) det(\mathbf{U})_n.$$

Once these computations have been made, bounds for the condition number κ are available at no extra cost because

$$\pi_n \sum_{i=1}^{n-1} \frac{\pi_i}{u_{ii}} = \frac{1}{(1 + \|\mathbf{x}\|_1)^2} \sum_{i=1}^{n-1} \frac{x_i}{u_{ii}} \leq \kappa < \frac{2\delta(n-1)}{(1 + \|\mathbf{x}\|_1) \, det(\mathbf{U})_n}$$

$$\leq \frac{2(n-1)}{(1 + \|\mathbf{x}\|_1) \, det(\mathbf{U})_n}.$$

The absence of small pivots (or the existence of a large determinant) is generally not a guarantee of a well-conditioned matrix. However, for Markov chains, the bounds in Theorem (4.1) allow the pivots to be used as condition estimators.

COROLLARY 4.2. *For an irreducible Markov chain whose transition matrix is $\mathbf{P}$, suppose that the LU factorization of $\mathbf{A} = \mathbf{I} - \mathbf{P}$ and the stationary distribution $\boldsymbol{\pi}^T$ have been computed as described in Theorem (4.1)*

- *If the pivots u_{ii} are large in the sense that $(1 + \|\mathbf{x}\|_1) \, det(\mathbf{U})_n$ is large relative to $2\delta(n-1)$, then the chain must be well-conditioned.*
- *If there are pivots u_{ii} which are small relative to $x_i / (1 + \|\mathbf{x}\|_1)^2$, then the chain must be ill-conditioned.*

Remark. Regardless of whether $\mathbf{A}$ or $\mathbf{A}^T$ is used, Gaussian elimination with finite-precision arithmetic can prematurely produce a zero pivot, and this can happen for well-conditioned chains. Consequently, practical implementation demands a strategy to deal with this situation—see [5] and [17] for a discussion of this problem along with possible remedies.

5. Using a QR Factorization. The utility of orthogonal triangularization is well-documented throughout the literature on matrix computations, and the use of a QR factorization to solve and analyze Markov chains is discussed in [6]. The following theorem shows that if a QR factorization is used to compute the stationary distribution, then the character of the chain is also available at very little extra cost, and consequently a low-cost estimate of κ can be obtained. This theorem also provides a theoretical explanation of the Golub-Meyer [6] computational observation that the condition of an irreducible chain is related to the reciprocals of the nonzero diagonal elements of the R-factor.

THEOREM 5.1. *For an irreducible Markov chain whose transition matrix is* $\mathbf{P}$, *the QR factorization of* $\mathbf{A} = \mathbf{I} - \mathbf{P}$ *is given by*

$$
\mathbf{A} = \mathbf{QR} = \begin{pmatrix} \mathbf{Q}_n & \mathbf{c} \\ \mathbf{d}^T & q_{nn} \end{pmatrix} \begin{pmatrix} \mathbf{R}_n & -\mathbf{R}_n \mathbf{e} \\ \mathbf{0} & 0 \end{pmatrix} = \begin{pmatrix} \mathbf{Q}_n \mathbf{R}_n & -\mathbf{Q}_n \mathbf{R}_n \mathbf{e} \\ \mathbf{d}^T \mathbf{R}_n & -\mathbf{d}^T \mathbf{R}_n \mathbf{e} \end{pmatrix}
$$

If $\mathbf{q} = \begin{pmatrix} \mathbf{c} \\ q_{nn} \end{pmatrix}$ *is the last column of* $\mathbf{Q}$, *then the stationary distribution is given by*

$$
\boldsymbol{\pi}^T = \frac{\mathbf{q}^T}{\sum_{i=1}^{n} q_{in}},
$$

and the character of the chain is

$$
\chi = \|\mathbf{q}\|_1 \, det(\mathbf{R})_n.
$$

Once these computations have been made, an upper bound for the condition number κ *is available at no extra cost because*

$$
\kappa < \frac{2\delta(n-1)}{\|\mathbf{q}\|_1 \, det(\mathbf{R})_n} \leq \frac{2(n-1)}{\|\mathbf{q}\|_1 \, det(\mathbf{R})_n}.
$$

Remark. A lower bound for κ analogous to the one in Theorem (4.1) does not seem to be readily available.

6. Summary.

The major conclusions of this article can be summarized as follows.

- Let $\mathbf{A} = \mathbf{I} - \mathbf{P}$ where $\mathbf{P}$ is the transition probability matrix for an n-state irreducible chain with eigenvalues $\{1, \lambda_2, \lambda_3, \cdots, \lambda_n\}$. If $\mathbf{A}^{\#}$ denotes the group inverse of $\mathbf{A}$, then the largest entry $\kappa = \max_{i,j} |a_{ij}^{\#}|$ is the condition number for the chain. The character of the chain is the real number $\chi = (1 - \lambda_2)(1 - \lambda_3) \cdots (1 - \lambda_n)$, and κ is bounded by

$$
\triangleright \quad \frac{1}{n \min_{\lambda_i \neq 1} |1 - \lambda_i|} \leq \kappa < \frac{2\delta(n-1)}{\chi} \leq \frac{2(n-1)}{\chi}
$$

where $\delta = \max_{i,j} \prod_{k \neq i,j} a_{kk} < 1$.

- If $\mathbf{A}$ (or $\mathbf{A}^T$) $= \mathbf{LU} = \begin{pmatrix} \mathbf{L}_n & \mathbf{0} \\ \mathbf{r}^T & 1 \end{pmatrix} \begin{pmatrix} \mathbf{U}_n & \mathbf{c} \\ \mathbf{0} & 0 \end{pmatrix}$, then

$$
\triangleright \quad \boldsymbol{\pi}^T = \frac{1}{1 + \|\mathbf{x}\|_1} (\mathbf{x}^T, 1) \quad \text{where} \quad \begin{cases} \mathbf{x}^T \mathbf{L}_n = -\mathbf{r}^T & \text{if } \mathbf{A}. \\ \mathbf{U}_n \mathbf{x} = -\mathbf{c} & \text{if } \mathbf{A}^T. \end{cases}
$$

$$\triangleright \quad \pi_n \sum_{i=1}^{n-1} \frac{\pi_i}{u_{ii}} \le \kappa < \frac{2\delta(n-1)\pi_n}{det(\mathbf{U})_n}.$$

$$\triangleright \quad \frac{1}{\left(1 + \|\mathbf{x}\|_1\right)^2} \sum_{i=1}^{n-1} \frac{x_i}{u_{ii}} \le \kappa < \frac{2\delta(n-1)}{\left(1 + \|\mathbf{x}\|_1\right) det(\mathbf{U})_n}.$$

These inequalities produce the following statements.

- If an irreducible chain is well-conditioned, then its subdominant eigenvalues must be well-separated from 1.
- If the subdominant eigenvalues of an irreducible chain are well-separated from 1 in the sense that χ is large relative to δ, then the chain must be well-conditioned.
- If an irreducible chain is well-conditioned, then the pivots u_{ii} which emerge during gaussian elimination cannot be small relative to $\pi_i \pi_n$.
- If there are no small pivots in the sense that $det(\mathbf{U})_n$ is large relative to $\delta\pi_n$, then the chain must be well-conditioned.

REFERENCES

[1] Barlow, J.L. Error bounds and condition estimates for the computation of null vectors with applications to Markov chains. Preprint, Computer Sci. Dept., Penn State U., pp.1–29 1991.

[2] Berman, A. and Plemmons, R.J., *Nonnegative Matrices In The Mathematical Sciences*, Academic Press, New York 1979.

[3] Campbell, S.L. and Meyer, C.D., *Generalized Inverses Of Linear Transformations*, Dover Publications, New York, (1979 edition by Pitman Pub. Ltd., London) 1991.

[4] Chatelin, F. and Miranker, W.L., Acceleration by aggregation of successive approximation methods *Linear Algebra and Its applications*, 43(1982), pp.17–47.

[5] Funderlic, R.E., and Meyer, C.D. (1986). Sensitivity of the stationary distribution vector for an ergodic Markov chain, *Linear Algebra and Its Applications*, 76(1986), pp.1–17.

[6] Golub, G.H. and Meyer, C.D., (1986). Using the QR factorization and group inversion to compute, differentiate, and estimate the sensitivity of stationary probabilities for Markov chains, *SIAM J. On Algebraic and Discrete Methods*, Vol. 7, No. 2, pp.273–281, 1986.

[7] Grassmann, W.K., Taksar, M.I., and Heyman, D.P., Regenerative analysis and steady state distributions for Markov chains em Operations Research, 33(1985), pp.1107–1116.

[8] Harrod, W.J. and Plemmons, R.J., Comparison of some direct methods for computing stationary distributions of Markov chains, *SIAM J. Sci. Statist. Comput.*, 5(1984), pp.453–469.

[9] Haviv, M., Aggregation/disaggregation methods for computing the stationary distribution of a Markov chain, *SIAM J. Numer. Anal.*, 22(1987), pp.952–966.

[10] Horn, R.A. and Johnson, C.R., em Topics In Matrix Analysis, Cambridge University Press, Cambridge, 1991.

[11] Kemeny, J.G. and Snell, J.L., *Finite Markov Chains,* D. Van Nostrand, New York 1960.

[12] Meyer, C.D., and Stewart, G.W., Derivatives and perturbations of eigenvectors, *SIAM J. On Numerical Analysis*, Vol. 25, No. 3, pp.679–691, 1988.

[13] Meyer, C.D., The role of the group generalized inverse in the theory of finite Markov chains, *SIAM Review*, Vol. 17, No. 3, pp.443–464, 1975.

[14] Meyer, C.D., The condition of a finite Markov chain and perturbation bounds for the limiting probabilities, SIAM J. On Algebraic and Discrete Methods, Vol. 1, No. 3, pp.273–283, 1980.

[15] Meyer, C.D., Sensitivity of Markov chains, NCSU Technical Report, October, 1991, pp.1–18.

[16] Schweitzer, P.J., Perturbation theory and finite Markov chains, *J. Appl. Prob.*, 5(1968), pp.401–413.

[17] Stewart, G.W. and Zhang, G., On a direct method for the solution of nearly uncoupled Markov chains, *Numerische Mathematik*, 59(1991), pp.1–11.

GAUSSIAN ELIMINATION, PERTURBATION THEORY, AND MARKOV CHAINS*

G. W. STEWART[†]

Abstract. The purpose of this paper is to describe the special problems that emerge when Gaussian elimination is used to determinin the steady-state vector of a Markov chain.

Key words. Gaussian elimination, Markov chain, rounding error, perturbation theory

1. Introduction. The purpose of this paper is to describe the special problems that emerge when Gaussian elimination is used to determinine the steady-state vector of a Markov chain. Although there are many iterative techniques for solving this problem, direct methods are appropriate when the problem is small or when it is sparse and unstructured. In such cases, Gaussian elimination, the simplest of the direct methods, is a natural candidate.

The analysis of direct methods for linear systems has traditionally combined rounding-error analysis with perturbation theory — the former to establish the stability of the algorithm in question and the latter to assess the accuracy of the solution. This fruitful interplay carries over to the solution of Markov chains, and will be one of the main themes of this paper.

The paper begins with a review of the basic facts about Gaussian elimination for general linear systems. We then turn to the application of the algorithm to general Markov chains. The theory here is quite satisfactory and justifies the algorithm for many, if not most, applications. However, there is an important class of chains for which the algorithm fails — the nearly completely decomposable (NCD) chains. Paradoxically, although the algorithm fails, the problem is well determined in the sense that all the information necessary to compute a solution is available at the outset of the computations. A closer study of this paradox motivates a variant of Gaussian elimination proposed by Grassmann, Taksar, and Heyman [6].

Throughout out this paper we will assume that the reader is familiar with Gaussian elimination and its relation to the LU decomposition. Treatments of this material can be found in most introductory books on numerical linear algebra (e.g., [5,12,17]). The symbol $\| \cdot \|$ denote a family

---

* This report is available by anonymous ftp from `thales.cs.umd.edu` in the directory `pub/reports`.

† Department of Computer Science and Institute for Advanced Computer Studies, University of Maryland, College Park, MD 20742. This work was supported in part by the Air Force Office of Scientific Research under Contract AFOSR-87-0188 and was done while the author was a visiting faculty member at the Institute for Mathematics and Its Applications, The University of Minnesota, Minneapolis, MN 55455.

of consistent matrix norms; i.e., one for which $\|AB\| \le \|A\|\|B\|$, whenever the product AB is defined.

2. Gaussian Elimination and Linear Systems. The need for a formal rounding-error analysis of Gaussian elimination became critical as the arrival of the electronic computer made it impossible for a human to monitor the individual numbers in a calculation. In an early analysis, the statistician Hotelling [8] arrived at the conclusion that the errors in Gaussian elimination could grow exponentially with the size of the matrix, which would have precluded its use for matrices of even modest size. A subsequent analysis by von Neumann and Goldstine [16] showed that the algorithm would solve positive definite systems accurately, provided they were what we now call well conditioned. Finally, in 1961 Wilkinson [18] provided a comprehensive error analysis of Gaussian elimination for general linear systems.

A key component of Wilkinson's treatment is the projection of the errors back onto the original problem, a procedure known as backward rounding-error analysis. Specifically, he showed that if Gaussian elimination is used to solve the system

$$Ax = b,$$

where A is of order n then the computed solution $\tilde{x}$ satisfies the equation

(2.1) $$(A + E)\tilde{x} = b,$$

where

(2.2) $$\|E\| \le \varphi(n)\gamma\epsilon_M.$$

Here $\varphi(n)$ is a slowly growing function of the size of the matrix that depends on the norm and details of the arithmetic used in the computations, and ϵ_M is the rounding unit for the arithmetic. The number γ is the largest matrix element encountered in the course of the elimination.

The analysis shows that provided γ is not large compared with A—i.e., there has been no undue growth of elements in the elimination—then the computed solution is the exact solution of a problem differing from the original by terms of order of magnitude of the rounding error. Since such errors are generally smaller than errors already present in the elements of A, the algorithm itself cannot be held responsible for inaccuracies in the solution. An algorithm with such a backward error analysis is said to be *stable*. By the above error analysis, partial pivoting (the practice of interchanging rows to bring the largest element in the column into the diagonal) is seen to be a device to ensure stability by limiting the growth of elements as reflected by γ.

Although stability is a compelling reason for using an algorithm, it is not sufficient for users who want to know the accuracy of their solutions.

Consequently, it is customary to supplement a rounding-error analysis with a perturbation theory for the problem being solved. The basic perturbation theory for linear systems is particularly simple. If $Ax = b$ and $\tilde{x}$ satisfies (2.1), then

$$\tilde{x} = x - A^{-1}Ex + O(\|E\|^2).$$

It follows that

(2.3)
$$\frac{\|\tilde{x} - x\|}{\|x\|} \lesssim \kappa(A)\frac{\|E\|}{\|A\|},$$

where $\kappa(A) = \|A\|\|A^{-1}\|$ is the *condition number* of A. The left-hand side of (2.3) is a relative error in the computed solution. The fraction on the right is the relative error in A due to the perturbation E. The number $\kappa(A)$, which is always greater than one, is a magnifying factor that says how much the error in A is magnified as it passes to an error in x.

Together the rounding-error analysis and the perturbation analysis form a neat summary of the properties of Gaussian elimination. The rounding error analysis shows that any inaccuracies in the solution must come from exceedingly small changes in the problem: the perturbation analysis gives us a means of assessing the errors in the solution due to these changes. We will now turn to a similar analysis of the use of Gaussian elimination to solve Markov chains.

3. Gaussian Elimination and Markov Chains. Now let P be the transition matrix of an irreducible Markov chain. Then up to normalization factors, P has unique, positive right and left eigenvectors corresponding to the eigenvector one. The right eigenvector is e; i.e., the vector whose components are all one. Our problem is to compute the left eigenvector, which we will denote by y^T.

The problem can be cast in a form that is more convenient for Gaussian elimination. Let

$$Q = I - P.$$

Then

$$y^T Q = y^T - y^T P = y^T - y^T = 0.$$

Thus the problem has been transformed from that of finding an eigenvector of P to that of finding a null vector of Q.[1] Note that $Qe = 0$; i.e., the row sums of Q are zero.

Gaussian elimination can be used to solve the equation $y^T Q = 0$ as follows.

---

[1] It is worth noting that this kind of reduction cannot be applied to the general eigenvalue problem, since we will not ordinarily know the eigenvalue a priori.

1. Use Gaussian elimination to decompose $Q^T = LU$, where L is unit lower triangular and U is upper triangular.

2. Partition

$$U = \begin{pmatrix} U_* & u \\ 0 & 0 \end{pmatrix}.$$

The matrix U_* will be upper triangular and nonsingular.

3. Compute

$$y^T = \frac{(-u^T U_*^{-T} \quad 1)}{\|(-u^T U_*^{-T} \quad 1)\|_1},$$

where $\|x\|_1 = e^T |x|$.

The third step of the algorithm is almost self-explanatory. Since L is nonsingular, any null vector of Q^T is a null vector of U and vice versa. The third step then amounts to computing a null vector of U by assuming its last component is one, solving the resulting triangular system, and normalizing.

The nonsingularity of U_*, which is not obvious, is closely bound up with the problem of pivoting. To see what is going on let us consider the first two columns of Q^T, which we write in the form

$$\begin{array}{cc} q_{11} & q_{12} \\ q_{21} & q_{22} \\ q_1 & q_2 \end{array}$$

The quantities q_{11} and q_{22} are nonnegative, while q_{21}, q_{12}, q_1, and q_2 are nonpositive. Moreover, since the components of the first column sum to zero, q_{11} can be zero only if q_{21} and q_1 are zero, in which case Q is reducible, contrary to assumption.

Thus the first step of Gaussian elimination can be performed on the second column, yielding a new column of the form

$$\bar{q}_{12} = 0$$
$$\bar{q}_{22} = q_{22} - \frac{q_{21} q_{12}}{q_{11}}$$
$$\bar{q}_2 = q_2 - \frac{q_1 q_{12}}{q_{11}}$$

By considering the signs of the quantities involved, we find that

1. $q_{22} \geq \bar{q}_{22}$,
2. $q_2 \geq \bar{q}_2$.

In other words, the diagonal element decreases while the off-diagonal elements increase in magnitude. Since the column sums are still zero, we have $q_{22} \geq \bar{q}_{22} \geq -\bar{q}_{i2}$ $(i > 2)$, which implies that all the elements in the reduced second column are bounded in magnitude by q_{22}; i.e., there is no net growth in the elements of the reduced column. Finally, the quantity $\bar{q}_{22}$ can be zero only if both q_1 and q_2 are zero, which contradicts irreducibility.

Since any column of Q^T can be symmetrically permuted into the second column, the above observations apply to any column. Thus the result of the first step of Gaussian elimination is an irreducible matrix with positive diagonal elements and nonpositive off-diagonal elements. By induction the same is true of the the subsequent steps, save the last, which must produce the single number $u_{nn} = 0$. Since there is no net growth in the elements, the reduction can be carried to completion, and the diagonal elements of U_*, which are the pivots in the elimination are positive. Since the growth factor γ in (2.2) is one, the algorithm is stable. This stability was first pointed out by Funderlic and Mankin [4].

Two further points. not only is there no need to pivot in the algorithm, but the usual form of partial pivoting is in some sense harmful. The reason is that interchanging two rows of Q^T (without interchanging the corresponding columns) destroys the properties that keep growth from occurring, and we are left with the (admittedly unlikely) possibility of instability. The second point is that symmetric pivoting, in which two rows and the same two columns are interchanged, does preserve the structure of Q and can be used with complete freedom. This fact is important in applications involving sparse matrices, in which pivoting is necessary to avoid fill-in [2].

Turning now to the perturbation theory of Markov chains, the basic theory goes back almost a quarter of a century and has been presented in a variety of ways [3,7,9,10]. Here we give it in a form that will be useful in the sequel. For proofs see [14].

It can be shown that the matrix P has the spectral decomposition

$$(3.1) \qquad P = ey^T + XBY^T,$$

where

$$\begin{pmatrix} y^T \\ Y^T \end{pmatrix} = (e\ X)^{-1}.$$

Alternatively,

$$(e\ X)^{-1}P(e\ X) = \operatorname{diag}(1, B);$$

that is, $(e\ X)$ transforms P via a similarity transformation into a block-diagonal matrix. It follows that the eigenvalues of B must be those of P other than one, and hence $I - B$ is nonsingular.

Now assume that $\tilde{P} = P + F$ is an irreducible stochastic matrix, and let $\tilde{y}^T$ be its steady-state vector. Then it can be shown that

$$(3.2) \qquad \tilde{y}^T \cong y^T + y^T FX(I - B)^{-1}Y^T,$$

from which it follows that

(3.3) $$\frac{\|\tilde{y}^{\mathrm{T}} - y^{\mathrm{T}}\|}{\|y^{\mathrm{T}}\|} \lesssim \|X(I - B)^{-1}Y^{\mathrm{T}}\|\|F\|.$$

This is the desired perturbation bound.

The matrix $(I - P)^{\#} \equiv X(I - B)^{-1}Y^{\mathrm{T}}$ is called the *Drazin pseudo-inverse* or the *group inverse* of the system. The bound (3.3) shows that the norm of the group inverse is a condition number for the steady-state vector of a Markov chain.

4. Nearly Completely Decomposable Chains. The results of the last section place Gaussian elimination for the solution of Markov chains on a par with Gaussian elimination for the solution of linear systems. The algorithms are backwards stable, and the problems have reasonable perturbation theories. However, in neither case does the analysis apply to matrices whose elements vary widely and systematically in size.

Consider, for example the following matrix:

$$Q = \begin{pmatrix} +0.75287 & -0.75283 & -0.00003 \\ -0.75283 & +0.75284 & -0.00001 \\ -0.00003 & -0.00001 & +0.00004 \end{pmatrix}.$$

Since the matrix is symmetric, the left and right eigenvector are both e, and there is no difference between applying Gaussian elimination to Q or Q^{T}.

Working in five decimal digits, we compute the correctly rounded multipliers for the first step of Gaussian elimination as

$$0.99996 = \mathrm{fl}(-0.75283/0.75287) \quad \text{and} \quad 0.39849{\cdot}10^{-4} = \mathrm{fl}(-0.00003/0.75287).$$

(Here fl denotes a floating point operation.) If we use these multipliers and make no further rounding errors, the matrix assumes the form.

(4.1) $$Q = \begin{pmatrix} +0.75287 & -0.75283 & -0.00003 \\ +0.0 & +0.000040113 & -0.000039999 \\ +0.0 & -0.000039999 & +0.000039999 \end{pmatrix}$$

This shows that the last two components of the computed null vector, which should be equal, will be in a ratio of 40113/39999. Thus, we have only two figures of accuracy.

Since Gaussian elimination is stable, the only way we can get an inaccurate answer is for the problem to be ill-conditioned. And indeed it is. For the problem is an example of a nearly completely decomposable (NCD) chain; that is, one which, after a suitable reordering of the states, is almost block diagonal. For the case of three blocks, such a chain has the form

$$P = \begin{pmatrix} P_{11} & E_{12} & E_{13} \\ E_{21} & P_{22} & E_{23} \\ E_{31} & E_{32} & P_{33} \end{pmatrix},$$

where the matrices E_{ij} are small. Such chains were introduced by Simon and Ando [11], and have been studied extensively since (e.g., see [1,13]).

Since the E_{ij} are small, each of the matrices P_{ii} has an eigenvalue near one. Consequently the entire matrix, in addition to an eigenvalue of one, has $k-1$ eigenvalues near one, where k is the number of blocks. Consequently, the matrix B in (3.1) has $k-1$ eigenvalues near one and the condition number $\|(I-P)^{\#}\| = \|X(I-B)^{-1}Y^{\mathrm{T}}\|$ will be large.

In spite of this ill-conditioning, the problem can be solved by the following aggregation-disaggregation technique (here ϵ is the norm of the matrix consisting of the off-diagonal blocks):

1. Compute $\hat{y}_i^{\mathrm{T}}$, the Perron eigenvector[2] of P_{ii} normalized so that $\hat{y}_i^{\mathrm{T}}\mathrm{e} = 1$.
2. Compute $\epsilon_{ij} = \hat{y}_i^{\mathrm{T}} E_{ij} \mathrm{e}$ and $\pi_{ii} = \hat{y}_i^{\mathrm{T}} P_{ii} \mathrm{e}$.
3. The *coupling matrix*

$$
C = \begin{pmatrix}
\pi_{11} & -\epsilon_{12} & -\epsilon_{13} \\
-\epsilon_{21} & \pi_{22} & -\epsilon_{23} \\
-\epsilon_{31} & -\epsilon_{32} & \pi_{33}
\end{pmatrix}
$$

   is easily seen to be an irreducible stochastic matrix. Compute its steady-state vector (ν_1, ν_2, ν_3).
4. Then

$$
y^{\mathrm{T}} = \left(\nu_1 \hat{y}_1^{\mathrm{T}}, \nu_2 \hat{y}_2^{\mathrm{T}}, \nu_3 \hat{y}_3^{\mathrm{T}}\right) + O(\epsilon).
$$

The solution provided by the algorithm has two components: the vectors $\hat{y}_i^{\mathrm{T}}$ and the *coupling coefficients* ν_i. Provided the diagonal blocks P_{ii} are well behaved, the former will be insensitive to perturbations in P. On the other hand, unless the perturbation in P is *small compared to the E_{ij}*, the elements of C and hence the coupling coefficients will be poorly determined. Since the E_{ij} are small compared to the P_{ii}, small relative perturbations in P can be large compared to the E_{ij} and harm the solution.

All this agrees with the perturbation theory for NCD Markov chains. It can be shown [14] that under suitable regularity conditions the matrix P has a spectral decomposition of the form

$$
(4.2) \qquad P = \mathrm{e}y^{\mathrm{T}} + X_{\mathrm{s}} B_{\mathrm{s}} Y_{\mathrm{s}}^{\mathrm{T}} + X_{\mathrm{f}} B_{\mathrm{f}} Y_{\mathrm{f}}^{\mathrm{T}},
$$

where

$$
\begin{pmatrix}
y^{\mathrm{T}} \\
Y_{\mathrm{s}}^{\mathrm{T}} \\
Y_{\mathrm{f}}^{\mathrm{T}}
\end{pmatrix} = \left(\mathrm{e}\ X_{\mathrm{s}}\ X_{\mathrm{f}}\right)^{-1}.
$$

---

[2] The Perron eigenvector of an irreducible nonnegative matrix is the positive eigenvector corresponding to the largest positive eigenvalue.

The matrix B_s is a perturbation of the identity, while the matrix B_f has eigenvalues whose eigenvalues are bounded in magnitude away from one.[3] In analogy with (3.2) the perturbed steady-state vector, due to a perturbation F in P, can be written

$$(4.3) \qquad \tilde{y} \cong y^T + y^T X_s F (I - B_s)^{-1} Y_s^T + y^T X_f F (I - B_f)^{-1} Y_f^T.$$

Since the eigenvalues of B_s are near one while those of B_f are not, the second term in (4.3) will dominate; i.e., the perturbations will tend to lie in the space spanned by the rows of Y_s^T. However, it can be shown that the row space of Y_s^T is essentially the same as the row space of

$$\begin{pmatrix} \hat{y}_1^T & 0 & 0 \\ 0 & \hat{y}_2^T & 0 \\ 0 & 0 & \hat{y}_3^T \end{pmatrix}.$$

Consequently, the components of $\tilde{y}^T$ will tend to lie, as they should, along the directions of the vector $\hat{y}_i^T$; however, their relative proportions, which correspond to the coupling coefficients, will change. Thus, it is the coupling coefficients that are sensitive to changes in P, a fact which agrees with our comments on the aggregation algorithm.

5. The GTH Algorithm. The analysis of the preceding section shows that unless we know the elements of the E_{ij} to high relative accuracy, the steady-state vector of the chain will be ill-determined. In situations where the E_{ij} must be determined empirically, this accuracy may be difficult to achieve, since their elements correspond to events that occur only infrequently. On the other hand, in parameter studies, where the behavior of a system is being modeled by a Markov chain, the E_{ij} can be taken as fully accurate. Thus it is reasonable to pose the following problem: How do we compute the steady-state vector of a NCD chain when the E_{ij} are known to high accuracy?

The obvious answer is to use the aggregation algorithm. However, this answer begs the question; for the coupling matrix C is itself nearly completely decomposable and hence ill-conditioned. For example, suppose that

$$C = \begin{pmatrix} 0.9999 & 0.1499e{-}3 \\ 0.2499e{-}3 & 0.9998 \end{pmatrix}.$$

---

[3] The subscripts "s" and "f", which stand for slow and fast, have the following origin. It is easy to see that

$$P^i = e y^T + X_s B_s^i Y_s^T + X_f B_f^i Y_f^T.$$

Since the eigenvalues of B_s are near one, B_s^i approaches zero more slowly that the ith power of B_f, whose eigenvalues are smaller. Consequently the decomposition (4.2) exhibits two transient behaviors: a fast transient associated with B_f and a slow transient associated with B_s. This behavior was noted by Simon and Ando.

Note that to four decimal digits, C is a correctly rounded stochastic matrix. However,

$$(5.1) \qquad I - C = 10^{-3} \begin{pmatrix} 0.1000 & -0.1499 \\ -0.2499 & 0.2000 \end{pmatrix}$$

has the positive eigenvector

$$y = (0.4968 \ 0.5032),$$

whereas the corresponding eigenvector of the exactly stochastic matrix

$$C_{\text{true}} = \begin{pmatrix} 0.9998501 & 0.1499e{-}3 \\ 0.2499e{-}3 & 0.9997501 \end{pmatrix}.$$

is

$$y_{\text{true}} = (0.5000 \ 0.5000).$$

Thus the aggregation algorithm will produce coupling coefficients that are inaccurate in the third figure.

Looking carefully at this example, we see that the problem lies with the diagonal elements of $I - C$, which are inaccurate. However, since the rows of $I - C$ must sum to zero, we can restore the accuracy by replacing $I - C$ with

$$10^{-3} \begin{pmatrix} 0.1499 & -0.1499 \\ -0.2499 & 0.2499 \end{pmatrix}.$$

In the general aggregation algorithm, this amounts to computing the coupling matrix in the form

$$I - C = \begin{pmatrix} \epsilon_{12} + \epsilon_{13} & -\epsilon_{12} & -\epsilon_{13} \\ -\epsilon_{21} & \epsilon_{21} + \epsilon_{23} & -\epsilon_{23} \\ -\epsilon_{31} & -\epsilon_{32} & \epsilon_{31} + \epsilon_{32} \end{pmatrix}.$$

This procedure of adjusting the diagonals restores the figures that could not be represented in the diagonals of the original matrix C.

The idea of diagonal adjustment can be applied to Gaussian elimination. For example, the trailing 2×2 principal submatrix of (4.1) should have zero column sums. Since it does not, we force them to be by replacing the diagonals with the sum of the off diagonals, to get the matrix

$$Q = \begin{pmatrix} +0.75287 & -0.75283 & -0.00003 \\ +0.0 & +0.000039999 & -0.000039999 \\ +0.0 & -0.000039999 & +0.000039999 \end{pmatrix},$$

which gives the right answer.

In general, after k steps of Gaussian elimination applied to Q^{T}, the matrix assumes the form

$$\begin{pmatrix} U_{11}^{(k)} & U_{12}^{(k)} \\ 0 & U_{22}^{(k)} \end{pmatrix},$$

where $U_{11}^{(k)}$ is of order k. Since the row sums of $U_{22}^{(k)}$ are known to be zero, instead of using Gaussian elimination to compute its diagonal elements, we use the alternative formula

$$(5.2) \qquad u_{jj}^{(k)} = - \sum_{\substack{i=k+1 \\ i \neq j}}^{i=n} u_{ij}, \qquad j = k+1, \ldots, n.$$

This, in essence, is the algorithm of Grassmann, Taksar, and Heyman [6] mentioned in the introduction. There are three points to be made about it.

In the first place, it is easy to implement and not very expensive. The sums (5.2) can be accumulated as the off-diagonal elements are generated during the elimination, which increases the work done in the inner loop by a single addition.[4]

Second, the numerical properties of the algorithm are not well understood. A possible justification is that the method never subtracts and hence cancellation cannot cause it to fail. Unfortunately, this line of reasoning, when formalized, is essentially the analysis of Hotelling mentioned in the introduction, and it leads to the same pessimistic conclusions. With G. Zhang, I have given a rounding-error analysis of a closely related algorithm [15]; however, the original algorithm remains unanalyzed. Nontheless, I believe that the GTH algorithm is stable and should be used routinely in the direct solution of Markov chains.

Third, the above discussion is a little unfair to Gaussian elimination, which is frequently asked to do the impossible: solve a problem that is not in the computer. The matrix (5.1) is a case in point. Owing to initial rounding errors, it is only approximately singular and Gaussian elimination will do a very good job of computing an approximate null vector. The fact that this vector is not what we want cannot be blamed on Gaussian elimination, which has no way of knowing it is dealing with a Markov chain. In this light, the GTH algorithm is seen to be an augmentation of Gaussian elimination that does know.[5]

---

[4] This assumes that the exterior product form of Gaussian elimination is used. If inner-product forms of the kind associated with the names of Crout and Doolittle are used, the algorithm is a little more complicated, though the amount of extra work stays the same.

[5] Some of the examples appearing in the literature purporting to show that Gaussian elimination fails are of the same nature: the damage is done in the initial rounding of problem.

REFERENCES

[1] P.-J. Courtois. *Decomposability.* Academic Press, New York, 1977.

[2] I. S. Duff, A. M. Erisman, and J. K. Reid. *Direct Methods for Sparse Matrices.* Clarendon Press, Oxford, 1986.

[3] R. E. Funderlic. Sensitivity of the stationary distribution vector for an ergodic Markov chain. *Linear Algebra and Its Applications,* 76:1–17, 1986.

[4] R. E. Funderlic and J. B. Mankin. Solution of homogeneous systems of linear equations arising from compartmental models. *SIAM Journal on Scientific and Statistical Computing,* 2:375–109, 1981.

[5] G. H. Golub and C. F. Van Loan. *Matrix Computations.* Johns Hopkins University Press, Baltimore, Maryland, 2nd edition, 1989.

[6] W. K. Grassmann, M. I. Taksar, and D. P. Heyman. Regenerative analysis and steady state distributions. *Operations Research,* 33:1107–1116, 1985.

[7] M. Haviv and L. van der Heyden. Perturbation bounds for the stationary probabilities of a finite Markov chain. *Advances in Applied Probability,* 16:804–818, 1984.

[8] H. Hotelling. Some new methods in matrix calculations. *Annals of Mathematical Statistics,* 14:1–34, 1943.

[9] C. D. Meyer. The condition of a Markov chain and perturbations bounds for the limiting probabilities. *SIAM Journal on Algebraic and Discrete Methods,* 1:273–283, 1980.

[10] P. J. Schweitzer. Perturbation theory and finite Markov chains. *Journal of Applied Probability,* 5:401–413, 1968.

[11] H. A. Simon and A. Ando. Aggregation of variables in dynamic systems. *Econometrica,* 29:111–138, 1961.

[12] G. W. Stewart. *Introduction to Matrix Computations.* Academic Press, New York, 1973.

[13] G. W. Stewart. On the structure of nearly uncoupled Markov chains. In G. Iazeolla, P. J. Courtois, and A. Hordijk, editors, *Mathematical Computer Performance and Reliability,* pages 287–302, Elsevier, New York, 1984.

[14] G. W. Stewart. Perturbation theory for nearly uncoupled Markov chains. In W. J. Stewart, editor, *Numerical Methods for Markov Chains,* pages 105–120, North Holland, Amsterdam, 1990.

[15] G. W. Stewart and G. Zhang. On a direct method for the solution of nearly uncoupled Markov chains. *Numerische Mathematik,* 59:1–11, 1991.

[16] J. von Neumann and H. H. Goldstine. Numerical inverting of matrices of high order. *Bulletin of the American Mathematical Society,* 53:1021–1099, 1947.

[17] D. S. Watkins. *Fundamentals of Matrix Computations.* John Wiley & Sons, New York, 1991.

[18] J. H. Wilkinson. Error analysis of direct methods of matrix inversion. *Journal of the ACM,* 8:281–330, 1961.

ALGORITHMS FOR PERIODIC MARKOV CHAINS

FRANÇOIS BONHOURE, YVES DALLERY* AND WILLIAM J. STEWART†

Abstract. Our concern in this paper is the development of algorithms for solving queueing network models whose underlying Markov chain is p-Cyclic. We illustrate how the period may be conviently determined and the transition matrix permuted to normal cyclic form. We show how numerical algorithms can take advantage of the periodic structure, both on the complete state space and on a reduced level. On area of particular interest occurs when the model is both periodic and nearly-completely- decomposable. In this case we show how iterative aggregation/disaggregation algorithms may be modified to take advantage of the periodicity property.

1. Introduction. Markov chains sometimes possess the property that the minimum number of transitions that must be made on leaving any state to return to that state, is a multiple of some integer $p > 1$. These models are said to be periodic of period p or *p-cyclic of index p*. Bonhoure and Dallery, [2], have shown that Markov chains that arise from queueing network models frequently possess this property. We shall present numerical algorithms to compute the stationary probability vector of such Markov chains. Not all of the details will be presented in this version of our paper. Rather our concern will be with presenting the main computational algorithms and we shall leave further details including proofs of theorems, certain aspects concerning implementation of preconditioning, etc. to the full paper, [2]. The discussion of iterative aggregation/disaggregation methods for Markov chains that are both nearly-completely-decomposable *and* cyclic which is presented in this version is not in the original version.

Consider, as an example, the queueing network model described below. All the stations contain a single exponential server except the second which consists of a two-phase Erlang server. The service rates are given by μ_i, $i = 1, 2, \cdots, 5$, and are as marked in the figure.

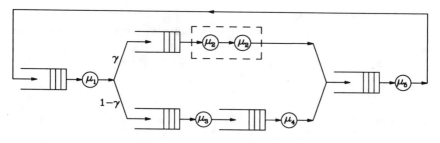

FIG. 1.1. *Sample Queueing Network*

* MASI, Université de Paris VI, France.

† North Carolina State University, Department of Computer Science. Research supported in part by NSF (DDM 8906248).

A state of the system may be completely described by the vector

$$(n_1, \ n_2, \ k, \ n_3, \ n_4, \ n_5)$$

where n_i describes the number of customers in station i and k denotes the service phase at the Erlang server. For example, when the number of customers in the network is given by $N = 2$, then the system contains 20 states as follows:

	n_1	n_2	k	n_3	n_4	n_5
1.	2	0	0	0	0	0
2.	1	1	1	0	0	0
3.	1	1	2	0	0	0
4.	1	0	0	1	0	0
5.	1	0	0	0	1	0
6.	1	0	0	0	0	1
7.	0	2	1	0	0	0
8.	0	2	2	0	0	0
9.	0	1	1	1	0	0
10.	0	1	1	0	1	0
11.	0	1	1	0	0	1
12.	0	1	2	1	0	0
13.	0	1	2	0	1	0
14.	0	1	2	0	0	1
15.	0	0	0	2	0	0
16.	0	0	0	1	1	0
17.	0	0	0	1	0	1
18.	0	0	0	0	2	0
19.	0	0	0	0	1	1
20.	0	0	0	0	0	2

The following graph displays all the transitions that are possible among these states. In this model, a return to any state is possible only in a number of transitions that is some positive (non-zero) multiple of 4.
For example, the following paths from state 5 back to state 5 are possible

```
5 ---> 6 ---> 1 ---> 4 ---> 5    of length 4,
5 --->10 --->13 --->19 --->20 ---> 6 ---> 1 ---> 4 ---> 5  of length 8,
etc.
```

From Figure 1.2, let us now group the states according to their generated distance from the initial state This is shown in figure 1.3, where the groups have been labelled $A, B, \cdots, G$. We have

$$
\begin{aligned}
A &= \{1\} \\
B &= \{2,4\} \\
C &= \{3,7,5,9,15\} \\
D &= \{6,8,10,12,16\}
\end{aligned}
$$

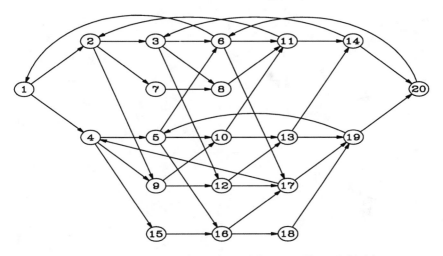

FIG. 1.2. *Possible State Transitions of Queueing Network Model.*

$$E = \{11, 13, 17, 18\}$$
$$F = \{14, 19\}$$
$$G = \{20\}$$

We shall use the term *pre-class* to designate a set of states A, B, $\cdots$, G.

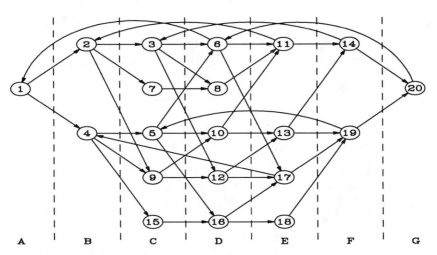

FIG. 1.3. *The Pre-classes of the Markov Chain.*

If we construct a graph which shows only the transitions among these pre-classes, we get the following:

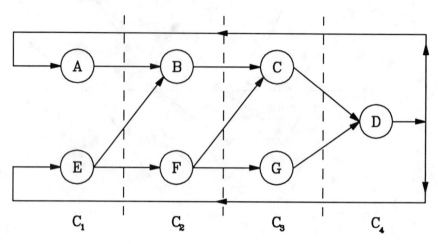

FIG. 1.4. *The Periodic-Classes of the Markov Chain.*

Finally, if we now consider only the transitions among these pre-classes, and rearrange the pre-classes as we did for the original states, we find the four *periodic classes* of the Markov chain, C_1, C_2, C_3, C_4.

$$
\begin{aligned}
C_1 &= A \cup E \\
C_2 &= B \cup F \\
C_3 &= C \cup G \\
C_4 &= D
\end{aligned}
$$

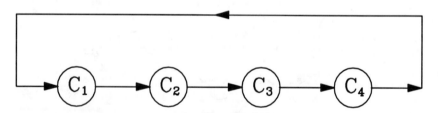

FIG. 1.5. *Transitions among the Periodic-Classes.*

Let us consider the structure of the infinitesimal generator when the states are ordered according to their periodic classes. The order of the states within any particular periodic class is unimportant.

New ordering: 1, 11, 13, 17, 18;   2, 4, 14, 19;   3, 7, 5, 9, 15, 20;   6, 8, 10, 12, 16.
Periodic class: C_1 C_2 C_3 C_4

We indicate an non-zero element of the infinitesimal generator by the symbol $\times$, a zero element by $\cdot$, and diagonal elements, which are equal to the negated sum of the off-diagonal elements, by an asterisk.

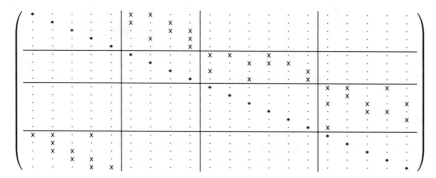

This structure shows that the matrix may be written in the form

$$
\begin{pmatrix}
D_1 & Q_1 & \cdot & \cdot \\
\cdot & D_2 & Q_2 & \cdot \\
\cdot & \cdot & D_3 & Q_3 \\
Q_4 & \cdot & \cdot & D_4
\end{pmatrix}
$$

in which the D_i are square diagonal submatrices.

In this paper we shall show how to make use of this structure to efficiently compute stationary probability vectors.

2. P-Cyclic Markov Chains

2.1. The Embedded Markov Chain. For every continuous time Markov chain, it is possible to define a discrete time chain at the instants of state departure. This is called the *embedded* Markov chain. Let S be the transition probability matrix for this discrete time Markov chain (DTMC). The probabilities s_{ij} of a transition from state i to state j are defined by:

$$
\begin{aligned}
s_{ij} &= \frac{q_{ij}}{-q_{ii}} \text{ for } i \neq j \\
s_{ij} &= 0 \text{ for } i = j
\end{aligned}
$$

Given the usual splitting of the infinitesimal generator Q:

$$
Q = D_Q - L_Q - U_Q
$$

where D_Q is the diagonal of Q, $-L_Q$ is the strictly lower triangular part of Q and $-U_Q$ is the strictly upper triangular part of Q, it is possible to characterize the matrix S as:

$$
(2.1) \qquad S = (I - D_Q^{-1}Q) = D_Q^{-1}(D_Q - Q) = D_Q^{-1}(L_Q + U_Q)
$$

Note that S is the Jacobi iteration matrix associated with the matrix Q, that it is stochastic and that its diagonal is zero. Since Q is irreducible,

S is also irreducible. The problem of determining the stationary probability vector of the original Markov chain is therefore that of finding the left-hand eigenvector x associated with the unique unit eigenvalue of S.

(2.2) $x = xS$ with $x > 0$ and $||x||_1 = 1$.

Indeed, the stationary solution of the CTMC may be obtained from:

(2.3) $$\pi = \frac{-xD_Q^{-1}}{||xD_Q^{-1}||_1}$$

2.2. Markov Chains with Periodic Graphs. For every Markov chain, whether continuous-time or discrete-time, we may associate a directed graph in which the nodes (or vertices) correspond to the states and the edges correspond to transitions between states. Note that, with the exception of loops, the graph of the embedded Markov chain is identical to that of the original continuous-time Markov chain. In what follows, we are interested in those cases in which the graph associated with the CTMC is periodic.

LEMMA 2.1. *If the graph of the original CTMC is periodic of period p, then the matrix S associated with the embedded Markov chain is also periodic of period p.*

Since S is periodic, it may be permuted to the following periodic form by means of a permutation matrix P:

$$S = \begin{pmatrix} 0 & S_1 & 0 & \cdots & 0 & 0 \\ 0 & 0 & S_2 & \cdots & 0 & 0 \\ 0 & 0 & 0 & \cdots & 0 & 0 \\ \vdots & \vdots & \vdots & \ddots & \vdots & \vdots \\ 0 & 0 & 0 & \cdots & 0 & S_{p-1} \\ S_p & 0 & 0 & \cdots & 0 & 0 \end{pmatrix}$$

Note that the permutation operation consists uniquely of re-ordering the states. Since the matrix is periodic, this re-ordering defines a partition of the set of states into p *classes* in such a way that the only transitions possible are from a state of class i to a state of the class $(i \bmod p) + 1$. With this re-ordering of the states, Q may be written as:

$$Q = \begin{pmatrix} D_1 & Q_1 & 0 & \cdots & 0 & 0 \\ 0 & D_2 & Q_2 & \cdots & 0 & 0 \\ 0 & 0 & D_3 & \cdots & 0 & 0 \\ \vdots & \vdots & \vdots & \ddots & \vdots & \vdots \\ 0 & 0 & 0 & \cdots & D_{p-1} & Q_{p-1} \\ Q_p & 0 & 0 & \cdots & 0 & D_p \end{pmatrix},$$

where the diagonal blocks D_i are themselves diagonal matrices. In what follows, we assume that the states of a CTMC with periodic graph are ordered so that the infinitesimal generator Q and the stochastic matrix S have the above structure. Let us partition π and x according to this periodic structure. That is:

$$\pi = (\pi_1, \pi_2, \cdots, \pi_i, \cdots, \pi_p),$$

$$x = (x_1, x_2, \cdots, x_i, \cdots, x_p).$$

Since the D_i's are diagonal, we have, from equations 2.1 and 2.3

$$(2.4) \qquad S_i \;=\; -D_i^{-1} Q_i,$$

$$(2.5) \qquad \pi_i \;=\; \frac{-x_i D_i^{-1}}{\|x D_Q^{-1}\|_1}.$$

2.3. Computation of the Periodicity. Let $G_M = (E, T)$ be the graph of the CTMC. The algorithm below constructs a graph $G_C = (F, V)$ whose period is the same as that of G_M and is easily computed.
Algorithm for the construction of G_C

$$F \leftarrow \emptyset, \quad V \leftarrow \emptyset, \quad k \leftarrow 1$$

Choose an arbitrary initial state $e_d \in E$

$$F_1 \leftarrow \{e_d\}, \quad F \leftarrow F_1$$

(1) $F_{k+1} \leftarrow \emptyset$

For each state $e \in F_k$

      For each state $e' \in E$ such that $(e, e') \in T$

          If $\exists\, k' \leq k$ such that $e' \in F_{k'}$ then

               $V \leftarrow V \cup \{(F_k, F_{k'})\}$

          else

               $F_{k+1} \leftarrow F_{k+1} \cup \{e'\}$

          endif

      end

end

If $F_{k+1} \neq \emptyset$ then

      $F \leftarrow F \cup F_{k+1}$

      $V \leftarrow V \cup \{(F_k, F_{k+1})\}$

      $k \leftarrow k + 1$

      goto (1)

else

      stop

endif

This algorithm constructs a graph G_C in which each node is a set of states of E. Each node of G_C is a *pre-class*. We have the following theorem.

THEOREM 2.2. *Let G_M be strongly connected. Then the graph G_C has the same periodicity as G_M.*

Thus, the computation of the periodicity of the graph G_M of a CTMC may be obtained from the graph G_C of identical periodicity. The construction of G_C corresponds to a pass through G_M from some initial state. Note that this pass is often that which is performed during the construction of the infinitesimal generator of the CTMC. A separate independent pass to compute G_C is unnecessary. The cost of constructing G_C is therefore small.

Furthermore, it is not necessary to compute G_C explicitly to compute its period. Indeed, it is only the lengths of the elementary cycles of G_C that are needed to compute its period, and all of the elementary cycles of G_C appear during the construction of the CTMC. In fact, if at stage k of the algorithm, state e' generated from state $e \in F_k, (e, e') \in T$, already belongs to an existing pre-class $F_{k'}$, we form an elementary cycle of G_C of length $k - k' + 1$. Since the gcd is an associative operation, it suffices to compute the gcd of these lengths to compute the period of G_C.

In conclusion, to compute the period of a graph associated with a CTMC we need to associate an integer with each state which represents the pre-class to which it belongs, and to compute a number, n_p, of gcd's of two elements. To compute the period of G_M directly from G_M, it is necessary to explicitly construct all the elementary cycles of G_M and this is extremely costly both in terms of memory requirements and computation time. The advantage of the proposed approach is precisely that it avoids the explicit construction of these cycles.

3. Numerical Methods Applied to P-Cyclic Matrices.

As always, we wish to compute the stationary probability vector π of a finite state, continuous-time Markov chain (CTMC), defined by an infinitesimal generator, Q. Let n be the number of states of the CTMC. Recall that q_{ij} represents the rate of transition from state i to state j, and that:

$$q_{ij} \geq 0 \text{ for } i \neq j; \quad q_{ii} = -\sum_{j \neq i} q_{ij}$$

We assume that Q is irreducible. In this case, the Markov chain is ergodic and the stationary probability vector π exists and is unique. The element π_i is the probability of being in state i when the system reaches statistical equilibrium and may be found by solving the system of linear homogeneous equations:

$$(3.1) \qquad \pi Q = 0 \text{ with } \pi > 0 \text{ and } ||\pi||_1 = 1$$

where $||\pi||_1 = \sum_i |\pi_i|$.

In this section, we discuss the use of the periodic structure of Q in calculating the stationary probability vector. Our aim is to compare the efficiency of methods applied when the matrix Q has been permuted to normal cyclic form with respect to that of the same method applied to Q in its original, non-permuted form.

Let us first note that there will be no difference when using methods that do not depend on the ordering of the states. This, in particular, is the case of the power method, the method of Arnoldi and *GMRES*. As a result, we only consider methods in which ordering is important. We shall successively consider direct *LU* decompositions, Gauss-Seidel, and preconditioned methods.

3.1. Direct Methods. Consider a direct *LU* decomposition of Q. The principle of the decomposition is to express Q as the product of two matrices L and U, i.e., $LU = Q$, L being a lower triangular matrix and U being an upper triangular matrix. Recall the form of the infinitesimal generator:

$$Q = \begin{pmatrix} D_1 & Q_1 & 0 & \cdots & 0 & 0 \\ 0 & D_2 & Q_2 & \cdots & 0 & 0 \\ 0 & 0 & D_3 & \cdots & 0 & 0 \\ \vdots & \vdots & \vdots & \ddots & \vdots & \vdots \\ 0 & 0 & 0 & \cdots & D_{p-1} & Q_{p-1} \\ Q_p & 0 & 0 & \cdots & 0 & D_p \end{pmatrix}$$

in which the diagonal matrices are themselves diagonal. Thus, with the exception of the last block row, this is already upper triangular. An *LU* decomposition must therefore have the following form:

$$\begin{pmatrix} I & 0 & \cdots & 0 & 0 \\ 0 & I & \cdots & 0 & 0 \\ \vdots & \vdots & \ddots & \vdots & \vdots \\ 0 & 0 & \cdots & I & 0 \\ L_1 & L_2 & \cdots & L_{p-1} & L_p \end{pmatrix} \begin{pmatrix} D_1 & Q_1 & \cdots & 0 & 0 \\ 0 & D_2 & \cdots & 0 & 0 \\ \vdots & \vdots & \ddots & \vdots & \vdots \\ 0 & 0 & \cdots & D_{p-1} & Q_{p-1} \\ 0 & 0 & \cdots & 0 & U_p \end{pmatrix} =$$

$$\begin{pmatrix} D_1 & Q_1 & \cdots & 0 & 0 \\ 0 & D_2 & \cdots & 0 & 0 \\ \vdots & \vdots & \ddots & \vdots & \vdots \\ 0 & 0 & \cdots & D_{p-1} & Q_{p-1} \\ Q_p & 0 & \cdots & 0 & D_p \end{pmatrix}$$

The *LU* decomposition is known once the blocks $L_1, L_2, \cdots, L_p$ and U_p are known. These are easily determined, since

$$L_1 = Q_p D_1^{-1}$$

and, for $i = 2, \cdots, p - 1$,

$$L_i = (-L_{i-1} Q_{i-1}) D_i^{-1}$$

and the only inverses needed so far are those of diagonal matrices. Finally, L_p and U_p may be obtained by performing a decomposition of the right-hand side of:

$$(3.2) \quad L_p U_p = (D_p - L_{p-1} Q_{p-1})$$
$$(3.3) \qquad\quad = D_p + (-1)^{p-1} Q_p (D_1^{-1} Q_1)(D_2^{-1} Q_2) \cdots (D_{p-1}^{-1} Q_{p-1})$$

In fact, the only block that is actually needed is U_p since it is the only one that is used in the backsubstitution. In certain cases, it may be advantageous to compute the right-hand side of 3.2 explicitly. This depends on the dimensions of the individual Q_i and on the order in which the multiplications are performed. Since the amount of fill-in generated in producing matrices L and U is likely to be much lower when Q is in periodic form, it is apparent that savings in time and memory will often result from using the periodic structure of Q as opposed to a structure resulting from an arbitrary ordering of the states.

Finally, note that from the definition of T_p in equation 4.6, equation 3.2 yields:

$$L_p U_p = T_p,$$

i.e., $L_p U_p$ is an LU decomposition of T_p, the matrix involved in the reduced scheme associated with the infinitesimal generator Q.

3.2. Gauss-Seidel applied to Q. We shall now apply the method of Gauss-Seidel to the infinitesimal generator, Q, of the CTMC when Q is in normal cyclic form. Gauss-Seidel consists of applying the following iterative scheme:

$$\pi^{(k+1)}(D_Q - U_Q) = \pi^{(k)} L_Q$$

where $\pi^{(k)} = (\pi_1^{(k)}, \pi_2^{(k)}, \cdots, \pi_p^{(k)})$ is the approximation to $\pi = (\pi_1, \pi_2, \cdots, \pi_p)$ at step k. Due to the special periodic structure of Q, this yields:

$$(3.4) \qquad \pi_1^{(k+1)} D_1 = -\pi_p^{(k)} Q_p$$
$$\pi_2^{(k+1)} D_2 = -\pi_1^{(k+1)} Q_1$$
$$\cdots = \cdots$$
$$\pi_p^{(k+1)} D_p = -\pi_{p-1}^{(k+1)} Q_{p-1}$$

from which we may deduce

$$(3.5) \qquad \pi_p^{(k+1)} D_p = -\pi_p^{(k)} Q_p (-D_1^{-1} Q_1) \cdots (-D_{p-1}^{-1} Q_{p-1})$$

Using equations 2.4 and 2.5, equation 3.5 may be written as:

$$(3.6) \qquad x_p^{(k+1)} = x_p^{(k)} S_p S_1 S_2 \cdots S_{p-1} = x_p^{(k)} R_p$$

Since R_p is a stochastic, irreducible and acyclic, convergence of the iterative scheme 3.6 and as a result that of Gauss-Seidel, is guaranteed.

4. Reduced Schemes

4.1. Reduced Schemes Associated with a Stochastic Matrix

We now show that, due to the special cyclic structure of matrices Q and S, it is possible to transform the original system of equations into a system of equations involving only a subset of the probability vector.

Let us first consider matrix S. S is a stochastic, irreducible and periodic matrix of period p. It has a unique left-hand eigenvector x associated with its unit eigenvalue. We have:

$$x = xS, \quad x > 0, \quad ||x_i||_1 = \frac{1}{p}$$

Furthermore, the pth roots of unity

$$\beta_k = e^{j \frac{2k\pi}{p}}, \quad k = 0, 1, \cdots, p-1$$

are eigenvalues of S with multiplicity one. Indeed all non-zero eigenvalues of S appear in p-tuples, each member of a tuple having the same multiplicity. In other words, if μ is a non-zero eigenvalue of S, then $\mu\beta_k$ $k = 1, 2, \cdots, p-1$, is also an eigenvalue of S with the same multiplicity as μ, [3].

Since $(\beta_k)^p = 1$ for $k = 0, 1, \cdots p-1$, the matrix $R = S^p$ has exactly p eigenvalues equal to one and no others of modulus one. It is thus a completely decomposable (block diagonal) stochastic matrix and may be written as:

$$R = \begin{pmatrix} R_1 & 0 & 0 & \cdots & 0 & 0 \\ 0 & R_2 & 0 & \cdots & 0 & 0 \\ \vdots & \vdots & \vdots & \ddots & \vdots & \vdots \\ 0 & 0 & 0 & \cdots & R_{p-1} & 0 \\ 0 & 0 & 0 & \cdots & 0 & R_p \end{pmatrix}$$

with

$$\begin{aligned} R_1 &= S_1 S_2 \cdots S_{p-1} S_p \\ R_2 &= S_2 S_3 \cdots S_p S_1 \\ &\cdots \qquad \cdots \\ R_p &= S_p S_1 \cdots S_{p-2} S_{p-1} \end{aligned}$$

Note that matrix R_i, $i = 1, \cdots, p$, is the matrix of p-step transition probabilities of the embedded Markov chain.

LEMMA 4.1. *Each R_i, $i = 1, \ldots, p$, is a stochastic, irreducible and acyclic matrix.*

Proof.

1. R_i is stochastic since R is stochastic and completely decomposable.
2. Since S is irreducible, there is, in particular, a path between any two states of a given aggregate. Since S is cyclic of period p, the length of this path is a multiple of p. Each transition in R corresponds to p consecutive steps in S. Therefore there is a path in R between the two states. This implies that each submatrix of R is irreducible.
3. The matrix R possesses only p eigenvalues of modulus one (in fact all equal to one), and one for each submatrix. Thus it is not possible for any submatrix to possess more than one eigenvalue of unit modulus. Since R_i is stochastic (with dominant eigenvalue equal to 1), this implies that R_i cannot be cyclic.

□

Now, due to the periodic structure of S, we have:

$$
\begin{aligned}
x_1 &= x_p S_p \\
x_2 &= x_1 S_1 \\
\cdots &= \cdots \\
x_p &= x_{p-1} S_{p-1}
\end{aligned}
$$

(4.1)

and thus we can deduce, for example that:

$$x_p = x_p S_p S_1 S_2 \cdots S_{p-1}$$

which leads to an equation of the form:

(4.2) $$x_p = x_p R_p$$

with

(4.3) $$R_p = S_p S_1 S_2 \cdots S_{p-1}$$

Equation 4.2 is called the *reduced scheme* associated with the stochastic matrix S, because it involves only a part of the original probability vector x. It is evident from equation 4.2 that x_p is an eigenvector corresponding to the unit eigenvalue of the stochastic matrix R_p. Note that it is sufficient to work with the reduced scheme since once x_p is obtained, the other vectors x_i may be determined from x_p by a simple matrix-vector multiplication according to equations 4.1.

4.2. Reduced Schemes Associated with an Infinitesimal Generator. Whereas equation 4.2 relates to a stochastic matrix, it is also possible to derive a similar equation using the infinitesimal generator. Recall that π is a solution of:

$$\pi Q = 0 \quad \text{with} \quad \pi > 0 \quad \text{and} \quad ||\pi||_1 = 1$$

Due to the periodic structure of Q, we have:

$$(4.4) \qquad \begin{aligned} \pi_1 D_1 &= -\pi_p Q_p \\ \pi_2 D_2 &= -\pi_1 Q_1 \\ \cdots &= \cdots \\ \pi_p D_p &= -\pi_{p-1} Q_{p-1} \end{aligned}$$

from which we may derive

$$\pi_p D_p = -\pi_p Q_p(-D_1^{-1})Q_1 \cdots (-D_{p-1}^{-1})Q_{p-1}.$$

This leads to an equation of the form

$$(4.5) \qquad \pi_p T_p = 0$$

with

$$(4.6) \qquad T_p = D_p + Q_p(-D_1^{-1}Q_1)\cdots(-D_{p-1}^{-1}Q_{p-1}).$$

Equation 4.5 is called the *reduced scheme* associated with the infinitesimal generator Q. It involves only a part of the original probability vector π. Again, it is sufficient to work with the reduced scheme since, once π_p is obtained, the other vectors π_i may be determined from π_p by a simple matrix-vector multiplication. Indeed, according to equations 4.4, we have:

$$\begin{aligned} \pi_1 &= -\pi_p Q_p D_1^{-1} \\ \pi_2 &= -\pi_1 Q_1 D_2^{-1} \\ \cdots &= \cdots \\ \pi_{p-1} &= -\pi_{p-2} Q_{p-2} D_{p-1}^{-1} \end{aligned}$$

It is worthwhile noticing that both reduced schemes are related. Indeed, matrices T_p and R_p are related by:

$$(4.7) \qquad T_p = D_p(I - R_p)$$

since equation 4.6 may be written as:

$$(4.8) \qquad T_p = D_p(I - (-D_p^{-1}Q_p)(-D_1^{-1}Q_1), \cdots, (-D_{p-1}^{-1}Q_{p-1})$$

which, using equations 2.4 and 4.3, leads to 4.7. Note that since R_p is stochastic and irreducible, it follows from 4.7 that T_p is an irreducible infinitesimal generator.

It may be of interest to use either of the reduced schemes (4.5 or 4.2) instead of the corresponding original equation (3.1 or 2.2). All of the blocks R_i have the same spectrum [3]. Since the eigenvalues of R are those of S raised to the power p, the spectrum of R_i is more suitable for iterative methods than that of S. Clearly the same is true for T_p.

4.3. Methods Based on the Reduced Scheme. We now show how
the reduced schemes may be used to reduce computational complexity.
We shall not consider any method that requires the matrix T_p (or R_p) in
explicit form since that would, in general, be too costly. In particular, this
eliminates LU decompositions and Gauss-Seidel iterations. On the other
hand, methods that only need to perform multiplication with the matrix
such as the power method, the method of Arnoldi or $GMRES$, may be used.

4.3.1. Iterative Methods. The power method on S consists of applying the following iterative scheme:

$$x^{(k+1)} = x^{(k)}S$$

For an arbitrary initial vector $x^{(0)}$, this will not converge because there are
p eigenvalues of unit modulus. There will, however, be convergence of the
$x_i^{(k)}$ towards the individual x_i, correct to a multiplicative constant. Since
we know that $||x_i||_1 = 1/p$, it is therefore possible to determine x with
the power method applied to the matrix S. Note that if $x^{(0)}$ is such that
$||x_i^{(0)}||_1 = 1/p$, then the power method will converge directly towards x.

We now show that it is possible to go p times faster. It is evident
from equation 4.2 that x_p is the eigenvector corresponding to the unit
eigenvalue of the stochastic matrix R_p. It is therefore possible to apply the
power method to this matrix to determine the vector x_p:

$$(4.9) \qquad x_p^{(k+1)} = x_p^{(k)}S_pS_1S_2\cdots S_{p-1} = x_p^{(k)}R_p$$

We shall call this the *accelerated* power method. Its convergence is guaranteed because the stochastic matrix R_p is irreducible and acyclic. It is
important not to compute the matrix R_p explicitly. Rather, the multiplications of 4.9 should be performed succcessively. In this way, the total cost
of a step is equal to the number of non-zero elements in the matrix S.

Let us now compare the regular power method on S and the accelerated
power method. The amount of computation per iteration step is identical
for both methods; it is equal to the number of non-zero elements in S.
On the other-hand, the application of one step of the accelerated power
method is identical to p steps of the regular power method. Therefore
the accelerated power method will converge p times faster with the same
amount of computation per iteration. This result may be interpreted as a
consequence of the better spectral properties of R_p with respect to those
of S.

Other iterative methods like Arnoldi, [1], $GMRES$, etc., whose only
involvement with a matrix involves multiplying it with a vector, can also
be applied to the reduced scheme 4.5. Again these methods will make use
of the better spectral properties of T_p with respect to those of Q. The
improvement obtained by these methods is the same order of magnitude
as that achieved by the power method. Moreover, the construction of the

orthonormal basis of the Krylov subspace requires less computation time since the dimension of the reduced scheme is of the order η/p where η is the number of states of the Markov chain. Note that since the spectrum of all matrices T_i is the same, it is advantageous to work with the block of smallest size (which cannot exceed η/p).

5. A/D Methods for NCD, P-Cyclic Markov Chains

5.1. Aggregation/Disaggregation (A/D) Methods

Aggregation/disaggregation (A/D) methods are iterative methods that have been used for the numerical solution of large Markov chains; (see [4,5] and the references therein). In this section, we discuss the use of the periodicity property when implementing A/D methods for solving CTMC's that are periodic.

In these methods, the state space is partitioned into a set of J blocks. Each iteration of an A/D procedure consists of two steps. In the first, an approximation to the stationary probability vector is used to compute an approximation to the conditional probability of each state of block j, $1 \leq j \leq J$, conditioned on the CTMC being in one of the states of block j. This is sometimes referred to as the "disaggregation step". In the second step, an approximation to the probability of being in each of the blocks is obtained and this used to remove the conditional from the probabilities obtained in the first step. This is sometimes called the "aggregation step" or "coupling step" and it yields a new approximation with which to begin a new iteration.

The part of the disaggregation step corresponding to block j , is obtained by solving the system of equations:

$$(5.1) \qquad \pi^j Q^j = b^j, \quad ||\pi^j||_1 = 1,$$

where π^j is an approximate vector of conditional probabilities of block j, Q^j is the submatrix of the infinitesimal generator, Q, corresponding to the states of block j, and b^j is a vector obtained from the previous disaggregation and aggregation steps. The coupling step involves generating the $J \times J$ stochastic matrix whose jk^{th} element is given by

$$\frac{\pi^j}{||\pi^j||_1} S_{jk} e$$

and then determining its stationary probability vector. When the number of blocks is small with respect to the number of states, and this is most often the case, the computational complexity of the coupling step is negligible compared to that of the disaggregation steps.

A/D methods have proven to be especially efficient when applied to *nearly completely decomposable (NCD)* Markov chains. A Markov chain is to be NCD if its states can be partitioned into groups such that the transition rates between states of different groups is small compared to the transition rates between states of the same group. When applying an A/D method to an NCD Markov chain, a block is associated with each group of the Markov chain. With this partitioning of the state space, A/D methods usually converge very rapidly.

5.2. State Space Ordering. We now show how to take advantage of the periodicity of the Markov chain when using A/D methods. Consider a Markov chain whose associated graph is periodic of period p. We assume that the set of states is partitioned into J blocks. Note that this partitioning of the state space into blocks is not at all related to the partitioning of the state space into aggregates according to the periodic structure of the Markov chain.

Consider the submatrix Q^j corresponding to block j. The graph associated with this submatrix, G^j, is a subgraph of the graph of the CTMC. As a result, the length of all elementary cycles of G^j is a multiple of p. Therefore, it is possible to order the states corresponding to block j so that the submatrix Q^j has the following form:

$$(5.2) \qquad Q^j = \begin{pmatrix} D_1^j & Q_1^j & 0 & \cdots & 0 & 0 \\ 0 & D_2^j & Q_2^j & \cdots & 0 & 0 \\ 0 & 0 & D_3^j & \cdots & 0 & 0 \\ \vdots & \vdots & \vdots & \ddots & \vdots & \vdots \\ 0 & 0 & 0 & \cdots & D_{p-1}^j & Q_{p-1}^j \\ Q_p^j & 0 & 0 & \cdots & 0 & D_p^j \end{pmatrix}$$

Notice that the diagonal blocks D_i^j are themselves diagonal matrices and that the structure of Q^j is identical to that of Q when Q is given in periodic form.

Given a partitioning of the state space into blocks, it is not difficult to permute the elements of each block so that it has the form 5.2. The periodic class of each state is determined as usual during the generation of the matrix, and it suffices to permute the states of each block accordingly. The net result is that the states are ordered, first with respect to the partitioning into blocks, and secondly within each block, according to the periodicity.

5.3. Application to P-Cyclic Matrices. When solving each system of equations, $\pi^j Q^j = b^j$, during each disaggregation step, it is possible to take advantage of the periodic structure of the coefficient matrices Q^j. This has been our theme through the entire paper. First, the methods described in section 3 when applied to equation 5.1 will also take advantage of the periodic form of Q^j. When using the LU decomposition method, the savings in computation time and memory requirements can be especially significant.

Secondly, as described in section 4 for the original system $\pi Q = 0$, it is also possible to work with a reduced scheme instead of equation 5.1.

Indeed, let us partition the vector of conditional probabilities of block j, π^j, according to the periodic structure of Q^j, i.e., $\pi^j = (\pi_1^j, \pi_2^j, \cdots, \pi_p^j)$. Equation 5.1 may then be decomposed into:

$$\pi_1^j D_1^j = -\pi_p^j Q_p^j + b_1^j$$
$$\pi_2^j D_2^j = -\pi_1^j Q_1^j + b_2^j$$
$$\cdots = \cdots$$
$$\pi_p^j D_p^j = -\pi_{p-1}^j Q_{p-1}^j + b_p^j$$

Consider for instance, the subvector π_p^j. From the above set of equations, it follows that:

$$
\begin{aligned}
\pi_p^j D_p^j = & -\pi_p^j Q_p^j (-D_1^j)^{-1} Q_1^j \cdots (-D_{p-1}^j)^{-1} Q_{p-1}^j \\
& + b_1^j (-D_1^j)^{-1} Q_1^j \cdots (-D_{p-1}^j)^{-1} Q_{p-1}^j \\
& + b_2^j (-D_2^j)^{-1} Q_2^j \cdots (-D_{p-1}^j)^{-1} Q_{p-1}^j \\
& + \cdots \\
& + b_{p-1}^j (-D_{p-1}^j)^{-1} Q_{p-1}^j \\
& + b_p^j
\end{aligned}
$$

which leads to:

$$
\begin{aligned}
\pi_p^j (D_p^j + Q_p^j (-D_1^j)^{-1} Q_1^j & \cdots (-D_{p-1}^j)^{-1} Q_{p-1}^j) \\
& = b_1^j (-D_1^j)^{-1} Q_1^j \cdots (-D_{p-1}^j)^{-1} Q_{p-1}^j \\
& + b_2^j (-D_2^j)^{-1} Q_2^j \cdots (-D_{p-1}^j)^{-1} Q_{p-1}^j \\
& + \cdots \\
& + b_{p-1}^j (-D_{p-1}^j)^{-1} Q_{p-1}^j + b_p^j.
\end{aligned}
$$

i.e., an equation of the type:

(5.3) $$\pi_p^j T_p^j = c_p^j.$$

This equation is the reduced scheme associated with equation 5.1. It involves only a part of the vector, π^j, of conditional probabilities, namely π_p^j. It is similar to equation 4.5 except that the right-hand term is non-zero. By using a form of nesting, the right-hand side, c_p^j, may be formed in a number of operations that is equal to the number of non-zero elements in the matrix Q^j. Despite this non-zero right-hand side, the methods discussed in section 4 can still be applied and the conclusions reached there concerning the saving in memory requirements and computation time are also applicable here.

In summary, it is still possible to take advantage of the periodicity property when using aggregation/disaggregation methods. Since usually, most of the complexity of A/D methods is due to the disaggregation steps, it should be expected that the savings achieved will be of the same order of magnitude as that obtained by a periodic method over a non-periodic method applied to the entire matrix, Q.

REFERENCES

[1] W.E. Arnoldi. The principle of minimized iteration in the solution of the matrix eigenvalue problem. *Quart. Appl. Math.*, 9:17–29, 1951.

[2] F. Bonhoure, Y. Dallery and W.J. Stewart. On the Use of Periodicity Properties for the Efficient Numerical Solution of Certain Markov Chains. Submitted to *Journal of Numerical Linear Algebra with Applications*

[3] K. Kontovassilis, R.J. Plemmons and W.J. Stewart. Block SOR for the computation of the steady state distribution of finite Markov chains with p-cyclic infinitesimal generators. To appear in *Linear Algebra and its Applications*, 1991.

[4] P.J. Schweitzer. A survey of aggregation/disaggregation in large Markov chains. *First International Workshop on the Numerical Solution of Markov Chains*, Raleigh, NC, January 1990.

[5] W.J. Stewart. Decomposition Methods. Technical report, Dept. of Computer Science, North Carolina State University, Raleigh, N.C. 27695-8206, 1990.

ITERATIVE METHODS FOR QUEUEING NETWORKS WITH IRREGULAR STATE-SPACES

RAYMOND H. CHAN*

Abstract. In this paper, we consider the problem of finding the steady-state probability distribution for Markovian queueing networks with overflow capacity. Our emphasis is on networks with non-rectangular state-spaces. The problem is solved by the preconditioned conjugate gradient method with preconditioners that can be inverted easily by using separation of variables. By relating the queueing problems with elliptic problems, and making use of results from domain decomposition for elliptic problems on irregular domains, we derive three different kinds of such separable preconditioners. Numerical results show that for these preconditioners, the numbers of iterations required for convergence increase very slowly with increasing queue size.

Key words. Preconditioned conjugate gradient method, queueing networks, domain decomposition, substructuring method, Additive Schwarz method, capacitance matrix method.

AMS(MOS) subject classifications. 65N20, 65F10, 60K25.

1. Introduction. In this paper, we are interested in finding steady-state probability distributions for Markovian queueing networks with overflow capacity. From the steady-state probability distribution, we can compute for examples, the blocking probability of the network, the probability of overflow from one queue to another and the waiting time for customers in various queues.

In matrix terms, finding the steady-state probability distribution is equivalent to finding a right null-vector $\mathbf{p}$ of a matrix A, called the *generating matrix* of the network. The null-vector $\mathbf{p} = (p_1, \cdots, p_N)^t$ that satisfies the probability constraints

$$(1.1) \qquad \begin{aligned} \sum_{j=1}^{N} p_j &= 1, \\ p_j &\geq 0, \end{aligned}$$

will be the required probability distribution vector for the network. The problem is challenging because conventional methods for finding eigenvectors will not be cost effective for such problem as the size N of the matrix A is usually very large. Typically, we have

$$N = \prod_{i=1}^{q} n_i ,$$

where q is the number of queues in the network and n_i is the number of buffers in the ith queue, $1 \leq i \leq q$. However, A possesses rich algebraic structures that one can exploit in finding $\mathbf{p}$.

---

* Department of Mathematics, University of Hong Kong, Hong Kong. Research supported in part by ONR contract no. N00014-90-J-1695 and DOE grant no. DE-FG03-87ER25037.

For the networks that we will discuss in this paper, the matrix A is irreducible, has zero column sum, positive diagonal entries and non-positive off-diagonal entries. Thus $(I - A\text{diag}(A)^{-1})$ is a stochastic matrix. From Perron and Frobenius theory, we then know that A has a one-dimensional null-space with a positive null-vector $\mathbf{p}$, see for instance Varga [16]. Another important feature of A is that it is sparse. Its matrix graph is the same as that of the q-dimensional discrete Laplacian. Thus each row of A has at most $2q + 1$ non-zero entries.

One usual approach to the problem is to consider the partition

$$A = \begin{bmatrix} B & \mathbf{d} \\ \mathbf{c}^t & \alpha \end{bmatrix} \quad \text{and} \quad \mathbf{p} = \begin{bmatrix} \mathbf{f} \\ 1 \end{bmatrix}$$

and solve the reduced $(N - 1)$-by-$(N - 1)$ nonsingular system

$$(1.2) \qquad\qquad\qquad B\mathbf{f} = -\mathbf{d}$$

by direct or iterative methods, see Funderlic and Mankin [6], Kaufman [9], O'Leary [10] and Plemmons [12]. Sparsity of the matrix B is usually exploited in these methods.

However, from numerical pde's, we know that with suitable domain of definition and compatible boundary conditions, the q-dimensional discrete Laplacian matrix can be inverted efficiently by using the separability of the Laplacian operator; and this approach will not work on any reduced system of size $(N - 1)$-by-$(N - 1)$. Thus in this paper, instead of reducing the size of the matrix by one, we consider N-by-N singular matrices as preconditioners for the system $A\mathbf{p} = \mathbf{0}$. These preconditioners will cancel the singularity of A in the sense that the resulting preconditioned systems are N-by-N nonsingular systems. By working on matrices of size N-by-N, we can exploit the fast inversion of separable components in A and these separable components will be used as building blocks for constructing preconditioners for A.

The outline of the paper is as follows. In §2, to consolidate the idea, we introduce our method for queueing networks with rectangular state-spaces first. We will illustrate the idea of using separable components in the generating matrix A in constructing preconditioners for A. We will also point out the relationship between queueing networks and elliptic problems which will be useful later in designing preconditioners for queueing problems with irregular state-spaces. Convergence analysis for 2-queue single-server networks with rectangular state-spaces is also given. In §3, we consider queueing networks with irregular state-spaces. We will illustrate how the results from domain decomposition can be used to construct separable preconditioners for these queueing problems. In particular, we will make use of the ideas from the substructuring method, the capacitance matrix method and the additive Schwarz method to derive three different preconditioners for these queueing problems. We remark that domain decomposition methods are most well-suited for these large queueing problems because they can be made parallel easily. Numerical results are given in §4. They show that our preconditioners work very well for the test problems. The numbers of iterations required for convergence increase very slowly as the queue size increases.

2. Networks with Rectangular State-spaces. Let us first introduce the notations that we will be using. Assume that the network has q queues receiving customers from q independent Poisson sources. In the ith queue, $1 \leq i \leq q$, there are s_i parallel servers and $n_i - s_i - 1$ spaces for customers. Arrival of customers is assumed to be Poisson distributed with rate λ_i and the service time of each server is exponential with mean $1/\mu_i$. To illustrate our method and what we mean by separable components in a generating matrix, we restrict ourselves to queueing networks with rectangular state-spaces in this section.

2.1. A 2-Queue Overflow Network. For simplicity, we begin with a simple 2-queue overflow network discussed in Kaufman [9]. Here customers entering the first queue can overflow to the second queue if the first queue is full and the second queue is not yet full. However, customers entering the second queue will be blocked and lost if the second queue is full, see Figure 2.1.

Let $p_{i,j}$ be the steady-state probability that there are i and j customers in queues 1 and 2 respectively. The Kolmogorov balance equations, i.e. the equations governing $p_{i,j}$, are given by:

$$\{\lambda_1(1 - \delta_{in_1-1}\delta_{jn_2-1}) + \lambda_2(1 - \delta_{jn_2-1}) + \mu_1 \min(i, s_1) + \mu_2 \min(j, s_2)\} p_{ij}$$
$$= \lambda_1(1 - \delta_{i0})p_{i-1,j} + \mu_1(1 - \delta_{in_1-1})\min(i + 1, s_1)p_{i+1,j}$$
$$+ (\lambda_1\delta_{in_1-1} + \lambda_2)(1 - \delta_{j0})p_{i,j-1} + \mu_2(1 - \delta_{jn_2-1})\min(j + 1, s_2)p_{i,j+1},$$

for $0 \leq i < n$, $0 \leq j < n_2$. Here δ_{ij} is the Kronecker delta. These equations can be expressed as a matrix equation $A\mathbf{p} = \mathbf{0}$. The matrix A is called the generating matrix of the network while the vector $\mathbf{p}$, after normalization by (1.1), is called the steady-state probability distribution vector of the network. The matrix A is known to be non-separable with no closed form solution for $\mathbf{p}$. However, we will show below that A can be partitioned into the sum of a separable matrix and a low rank matrix.

To derive the separable matrix, let us assume for the moment that overflow of customers from queue 1 to queue 2 is not allowed. Thus the two queues are independent of each other. Such network is said to be *free*. Let A_0 be the generating matrix for such 2-queue free network. Then A_0 is separable and is given by

$$(2.1) \qquad\qquad A_0 = G_1 \otimes I_2 + I_1 \otimes G_2,$$

where I_i is the identity matrix of size n_i, $\otimes$ is the Kronecker tensor product

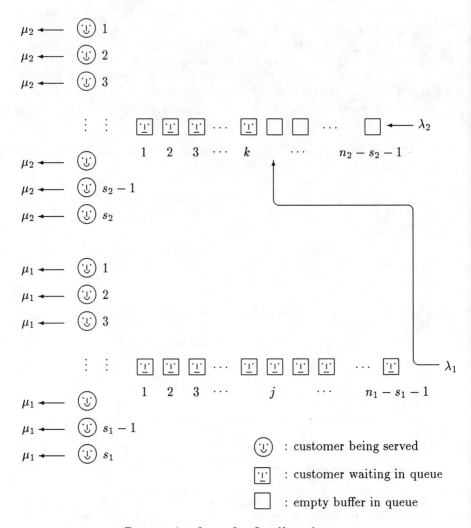

FIG. 2.1. *A 2-Queue Overflow Network.*

and G_i, $i = 1, 2,$ are n_i-by-n_i tridiagonal matrix given by

$$
(2.2) \quad \begin{bmatrix}
\lambda_i & -\mu_i & & & & & \\
-\lambda_i & \lambda_i + \mu_i & -2\mu_i & & & & \\
& \ddots & \ddots & \ddots & & 0 & \\
& & -\lambda_i & \lambda_i + s_i\mu_i & -s_i\mu_i & & \\
& & & \ddots & \ddots & \ddots & \\
& 0 & & & -\lambda_i & \lambda_i + s_i\mu_i & -s_i\mu_i \\
& & & & & -\lambda_i & s_i\mu_i
\end{bmatrix}.
$$

We note that G_i is just the generating matrix of the ith queue and (2.1) reflects the fact that the queues are independent of each other.

The difference $R = A - A_0$, which corresponds to the addition of the overflow queueing discipline from queue 1 to queue 2, is given by

$$
(2.3) \quad R = \left(\mathbf{e_{n_1}} \mathbf{e_{n_1}}^t\right) \otimes \lambda_1 \cdot R_2,
$$

where $\mathbf{e_j}$ is the jth unit vector and

$$
(2.4) \quad R_i = \begin{bmatrix}
1 & & & & \\
-1 & 1 & & 0 & \\
& \ddots & \ddots & & \\
& & \ddots & & \\
0 & & -1 & 1 & \\
& & & -1 & 0
\end{bmatrix},
$$

is an n_i-by-n_i matrix. Clearly R is a matrix of rank n_2.

2.2. Properties of the Preconditioner. Our method is based on the idea of partitioning A into $A_0 + R$ and using A_0 as a preconditioner. In this subsection, we show that the generalized inverse A_0^+ of A_0 can be obtained easily by using the separability of A_0 and we also discuss some of the properties of A_0^+.

Since the upper and lower diagonals of the matrix G_i given in (2.2) are nonzero and of the same sign, there exists a diagonal matrix D_i such that $D^{-1}G_iD_i$ is symmetric. In fact $D_i = \operatorname{diag}(d_1^i, \cdots, d_{n_i}^i)$ with

$$
(2.5) \quad d_j^i = \begin{cases}
1 & j = 1, \\
\prod_{k=1}^{j-1}\left(\frac{\lambda_i}{\min(k, s_i)\mu_i}\right)^{\frac{1}{2}} & 1 < j \le n_i.
\end{cases}
$$

As $D_i^{-1}G_iD_i$ is symmetric and tri-diagonal, we can find in $O(n_i^2)$ operations an orthogonal matrices Q_i such that

$$
Q_i^t D_i^{-1} G_i D_i Q_i = \Lambda_i
$$

is diagonal, see Golub and van Loan [7]. Thus by (2.1), we see that

$$
(Q_1 \otimes Q_2)^t (D_1 \otimes D_2)^{-1} A_0 (D_1 \otimes D_2)(Q_1 \otimes Q_2) = \Lambda_1 \otimes I_2 + I_1 \otimes \Lambda_2 \equiv \Lambda,
$$

where Λ is a diagonal matrix. Therefore we can define a generalized inverse A_0^+ of A_0 as

$$(2.6) \qquad A_0^+ = (D_1 \otimes D_2)(Q_1 \otimes Q_2)\Lambda^+(Q_1 \otimes Q_2)^t(D_1 \otimes D_2)^{-1}.$$

Since G_i has zero column sum and $D_i^{-1}G_iD_i$ is symmetric, it follows that

$$(2.7) \qquad\qquad G_iD_i^2\mathbf{1_i} = D_i^2G_i^t\mathbf{1_i} = 0,$$

where $\mathbf{1_i}$ is the n_i-vector of all ones. From (2.1), we thus see that a right null-vector of A_0 is given by

$$(2.8) \qquad\qquad \mathbf{p_0} = (D_1^2 \otimes D_2^2)(\mathbf{1_1} \otimes \mathbf{1_2}),$$

which after normalization by (1.1), will give us the steady-state probability distribution of the free network.

Using the spectral decomposition of A_0 in (2.6) and the fact that

$$(2.9) \qquad\qquad Im(A_0) = Im(A_0^+) = \langle\mathbf{1}\rangle^\perp,$$

we can easily verify the following property of A_0.

LEMMA 1 (CHAN [3]). *Let A_0^+ be the generalized inverse of A_0 as defined in (2.6). Then*
   (i) $\mathbf{R}^N = \langle\mathbf{p_0}\rangle \oplus Im(A_0)$.
   (ii) For all $\mathbf{x} \in Im(A_0^+)$, there exists a unique $\mathbf{y} \in Im(A_0)$ such that $A_0^+\mathbf{y} = \mathbf{x}$.
   (iii) For all $\mathbf{x} \in Im(A_0)$, we have $A_0A_0^+\mathbf{x} = A_0^+A_0\mathbf{x} = \mathbf{x}$.

Note that by (2.6), given any vector $\mathbf{y}$, the matrix-vector multiplication $A_0^+\mathbf{y}$ can be computed in $O(n_1n_2(n_1 + n_2))$ operations. Moreover, there is no need to generate A_0^+ explicitly, all we need are storages for each individual Q_i, D_i and Λ_i, $i = 1, 2$. We emphasize that if $s_i = 1$, then G_i has constant upper and lower diagonals and hence for any vector $\mathbf{x}$, the product $Q_i\mathbf{x}$ can be computed by using Fast Fourier Transform in $O(n_i \log n_i)$ operations, see Chan [3]. In particular, if $s_1 = s_2 = 1$, then $A_0^+\mathbf{y}$ can be obtained in $O(n_1n_2 \log(n_1n_2))$ operations.

2.3. The Method. Let us now go back to the problem of finding the null-vector $\mathbf{p}$ for A. By Lemma 1, we see that for any null-vector $\mathbf{p}$ of A, there exist unique real number α, $\mathbf{x} \in Im(A_0^+)$ and $\mathbf{y} \in Im(A_0)$ such that

$$\mathbf{p} = \alpha\mathbf{p_0} + \mathbf{x} = \alpha\mathbf{p_0} + A_0^+\mathbf{y}.$$

Since $\mathbf{1}^t\mathbf{p} \neq 0$ and $\mathbf{1}^t\mathbf{x} = 0$, we see that $\alpha \neq 0$.

For the moment, let us concentrate on finding the null-vector $\mathbf{p}$ with $\alpha = 1$, i.e.

$$(2.10) \qquad\qquad \mathbf{p} = \mathbf{p_0} + A_0^+\mathbf{y}.$$

Putting this into the expression $A\mathbf{p} = \mathbf{0}$, we get $(A_0 + R)(\mathbf{p_0} + A_0^+\mathbf{y}) = \mathbf{0}$. After simplification, we then have

$$A_0 A_0^+\mathbf{y} + RA_0^+\mathbf{y} = -R\mathbf{p_0}.$$

Since $\mathbf{y} \in Im(A_0)$, by Lemma 1(iii), we have $A_0 A_0^+\mathbf{y} = \mathbf{y}$. Therefore the above equation reduces to

$$(2.11) \qquad\qquad (I + RA_0^+)\mathbf{y} = -R\mathbf{p_0}.$$

Note that the equation has unique solution $\mathbf{y}$ in $Im(A_0)$. In particular, the mapping $(I + RA_0^+)$ is invertible on $Im(A_0)$. In fact, we can show further that the matrix $(I + RA_0^+)$ is indeed invertible on $\mathbf{R}^N$, see Chan [3]. Hence the system (2.11) can be solved by iterative methods or even direct methods without any restriction onto the subspace $Im(A_0)$. Thus we see that by preconditioning A from the left by A_0, we have basically cancel the singularity in A and reduce the singular system $A\mathbf{p} = \mathbf{0}$ to a nonsingular system (2.11). Once we have the solution $\mathbf{y}$ from (2.11), the null-vector $\mathbf{p}$ can be obtained by (2.10) and by using the normalization constraints in (1.1), we then have the steady-state probability distribution.

Because of the sparsity of R, we see that the matrix $(I + RA_0^+)$ is also sparse and has at most n_2 eigenvalues different from 1. Thus the matrix is an ideal candidate for the conjugate gradient type methods whose convergence rate depends on how clustered the spectrum of the matrix is, see Golub and van Loan [7]. Notice that the cost of forming the matrix-vector product $(I + RA_0^+)\mathbf{x}$, which is required in every iteration of the conjugate gradient method, can be further reduced if sparsity of R is exploited. Since the matrix $(I + RA_0^+)$ is nonsymmetric, one can apply the conjugate gradient method to the normal equation

$$(2.12) \qquad (I + RA_0^+)^t(I + RA_0^+)\mathbf{y} = -(I + RA_0^+)^t R\mathbf{p_0}$$

or apply other generalized conjugate gradient type methods, see for instance Young and Jea [17] or Saad and Schultz [15].

2.4. Convergence Analysis. Clearly, the total cost of our method depends on its convergence rate. In this section, we give the convergence rate of our method in the single server case, i.e. $s_1 = s_2 = 1$. Notice that in analyzing the convergence rate, which is a function of N, we need to know the relationship between the parameters of the queues as a function of N. To this end, let us consider a typical 1-queue, single server, Markovian network. It can be seen from (2.7) that in this case, the steady state probabilities are given by

$$p_j = \frac{\rho - 1}{\rho^n - 1}\rho^j, \quad 0 \le j < n,$$

where $\rho = \lambda/\mu$ is the traffic density of the queue. Thus as we increase the queue size n, in order that the probabilities p_j for $j \approx n$ are not

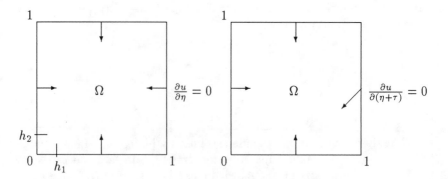

FIG. 2.2. *Neumann and Oblique Problems on Unit Square.*

exponentially small, one possible limit to consider is $\rho = 1 + cn^{-\alpha}$ for some constants $c \in \mathbf{R}$ and $\alpha \geq 1$.

We remark that we can obtain the same limit if we consider the continuous analogue of queueing networks. We begin by noting that in the 2-queue single-server case, when $\lambda_i = \mu_i = 1$, the matrix A_0 is exactly equal to the two-dimensional discrete Laplacian matrix on the unit square Ω with Neumann boundary conditions on every sides and mesh size $h_i = 1/(n_i-1)$, see Figure 2.2. When $\lambda_i \neq \mu_i$, A_0 resembles the finite difference approximation of the second order elliptic equation

$$(\lambda_1 + \mu_1)p_{xx} + (\lambda_2 + \mu_2)p_{yy}$$
$$(2.13) \qquad + \ 2(n_1 - 1)(\mu_1 - \lambda_1)p_x + 2(n_2 - 1)(\mu_2 - \lambda_2)p_y = 0$$

with the equation defined on the same unit square Ω and having the same Neumann boundary conditions on every sides. Moreover, we observe that R_2 in (2.4) is a first order differencing matrix. Hence the matrix R in (2.3) resembles a tangential operator on the edge $x = 1$ where overflow occurs. Compared to A_0, we see that the generating matrix A of the overflow problem resembles the same second order elliptic equation on the unit square with the Neumann boundary condition on the edge $x = 1$ being replaced by an oblique derivative, see Figure 2.2.

From (2.13), we see that if for large n_i, λ_i and μ_i are related by $\lambda_i/\mu_i = 1 + c_i n_i^{-\alpha_i}$ for some constants $c_i \in \mathbf{R}$ and $\alpha_i \geq 1$, then the second order terms in (2.13) are the dominant terms. In the following, we will analyze the convergence rate of our method under such limit. We note that if $\alpha_i < 1$ and $\lambda_i < \mu_i$, then $\|A\mathbf{p_0}\|_2$ tends to zero exponentially fast as n_i increases, see Chan [3]. In particular, $\mathbf{p_0}$ will already be a good estimate for $\mathbf{p}$ in such cases.

Notice that because of the low rank of R, the actual number of unknowns in the vector $\mathbf{y}$ in (2.11) is equal to n_2. Let the last n_2-by-n_2 principal submatrix of $(I + RA_0^+)$ be denoted by S. For ease of presentation, we assume in the following Lemma and Theorem that the queueing parameters λ_i, μ_i, s_i and n_i are the same for both queues with $n_1 = n_2 = n$. Full version and proof of the Lemma and Theorem are given in Chan [3].

LEMMA 2. *Assume that for both queues, $s_i = 1$ and $\mu_i = \lambda_i + cn^{-\alpha}$ for some constants $c \in \mathbf{R}$ and $\alpha \geq 1$. Then for sufficiently large n, $\|S^{-1}\|_2 < O(n^3)$ and*

$$S^t S = 2I + L + U,$$

where $\|U\|_2 < O(n^{1-\alpha}/\log n)$ and rank $L = O(\log n)$.

The Lemma states that the singular values of S are clustered around $\sqrt{2}$ except for at most $O(\log n)$ outlying ones. Applying standard error analysis of conjugate gradient method to the above results on S, we obtain the following Theorem.

THEOREM 1. *Assume that for both queues, $s_i = 1$ and $\mu_i = \lambda_i + cn^{-\alpha}$ for some constants $c \in \mathbf{R}$ and $\alpha \geq 1$. Then for large n, the conjugate gradient method applied to the normal equation (2.12) will converge within $O(\log^2 n)$ steps.*

The proof of the above Lemma involves purely linear algebra. However, the same result is anticipated if we look at the continuous analogue of the overflow problem. We recall that the matrix $(I + RA_0^+) \approx AA_0^+$ represents the mapping that maps the Neumann boundary data to the oblique boundary data. By using regularity theorem and trace theorem in elliptic theory, we have the following result on this mapping.

THEOREM 2 (CHAN [2]). *Let Ω be a bounded region in $\mathbf{R}^2$ with a smooth boundary $\partial\Omega$. Let*

$$E \equiv \left\{ g \in H^{-\frac{1}{2}}(\partial\Omega) \mid \int_{\partial\Omega} g d\tau = 0 \right\}$$

be equipped with the Sobolev $H^{-\frac{1}{2}}(\partial\Omega)$ norm. Let T be the Neumann-to-oblique mapping that maps g_1 in E to g_2 where g_1 and g_2 are boundary values of the problems

$$(N): \begin{cases} \Delta u = 0 & \text{in } \Omega, \\ \dfrac{\partial u}{\partial \eta} = g_1 & \text{on } \partial\Omega, \end{cases} \quad \text{and} \quad (O): \begin{cases} \Delta u = 0 & \text{in } \Omega, \\ \dfrac{\partial u}{\partial (\eta + \tau)} = g_2 & \text{on } \partial\Omega, \end{cases}$$

with both problems normalized by $\int_\Omega u = 0$. Here η and τ are the normal and tangential vectors respectively. Then T is a one-one onto mapping on E and satisfies

$$c\|Tg\|_{H^{-\frac{1}{2}}(\partial\Omega)} \leq \|g\|_{H^{-\frac{1}{2}}(\partial\Omega)} \leq C\|Tg\|_{H^{-\frac{1}{2}}(\partial\Omega)}$$

for some constants $c = c(\Omega) > 0$ and $C = C(\Omega) > 0$.

Thus the Neumann-to-oblique mapping T is well-conditioned. Hence, we expect the matrix $(I + RA_0^+)$ to be also well-conditioned for large n and if the conjugate gradient method is used, we expect fast convergence.

In the multi-server case, i.e. $s_i > 1$, instead of (2.13), the underlying continuous equation is of the form

$$(\lambda_1 + s_1\mu_1)p_{xx} + (\lambda_2 + s_2\mu_2)p_{yy}$$
$$(2.14) \quad + \quad 2(n_1 - 1)(s_1\mu_1 - \lambda_1)p_x + 2(n_2 - 1)(s_1\mu_2 - \lambda_2)p_y = 0$$

in the region where the states (i, j) satisfy $s_1 \leq i < n_1$ and $s_2 \leq j < n_2$. In other part of the rectangular state-space, the equation will be one with variable coefficients, with the coefficients of the second order terms decreasing in magnitude with decreasing i and j. Hence for large n_i, a reasonable limit to consider is

$$(2.15) \qquad \frac{\lambda_i}{s_i\mu_i} = 1 + c_i n_i^{-\alpha}$$

for some constants $c_i \in \mathbf{R}$ and $\alpha \geq 1$. We note however that under such limit, by Stirling's formula, the last entry $d_{n_i}^i$ of D_i in (2.5) is given

$$(2.16)\ d_{n_i}^i = \left[\frac{1}{s_i!}\left(\frac{\lambda_i}{\mu_i}\right)^{s_i}\left(\frac{\lambda_i}{s_i\mu_i}\right)^{n_i-s_i-1}\right]^{\frac{1}{2}} \approx \left[\frac{s_i^{s_i}}{s_i!}\right]^{\frac{1}{2}} \approx (2\pi s_i)^{-1/4}e^{s_i/2}.$$

Thus for large s_i, the matrix D_i will be ill-conditioned and hence the spectral decomposition of A_0^+ in (2.6) will be unstable. Experimental results show that our method works very well under limit (2.15) for small s_i but will break down when s_i becomes large, see Chan [3] and also §4.

2.5. General Overflow Queueing Networks with Rectangular State-space. The above method for finding $\mathbf{p}$ in $A\mathbf{p} = 0$ can be readily generalized to overflow queueing networks with more than two queues, provided that all overflows occur when and only when the queue is full. In this case, the state-space of the problem is the q-dimensional unit cube and one can automate the whole procedure for finding $\mathbf{p}$. More precisely, one can write a program that accepts queue parameters and overflow disciplines as input and outputs the steady-state probability distribution.

To see how this can be done automatically, we first note that the generating matrix for any q-queue free network is given by

$$A_0 = G_1 \otimes I_2 \otimes \cdots \otimes I_q + I_1 \otimes G_2 \otimes \cdots \otimes I_q + \cdots + I_1 \otimes I_2 \otimes \cdots \otimes G_q,$$

where G_i are given by (2.2). Clearly A_0 is still separable with a null-vector given by

$$\mathbf{p_0} = (D_1^2 \otimes \cdots \otimes D_q^2)(\mathbf{1_1} \otimes \cdots \otimes \mathbf{1_q}).$$

Any addition of overflow queueing disciplines to this free network corresponds to addition of matrices of the form (2.4) to the matrix $R = A - A_0$. With R and A_0 known, we can apply our method to obtain the system (2.11) which can then be solved by conjugate gradient type methods.

TABLE 2.1
Cost Comparison

Method	Operations	Storage
Normal equation	$O(n^{q+1})$ per iteration	$O(n^q)$
Point SOR	$O(n^q)$ per iteration	$O(n^q)$
Block SOR	$O(n^{q+1})$ per iteration	$O(n^q)$
Band solver	$O(n^{3q-2})$	$O(n^{2q-1})$

As an example, consider a 3-queue network where customers from queue i, $i = 1, 2$, can overflow to queue $i + 1$ provided that queue i is full. Then the matrix $R = A - A_0$ is given by

$$(2.17) \quad R = \left(\mathbf{e_{n_1}} \mathbf{e_{n_1}}^t\right) \otimes \lambda_1 \cdot R_2 \otimes I_3 + I_1 \otimes \left(\mathbf{e_{n_2}} \mathbf{e_{n_2}}^t\right) \otimes \lambda_2 \cdot R_3$$
$$+ \left(\mathbf{e_{n_1}} \mathbf{e_{n_1}}^t\right) \otimes \left(\mathbf{e_{n_2}} \mathbf{e_{n_2}}^t\right) \otimes \lambda_1 \cdot R_3,$$

with R_i given by (2.4). The last term above represents the flow from queue 1 to queue 3 when both queues 1 and 2 are full. We further note that one does not have to generate R nor the terms in R explicitly. For in conjugate gradient type methods, one only needs to compute the product $R\mathbf{x}$ which can be computed term-wise and the product for each term can be computed easily without forming each individual term explicitly.

The cost per iteration of our method will mainly depend on the cost of computing the matrix-vector multiplication $A_0^+\mathbf{y}$. However, in view of (2.6), this can be done in

$$O(N(n_1 + n_2 + \cdots + n_q))$$

operations where $N = \prod_{i=1}^q n_i$. In the case of single-server, i.e. $s_1 = s_2 = \cdots = s_q = 1$, the cost is reduced to $O(N \log N)$ by using FFT. Storage for few N-vectors and the small dense matrices Q_i will be required and will be of order $O(N)$.

Table 2.1 below compares the cost of our method as applied to the normal equation (2.12) with the cost of other methods as applied to system (1.2). For simplicity, we let $n_i = n$ for all $i = 1, \cdots, q$. We emphasize again that FFT can be used to speed up the computation of $A_0^+\mathbf{x}$ if any one of the s_i is equal to 1.

2.6. Some Remarks on the Method. We remark that our method is a generalization of the method in (1.2). In the simplest case if we partition $A = A_0 + R$ with A_0 given by

$$A_0 = I_N - \mathbf{e_N} \mathbf{e_N}^t$$

we then have $\mathbf{p_0} = \mathbf{e_N}$ and equation (2.11) is reduced basically to (1.2). If we choose

$$A_0 = \text{diag}(A) - (\mathbf{e_N}^t A \mathbf{e_N}) \cdot \mathbf{e_N} \mathbf{e_N}^t$$

then (2.11) is similar to the Jacobi method applied to (1.2). The main difference between our method and that of (1.2) is that our preconditioner A_0 is an N-by-N matrix and any preconditioners for (1.2) will be of size $(N-1)$-by-$(N-1)$. The ability of using N-by-N matrices as preconditioners enables us to exploit the separable components of the original generating matrix A and hence speed up the inversion of the preconditioners.

We finally note that one can also precondition the system $A\mathbf{p} = \mathbf{0}$ from the left by A_0^+. More precisely, we expand $\mathbf{p}$ as $\mathbf{p} = \beta\mathbf{p_0} + \mathbf{y}$ where $\mathbf{y} \in Im(A_0)$. By (1.1), $\beta \neq 0$ and by Lemma 1(iii), the system $A\mathbf{p} = \mathbf{0}$ is reduced to

$$(2.18) \qquad A_0^+ A\mathbf{y} = (I + A_0^+ R)\mathbf{y} = -A_0^+ R\mathbf{p_0}.$$

One can also prove that the matrix $(I + A_0^+ R)$ is nonsingular. We note further that in view of (2.9), $\mathbf{p}$ can be expanded as $\mathbf{p} = \gamma\mathbf{1} + \mathbf{x}$ where $\mathbf{x} \in Im(A_0)$. In particular, we can replace the right hand side of (2.18) by $-A_0^+ A\mathbf{1}$. However, we note that if A_0 is expected to be a close approximation to A, then we also expect $\mathbf{p_0}$ to be a better approximation to $\mathbf{p}$ than $\mathbf{1}$.

3. Networks with Irregular State-spaces. In this section, we consider queueing networks where the state-spaces are no longer rectangular and we will make use of the results from elliptic solvers for irregular domains to design preconditioners for these networks. To illustrate the idea, we consider the 2-queue one-way overflow network in §2.1 again and assume additionally that customers waiting for service in queue 1 must be served at queue 2 once servers in queue 2 become available. This network is almost the same as the one discussed in Kaufman et. al. [8], except now that we also allow the one-way overflow of customers from queue 1 to queue 2 when queue 1 is full. The Kolmogorov equations of the network are given in Chan [4].

According to the overflow discipline, we see that

$$p_{ij} = 0, \qquad s_1 < i < n_1, \quad 0 \le j < s_2 .$$

Hence the state-space of the network is no longer rectangular, but is given by an L-shaped region, see Figure 3.1. The subregions Ω_1 and Ω_2 are defined to be the set of states (i, j) where $0 \le i \le s_1$ and $s_1 < i < n_1$ respectively. They are separated by the interface $\tau = \{(s_1, j)\}_{j=s_2}^{n_2-1}$.

The generating matrix A of the network is in the block form

$$A = \begin{bmatrix} T_1 & D_{11} & 0 \\ D_{12} & C & D_{22} \\ 0 & D_{21} & T_2 \end{bmatrix}$$

where T_k, $k = 1, 2$, correspond to the interactions between the states within the subregions $\Omega_k \setminus \tau$, C corresponds to states in τ and D_{kl} correspond to

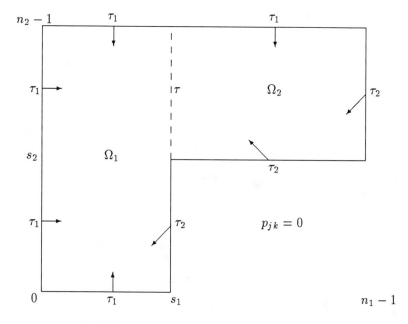

FIG. 3.1. *The State-space of the Queueing Network.*

the interactions between the states in the subregion Ω_k and the states in τ. Notice that there is no interaction between the states in $\Omega_1 \setminus \tau$ and states in Ω_2. Since A is no longer of size $n_1 n_2$, A_0 cannot be used as preconditioner.

The continuous analogue of this problem is a second order elliptic equation on the L-shaped region with oblique derivatives on boundary τ_2 and Neumann boundary conditions on τ_1. There are many domain decomposition techniques for solving elliptic equations on irregular regions. In the following subsections, we will apply some of the ideas there to design preconditioners for our queueing networks.

3.1. Substructuring Method. For elliptic problem on an L-shaped domain as in Figure 3.1, the theory of substructuring suggests the following Dirichlet-Neumann map as a preconditioner:

1. solve the problem defined in Ω_1 with Neumann boundary conditions on every side,
2. then use the value on the interface τ to solve a Dirichlet-Neumann problem in Ω_2,

see for instance Bjørstad and Widlund [1].

In term of the matrix A, that means we write $A = A_1 + R$, where

$$A_1 = \begin{bmatrix} T_1 & D_{11} & 0 \\ D_{12} & \tilde{C} & 0 \\ 0 & D_{21} & \tilde{T}_2 \end{bmatrix}.$$

The matrix $\tilde{C}$ is chosen such that the submatrix

$$W \equiv \begin{bmatrix} T_1 & D_{11} \\ D_{12} & \tilde{C} \end{bmatrix}$$

corresponds to a Neumann problem in subregion Ω_1. In terms of the queues, W will be the generating matrix of a 2-queue free network with s_1 spaces in the first queue and $n_2 - 1$ spaces in the second queue. Hence W will be separable and has a 1-dimensional null-space.

The submatrix $\tilde{T}_2$ in A_1 above corresponds to a Dirichlet-Neumann problem in subregion Ω_2, with Dirichlet data being transported from τ by D_{21}. More precisely, we have

$$\tilde{T}_2 = V_1 \otimes I_{n_2-s_2} + I_{n_1-s_1-1} \otimes V_2$$

with

$$(3.1) \quad V_1 = \mathrm{tridiag}(-\lambda_1, \lambda_1 + s_1\mu_1, -s_1\mu_1) - \lambda_1 \cdot \mathbf{e_{n_1-s_1-1}}\mathbf{e_{n_1-s_1-1}}^t,$$

and

$$V_2 = \mathrm{tridiag}(-\lambda_2, \lambda_2+s_2\mu_2, -s_2\mu_2)-s_2\mu_2\cdot\mathbf{e_1}\mathbf{e_1}^t-\lambda_2\cdot\mathbf{e_{n_2-s_2}}\mathbf{e_{n_2-s_2}}^t.$$
$$(3.2)$$
Thus $\tilde{T}_2$ will also be separable but it will be nonsingular.

Since A_1 is in block lower-triangular form, we see that A_1 is singular with a one-dimensional null-space and a null-vector

$$\mathbf{p_1} = \begin{bmatrix} \mathbf{w} \\ -\tilde{T}_2^{-1}(0, \ D_{21})\mathbf{w} \end{bmatrix},$$

where $\mathbf{w}$ is a null-vector of W. We can easily check that A_1 also satisfies the properties listed in Lemma 1 for A_0. Thus we can again expand $\mathbf{p} = \alpha\mathbf{p_1}+A_1^+\mathbf{y}$ where $\mathbf{y} \in Im(A_1)$ and reduce the homogeneous system $A\mathbf{p} = 0$ to a nonsingular system

$$(3.3) \qquad\qquad (I + RA_1^+)\mathbf{y} = -R\mathbf{p_1},$$

where $R = A - A_1$, see Chan [4].

We emphasize that the matrix R is sparse. In fact, $\tilde{C}$ and $\tilde{T}_2$ differ from C and T_2 only on the rows that correspond to those states on the interface τ and on the edges τ_2; and on such rows, they can only differ by at most three entries. The matrix-vector multiplication $A_0^+\mathbf{x}$ can be done efficiently by using the separability of W and $\tilde{T}_2$. We emphasize that for all arbitrary values of s_i, the submatrix $\tilde{T}_2$ can be inverted easily by using FFT, for it corresponds to a constant-coefficient Dirichlet-Neumann problem in Ω_2, see (2.14). In fact, we see from (3.1) and (3.2) that both matrices V_1 and V_2 have constant upper and lower diagonals.

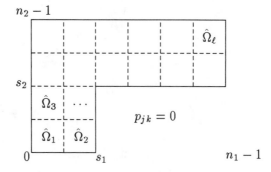

FIG. 3.2. *Partition into Many Subdomains.*

The idea of using two subregions can easily be extended to many subregions. Consider the domain in Figure 3.1 being partitioned into many subregions, see Figure 3.2. In $\hat{\Omega}_1$, we solve a Neumann problem. With the values obtained on the boundary $\partial\hat{\Omega}_1$, we solve mixed problems in $\hat{\Omega}_2$ and $\hat{\Omega}_3$ and so on.

The preconditioner will be of the block lower-triangular form:

$$A_1 = \begin{bmatrix} T_{11} & & & 0 \\ D_{21} & T_{22} & & \\ & \ddots & & \ddots \\ D_{\ell 1} & \cdots & D_{\ell(\ell-1)} & T_{\ell\ell} \end{bmatrix}$$

The submatrix T_{11} corresponds to a Neumann problem on $\hat{\Omega}_1$, the submatrices T_{ii}, $2 \le i \le \ell$ correspond to mixed problems in $\hat{\Omega}_i$ and the off-diagonal block matrices D_{ij} will transport the required boundary data from one subregion to another. Notice that for each $i = 1, \cdots, \ell$, only two D_{ij} will be non-zero, and for those $\hat{\Omega}_i$ that lie inside Ω_2, their corresponding T_{ii} can be inverted by using FFT.

3.2. Capacitance Matrix Method. Another method of solving elliptic problem on irregular domain is to embed the whole domain into a rectangular domain and use the preconditioners on the rectangular domain as preconditioners for the embedded system, see Proskurowski and Widlund [13]. To illustrate the idea, let us order the states in the L-shaped domain first and denote A_0 to be the generating matrix of the 2-queue free network on the rectangular domain $[0, n_1 - 1] \times [0, n_2 - 1]$. We then partition A_0 as

$$A_0 = \begin{bmatrix} A_{11} & A_{12} \\ A_{21} & A_{22} \end{bmatrix}$$

where A_{22} gives the interactions between states that are both in the region $[s_1 + 1, n_1 - 1] \times [0, s_2 - 1]$. We note that A_{22} corresponds to a mixed problem in that region and can easily be proved to be nonsingular.

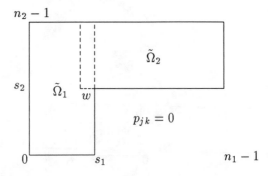

FIG. 3.3. *Partition into Overlapping Subdomains.*

We now embed the generating matrix A for the L-shaped domain into the whole rectangular domain as

$$A_e = \begin{bmatrix} A & A_{12} \\ 0 & A_{22} \end{bmatrix}$$

Since A_{22} is nonsingular, we see that A_e has a one-dimensional null-space and $\mathbf{p_e} = [\mathbf{p}, 0]^t$ is a null vector to A_e. Again we expand $\mathbf{p_e}$ as $\mathbf{p_e} = \mathbf{p_0} + A_0^+ \mathbf{y}$ with $\mathbf{y} \in Im(A_0)$. Then the vector $\mathbf{y}$ can be obtained by solving

(3.4) $(I + RA_0^+)\mathbf{y} = -R\mathbf{p_0},$

where $R = A_e - A_0$. Once again the matrix $(I + RA_0^+)$ can be shown to be nonsingular, see Chan [4].

3.3. Additive Schwarz Method. One of the basic ideas of the additive Schwarz method is to extract easily invertible components from the matrix A, inverse each of them individually and then add their inverses together to form an approximate inverse of the matrix A, see Dryja and Widlund [5]. To illustrate the idea, let us partition the L-shaped domain in Figure 3.1 into two overlapping subdomains, see Figure 3.3. For $i = 1, 2$, let B_i be the matrix that corresponds to the Neumann problem in $\tilde{\Omega}_i$ and zero elsewhere. Clearly, both matrices are singular and separable. Similar to (2.9), we can prove that

$$\mathbf{1}^t B_1^+ = \mathbf{1}^t B_2^+ = \mathbf{1}^t (B_1^+ + B_2^+) = \mathbf{0},$$

or equivalently,

(3.5) $Im(B_1^+ + B_2^+) \subseteq \langle \mathbf{1} \rangle^{\perp}.$

The matrix $(B_1^+ + B_2^+)$ will be used to precondition our system. Notice that it is in general difficult to find a null-vector for the preconditioner. Thus to find a null-vector $\mathbf{p}$ for A, we expand $\mathbf{p} = \alpha \mathbf{1} + \mathbf{x}$ where $\mathbf{x} \in \langle \mathbf{1} \rangle^{\perp}$.

By the normalization constraints on $\mathbf{p}$, we see that $\alpha \neq 0$. Therefore we can set $\alpha = 1$ and rewrite $A\mathbf{p} = \mathbf{0}$ as

$$(3.6) \qquad\qquad A\mathbf{x} = -A\mathbf{1}.$$

The system is then preconditioned by $(B_1^+ + B_2^+)$ to form

$$(3.7) \qquad\qquad (B_1^+ + B_2^+)A\mathbf{x} = -(B_1^+ + B_2^+)A\mathbf{1}.$$

Notice that unlike the methods mentioned in previous sections, the preconditioned system in this case is still singular. However, because of (3.5), if the above equation is solved by conjugate gradient type methods with initial guess in $\langle \mathbf{1} \rangle^\perp$, then each subsequent iterant will automatically be in $\langle \mathbf{1} \rangle^\perp$ again.

Since the matrix $(B_1^+ + B_2^+)$ is singular, any matrix-vector product of the form $(B_1^+ + B_2^+)\mathbf{x}$ will have no component along the null vector of $(B_1^+ + B_2^+)$, which is in general difficult to find. To partially remedy this, we add to the product a vector of the form $\beta\mathbf{q_1} + \gamma\mathbf{q_2}$, where $\mathbf{q_i}$ are the normalized null-vectors of B_i^+, $i = 1,2$ (extended by zeros outside their respective domains of definition). By the separability of B_i^+, $\mathbf{q_i}$ can be obtained easily. In order that the vector

$$(B_1^+ + B_2^+)\mathbf{x} + \beta\mathbf{q_1} + \gamma\mathbf{q_2}$$

is still in $\langle \mathbf{1} \rangle^\perp$, we need $\gamma = -\beta$. The remaining degree of freedom can be determined by imposing extra conditions on the vector at the intersection of the subdomains. One successful choice is to equate the mean values of the solutions of the subdomains at the intersection, i.e.

$$(3.8) \qquad \sum_{k \in \tilde{\Omega}_1 \cap \tilde{\Omega}_2} (B_1^+\mathbf{x} + \beta\mathbf{q_1})_k = \sum_{k \in \tilde{\Omega}_1 \cap \tilde{\Omega}_2} (B_2^+\mathbf{x} + \gamma\mathbf{q_2})_k,$$

see Mathew [11]. Our numerical results show that the addition of the vector $\beta\mathbf{q_1} + \gamma\mathbf{q_2}$ does improve the convergence rate especially when the size of the overlapping is small, see §4.

Obviously, because of the overlapping, the cost per iteration is higher than that in the substructuring method, but this is usually compensated by the advantage that $B_1^+A\mathbf{y}$ and $B_2^+A\mathbf{y}$ can be computed in parallel. Clearly the idea can be generalized to the case of many subdomains if more processors are available. However, it is already noted in domain decomposition literature that unless a coarse grid component is added, the additive Schwarz method will not converge as fast as the other domain decomposition preconditioners. This fact is also verified in the numerical results in §4 for queueing networks. It is an interesting project to formulate and implement a suitable coarse grid structure to the queueing problem. One possible way is to aggregate the states together to form superstates and use the balance equations for the superstates to derive a coarse grid

Comparison of Cost for Different Methods.

Method	Operations
Substructuring	$O(sn^2 + (n-s)^2 \log(n-s))$ per iteration
Capacitance Matrix	$O(n^3)$ per iteration
Additive Schwarz	$O(sn^2 + (n-s+w)^2(n-s))$ per iteration
Point-SOR	$O(ns + (n-s)^2)$ per iteration
Band Solver	$O([ns + (n-s)^2]n^2)$

formulation for the queueing problem. Another possible approach is to use the idea of algebraic multigrid method to construct the coarse grid matrix directly from the given generating matrix, see for instance, Ruge and Stüben [14].

We conclude this section by listing the costs of different methods in Table 3.1. For simplicity, we assume that $n_1 = n_2 = n$ and $s_1 = s_2 = s$. The variable w in the additive Schwarz method denotes the width of the overlapping region, see Figure 3.3.

4. Numerical Results. In this section, we apply our method to the queueing networks considered in previous sections. The parameters of the queues are assumed to be the same for all queues and are related by

$$s_i \mu_i = \lambda_i + (n_i - 1)^{-\alpha}, \quad i = 1, 2, \cdots, q,$$

with $\lambda_i \equiv 1$. In the examples, the preconditioned systems were solved by a generalized conjugate gradient method, called the Orthodir method which does not require the formation of the normal equation, see Young and Jea [17]. We chose zero vector to be the initial guess and the stopping criterion to be $||r_k||_2/||r_0||_2 \leq 10^{-6}$, where r_k is the residual vector after k iterations. Example 1 is taken from Chan [4] to illustrate the effectiveness of our method over the point-SOR method. Computations in Example 2 were done by Matlab on SPARC stations at UCLA.

Example 1: We consider the 3-queue network with overflow queueing disciplines $q_1 \rightarrow q_2$, $q_2 \rightarrow q_3$ and $q_1 \rightarrow q_2 \rightarrow q_3$, see (2.17). Table 4.1 shows the number of iterations required for convergence for our preconditioned system (2.11). Table 4.2 compares the performance of our method with that of the point-SOR method as applied to (1.2) with $\mathbf{p_0}$ as the initial guess. The optimal relaxation factors ω^* were obtained numerically. We see that our method performs much better than the point-SOR method especially for small s_i. We note that in the test, we have not used FFT to speed up our algorithm in the case where all $s_i = 1$.

Example 2: We consider solving the 2-queue network with line-jumping in §3 by the different techniques we mentioned there. For the additive Schwarz

TABLE 4.1
Numbers of Iterations required for Convergence.

n_i	N	s_i	α 1	α 2	α 3	s_i	α 1	α 2	α 3	s_i	α 1	α 2	α 3
4	64	1	10	10	10	3	9	9	9	3	9	9	9
8	512	1	14	14	14	3	14	14	14	6	13	13	13
16	4096	1	18	18	18	3	18	18	18	9	17	17	17

TABLE 4.2
Comparison with point-SOR method when $\alpha = 2$.

n_i	s_i	N	ω^*	number of iterations point-SOR	number of iterations orthodir	time in seconds point-SOR	time in seconds orthodir
4	1	64	1.700	69	10	1.176	0.461
4	3	64	1.593	30	9	0.529	0.420
8	1	512	1.831	242	14	31.282	3.815
8	7	512	1.715	49	12	5.997	3.274

method, two kinds of overlapping were tested: the maximum one with $\tilde{\Omega}_2 = [0, n_1-1] \times [s_2, n_2-1]$ and the minimum one with $\tilde{\Omega}_2 = [s_1-1, n_1-1] \times [s_2, n_2-1]$, see Figure 3.3. No coarse grid components were added however. Table 4.3 shows the numbers of iterations required for convergence for three different sets of queueing parameters. In the table, N denotes the total number of states in the L-shaped domain, the symbols No, DN, CM, ASmax and ASmin stand for no-preconditioning (see (3.6)), preconditioning by Dirichlet-Neumann preconditioner (3.3), capacitance matrix method (3.4) and additive Schwarz method (3.7) with maximum and minimum overlap respectively. In both additive Schwarz methods, we have added the null-vectors according to (3.8). As a comparison, we also tested the case of minimum overlapping without adding the null-vector components (i.e. $\beta = \gamma = 0$) and the results are shown under the column ASmin*. To check the accuracy of our computed solution $\mathbf{p_c}$, we have computed $\|A\mathbf{p_c}\|_2$ and found it to be less than 10^{-7} in all the cases we tested.

From the numerical results, we see that the numbers of iterations for the non-preconditioned systems grow linearly with the queue size, a well-known phenomenon for second order elliptic problems. However, for the Dirichlet-Neumann preconditioner and the capacitance matrix preconditioner, the numbers of iterations grow very slowly with increasing queue size. Notice that when s_i are smaller, the L-shaped domain is more rectangular. Thus it is not surprising to see that the matrix capacitance method is a better choice for networks with small s_i.

For the additive Schwarz method, as already mentioned, it will not be as competitive as the other preconditioners if coarse grid components are not added. We note that adding the null-vectors $\mathbf{q_1} - \mathbf{q_2}$ speeds up the

TABLE 4.3
Numbers of Iterations for Convergence.

n_i	s_i	α	N	No	DN	CM	ASmax	ASmin	ASmin*
10	2	1	86	41	12	12	16	18	19
20	4	1	340	81	16	15	19	25	27
40	8	1	1352	153	18	18	25	35	39
80	16	1	5392	>200	21	21	31	48	56
10	5	2	80	37	11	10	15	16	17
20	5	2	330	78	16	15	20	24	27
40	5	2	1430	161	19	19	25	34	38
80	5	2	6030	>200	23	21	30	49	54
10	4	3	80	39	12	10	15	16	18
20	8	3	312	74	14	14	19	23	26
40	16	3	1232	147	17	18	24	31	36
80	32	3	4896	>200	19	20	>60	>60	>60

convergence rate significantly. However, the speed-ups are less significant in the maximum overlapping case where the numbers of iterations with or without adding the null-vectors differ by only one iteration in all the cases we tested. Also the method shows instability in the last test problem when $s_i = 32$. We have tried GMRES(40), see Saad and Schultz [15], and it still did not converge after 30 iterations. The instability may partly due to the fact that the matrix D_i in (2.6) is ill-conditioned with condition number about 10^6 in this case, see (2.16). Another possible reason is that we are not able to recover exactly the components of $(B_1^+ + B_2^+)\mathbf{x}$ along the null vector of $(B_1^+ + B_2^+)$, see §3.3. Further research will be carried out in this direction.

Acknowledgments: I would like to thank Professors Olof Widlund, Tony Chan and Tarek Mathew for their guidance in the preparation of this paper.

REFERENCES

[1] P. Bjørstad and O. Widlund, *Iterative Methods for the Solutions of Elliptic Problems on Regions Partitioned into Substructures*, SIAM J. Numer. Anal., V23, pp. 1097-1120 (1986).

[2] R. Chan, *Iterative Methods for Overflow Queueing Models*, NYU Comput. Sci. Dept., Tech Report No. 171, 1985.

[3] R. Chan, *Iterative Methods for Overflow Queueing Models I*, Numer. Math. V51, pp.143-180 (1987).

[4] R. Chan, *Iterative Methods for Overflow Queueing Models II*, Numer. Math. V54, pp.57-78 (1988).

[5] M. Dryja and O. Widlund, *Some Domain Decomposition Algorithms for Elliptic Problems*, Proceedings of the Conference on Iterative Methods for Large Linear Systems, Austin, Texas, October, 1989, Academic Press.

[6] R. Funderlic and J. Mankin, *Solution of Homogeneous Systems of Linear Equations Arising from Compartmental Models*, SIAM J. Sci. Statist. Comput., V2, pp.375-383, (1981).

[7] G. Golub and C. van Loan, *Matrix Computations*, 2nd Ed., The Johns Hopkins University Press, Maryland, 1989.

[8] L. Kaufman, J. Seery and J. Morrison, *Overflow Models for Dimension PBX Feature Package*, Bell Syst. Tech. J., V60, pp. 661-676 (1981).

[9] L. Kaufman, *Matrix Methods for Queueing Problems*, SIAM J. Sci. Stat. Comput., V4, pp. 525-552 (1982).

[10] D. O'Leary, *Iterative Methods for Finding the Stationary Vector for Markov Chains*, Proceedings of the IMA Workshop on Linear Algebra, Markov Chains and Queueing Models, Minneapolis, Minnesota, January 1992, Springer-Verlag.

[11] T. Mathew, *Schwarz Alternating and Iterative Refinement Methods for Mixed Formulations of Elliptic Problems, Part I: Algorithms and Numerical Results*, UCLA CAM Report 92-04, January 1992.

[12] R. Plemmons, *Matrix Iterative Analysis for Markov Chains*, Proceedings of the IMA Workshop on Linear Algebra, Markov Chains and Queueing Models, Minneapolis, Minnesota, January 1992, Springer-Verlag.

[13] W. Proskurowski and O. Widlund, *A Finite Element-Capacitance Matrix Method for the Neumann Problem for Laplace's Equation*, SIAM J. Sci. Statist. Comput., V1, pp. 410-425 (1980).

[14] J. Ruge and K. Stüben, *Algebraic Multigrid* in Multigrid Methods, Frontier in Applied Mathematics, ed. S. McCormick, Philadelphia, SIAM, 1987.

[15] Y. Saad and M. Schultz, *GMRES: A Generalized Minimal Residual Algorithm for Solving Nonsymmetric Linear Systems*, SIAM J. Sci. Statist. Comput., V7, pp. 856-869 (1986).

[16] R. Varga, *Matrix Iterative Analysis*, New Jersey, Prentice-Hall, 1962.

[17] D. Young and K. Jea, *Generalized Conjugate Gradient Acceleration of Nonsymmetric Iterative Method*, Linear Algebra Appl, V34, pp.159-194 (1980).

ANALYSIS OF P–CYCLIC ITERATIONS FOR MARKOV CHAINS

APOSTOLOS HADJIDIMOS* AND ROBERT J. PLEMMONS†

Abstract. We consider the convergence theory of the Successive Overrelaxation (SOR) iterative method for the solution of linear systems $Ax = b$, when the matrix A has block a $p \times p$ partitioned p–cyclic form. Our purpose is to extend much of the p–cyclic SOR theory for nonsingular A to consistent singular systems and to apply the results to the solution of large scale systems arising, *e.g.*, in queueing network problems in Markov analysis. Markov chains and queueing models lead to structured singular linear systems and are playing an increasing role in the understanding of complex phenomena arising in computer, communication and transportation systems.

For certain important classes of singular problems, we develop a convergence theory for p–cyclic SOR, and show how to repartition for optimal convergence. Recent results by Kontovasilis, Plemmons and Stewart on the new concept of convergence of SOR in an *extended* sense are further analyzed and applied to the solution of periodic Markov chains.

AMS(MOS) subject classifications. 65F10. CR Catagory: 5.14

Key words. iterative methods, p–cyclic matrices, singular linear systems, optimal successive overrelaxation, Markov chains, queueing models.

1. Introduction. Markov chains sometimes possess the block partitioning property that the minimum number of transitions that must be made on leaving any state to return to that state, is a multiple of some integer $p > 1$. These models are said to be periodic of period p, or p–cyclic of index p. Bonhoure, Dallery and Stewart [2] have shown that Markov chains that arise from queueing network models frequently have this important property. These models usually possess extremely large numbers of states and iterative methods are normally used in their analysis. The problem of computing the stationary probabilities of such chains leads to linear systems of equations where one should take advantage of the resulting block cyclic structure of the coefficient matrix A in the iterations. With this in mind, we proceed to examine the application of block p–cyclic SOR iterative methods to the analysis of Markov chains.

More generally, block iterative methods are suitable for the solution of large and sparse systems of linear equations having matrices that possess a special structure. First we consider block p–cyclic SOR for arbitrary

* Department of Computer Science, Purdue University, West Lafayette, IN 47907. Research supported in part by NSF grant CCR-86-19817 and by the US Air Force under grant no. AFOSR-88-10243.

† Institute for Mathematics and its Applications, University of Minnesota, Minneapolis, MN 55455, and Department of Mathematics and Computer Science, Wake Forest University, P.O. Box 7388, Winston-Salem, NC 27109. Research supported by the US Air Force under grant no. AFOSR-91-0163.

consistent systems of linear equations. We are given

(1.1) $$Ax = b, \quad A \in \Re^{n \times n}, \quad x, b \in \Re^n$$

and the usual block decomposition

(1.2) $$A = D - L - U$$

where D, L and U are block diagonal, lower and upper triangular matrices respectively, and D is non-singular. The block SOR method for any $\omega \neq 0$ is defined as:

(1.3) $$Dx^{(m)} = Dx^{(m-1)} + \omega(Lx^{(m)} - Dx^{(m-1)} + Ux^{(m-1)} + b), \quad m = 1, 2, \ldots$$

The method can be equivalently described as

(1.4) $$x^{(m)} = \mathcal{L}_\omega x^{(m-1)} + c, \quad m = 1, 2, \ldots$$

where

(1.5) $$\mathcal{L}_\omega = (D - \omega L)^{-1}[(1 - \omega)D + \omega U)], \quad c = \omega(D - \omega L)^{-1}.$$

It is well known that, for nonsingular systems (1.1), SOR converges iff $\rho(\mathcal{L}_\omega) < 1$. The associated spectral convergence factor is then $\rho(\mathcal{L}_\omega)$.

For arbitrary systems (1), very little is known about the optimal relaxation parameter ω which minimizes $\rho(\mathcal{L}_\omega)$ as a function of ω. However, considerable information is known for the situations where the matrix A has a special block cyclic structure. For the important case of matrices with "Property A", Young [13] (see also [14]) discovered his famous result on the optimum ω. Here A in (1) has a special two–by–two cyclic block form. Young's result was generalized by Varga [11] (see also [12]) to consistently ordered p–cyclic matrices. In this case it is assumed (without loss of generality) that A has the partitioned block form

(1.6) $$A = \begin{pmatrix} A_1 & 0 & 0 & \cdots & B_1 \\ B_2 & A_2 & 0 & \cdots & 0 \\ 0 & B_3 & A_3 & \cdots & 0 \\ \vdots & \vdots & \ddots & \ddots & \vdots \\ 0 & 0 & \cdots & B_p & A_p \end{pmatrix}$$

where each diagonal submatrix A_i is square and nonsingular. With D in (1.2) defined by $D \equiv \mathrm{diag}(A_1, A_2, \ldots, A_p)$, the associated block Jacobi matrix J_p defined by $J_p \equiv I - D^{-1}A$, has the form

(1.7) $$J_p = \begin{pmatrix} 0 & 0 & 0 & \cdots & C_1 \\ C_2 & 0 & 0 & \cdots & 0 \\ 0 & C_3 & 0 & \cdots & 0 \\ \vdots & \vdots & \ddots & \ddots & \vdots \\ 0 & 0 & \cdots & C_p & 0 \end{pmatrix}$$

where $C_i \equiv -A_i^{-1} B_i$, $1 \le i \le p$.

The matrix J_p is weakly cyclic of index p if there exists a permutation matrix P such that $P J_p P^T$ has the form (1.7), where the null diagonal blocks are square. When J_p is already in the form (1.7) it is called *consistently ordered*. For such matrices Varga [11] proved the important relationship

$$(1.8) \qquad\qquad (\lambda + \omega - 1)^p = \lambda^{p-1} \omega^p \mu^p$$

between the eigenvalues μ of J_p and λ of $\mathcal{L}_\omega$. Assuming further that all eigenvalues of J_p^p satisfy

$$0 \le \mu^p \le \rho(J_p^p) < 1,$$

he showed that the optimum ω value ω_{opt} is the unique positive solution of the equation

$$(1.9) \qquad\qquad [\rho(J_p)\omega]^p = p^p (p-1)^{1-p} (\omega - 1)$$

in the interval $(1, \frac{p}{p-1})$. This ω_{opt} yields a convergence factor equal to

$$(1.10) \qquad\qquad \rho(\mathcal{L}_\omega) = (p-1)(\omega_{opt} - 1).$$

Similar results have been obtained (see, e.g., [5]) for the case where the eigenvalues of J_p^p are non-positive, that is,

$$-\left(\frac{p}{p-2}\right)^p < -\rho(J_p^p) \le \mu^p \le 0.$$

Few applications of p–cyclic SOR have been found for $p > 2$ (see, e.g., Berman and Plemmons [1]), but it turns out that, for example, ieast squares computations lead naturally to a 3–cyclic SOR iterative scheme. Markham, Neumann and Plemmons [9] have described a formulation of the problem leading to a 2–cyclic SOR method, obtained by repartitioning the 3–cyclic coefficient matrix into a 2–cyclic form. They showed that this method always converges for sufficiently small SOR parameter ω, in contrast to the 3–cyclic formulation, and that the 2–cyclic approach is asymptotically faster. Recently, Pierce, Hadjidimos and Plemmons [10] have generalized and extended the technology of p–cyclic SOR by showing that if the spectrum of the p-th power, J_p^p, of the block Jacobi matrix given in (1.7) is either nonpositive or nonnegative, then repartitioning a block p–cyclic matrix into a block q–cyclic form, $q < p$, results in asymptotically faster SOR convergence for the same amount of work per iteration. As a consequence, 2–cyclic SOR is asymptotically optimal for SOR under these conditions. In particular, it follows that 2–cyclic is optimal for SOR applied to linear equality constrained least squares in the Kuhn–Tucker formulation since here the spectrum of J_3^3 is nonpositive.

In general, the requirement that the spectrum of J_p^p be either non-positive or nonnegative is critical. Eiermann, Niethammer and Ruttan [3] have shown by considering experimental and theoretical counterexamples that, without this requirement, 2–cyclic SOR is not always superior to p–cyclic SOR, $p > 2$. Galanis and Hadjidimos [4] have now generalized all of this work for nonsingular systems by showing how to repartition a block p–cyclic consistently ordered matrix for optimal SOR convergence for the general real case of the eigenvalues of J_p^p. One must keep in mind that all these convergence results hold only in an asymptotic sense. Golub and de Pillis [6] have pointed out that short–term convergence, i.e., error reduction in the early iterations, may only be controlled by reducing the spectral norm of the iteration matrix, while long–term or asymptotic convergence is generally improved by minimizing the spectral radius, as described above.

All results mentioned thus far consider only *nonsingular* systems of equations of the form (1.1). For many applications, for example to Markov chains, the coefficient matrix A will be singular. Hadjidimos [7] has examined the singular case ($\det(A) = 0$ and $b \in \mathcal{R}(A)$), under the assumptions that: (1) the Jacobi matrix $J = J_p$ is weakly cyclic of index p, (2) the eigenvalues of J_p^p are nonnegative with $\rho(J) = 1$, and (3), that J has either a simple unit eigenvalue or a multiple one associated with 1×1 Jordan blocks. Hadjidimos proved, among other results, that ω_{opt} is the unique root of (1.9) (in the same interval as in the nonsingular case), where $\rho(J_p)$ has to be replaced by $\gamma(J_p)$, the maximum of the moduli of the eigenvalues of J_p, excluding those that have modulus 1, viz.,

$$\gamma(J_p) \equiv \max\{|\lambda|,\ \lambda \in \sigma(J),\ |\lambda| \neq 1\}.$$

Recall now that, if A is a singular irreducible M–matrix then $1 \in \sigma(\mathcal{L}_\omega)$ for all ω and, the conditions for *semiconvergence* (see, e.g., [1]) become:
- $\rho(\mathcal{L}_\omega) = 1$.
- Elementary divisors associated with 1 are linear, i.e., $\text{rank}(I - \mathcal{L}_\omega)^2 = \text{rank}(I - \mathcal{L}_\omega)$.
- If $\lambda \in \sigma(\mathcal{L}_\omega)$ with $|\lambda| = 1$, then $\lambda = 1$, i.e., $\gamma(\mathcal{L}_\omega) < 1$.

For consistency, will use the term convergence to mean semiconvergence in the singular case.

The results we will obtain here on optimal p–cyclic SOR for consistent linear systems $Ax = b$ have applications to discrete ergodic Markov Chain problems with a transition probability matrix P being cyclic with period p, as discussed in [8]. In particular, Markov chains sometimes possess the property that the minimum number of transitions that must be made on leaving any state to return to that state, is a multiple of some integer $p > 1$. These models are said to be periodic of period p, or p–cyclic of index p. In a recent interesting paper, Bonhoure, Dallery and Stewart [2] have shown that Markov chains that arise from queueing network models frequently possess this property.

Indeed, in the discrete case, the problem to be solved is

(1.11) $$\pi^T P = \pi^T, \quad \|\pi\|_1 = 1$$

or, equivalently,

$$(I - P^T)\pi = 0, \quad \|\pi\|_1 = 1,$$

where the element π_i is the probability of being in state i when the system reaches statistical equilibrium. It is immediate that, setting $A \equiv I - P^T$, and noting that if P is a cyclic stochastic matrix with transpose of the form (1.7), the corresponding homogeneous problem has a matrix that is of the form (1.6) and the associated Jacobi matrix is $J_p = P^T$. Therefore, all the results of this paper carry over to p–cyclic Markov Chains, simply by replacing J_p with P^T. In particular, the matrix A is a singular M–matrix and is irreducible when the chain is ergodic. Thus the conditions for semiconvergence described earlier apply to this Markov chain application.

For homogeneous continuous time p–cyclic Markov chains with infinitesimal generator Q, considered in [2,8], we are interested in solving

(1.12) $$\pi^T Q = 0, \quad \|\pi\|_1 = 1.$$

Equation (1.12) may also be written in the form (1.11), where

$$P = Q\Delta t + I,$$

if Δt is sufficiently small. In the p–cyclic case, the infinitesimal matrix Q in (1.12) is such that Q^T has the block form (1.6).

Markov chains and queueing models thus lead to structured singular, irreducible linear systems of the type considered in this paper. Queueing models are playing an increasing role in the understanding of complex phenomena arising in computer, communication and transportation systems.

Our purpose in this paper is to extend the p–cyclic SOR theory to the singular case and to apply the results to the solution of large scale singular systems arising, *e.g.*, in queueing network problems in Markov analysis. For certain important classes of singular problems, we provide a convergence theory for p–cyclic SOR in §2, and show in §3 how to repartition for optimal convergence. Recent results by Kontovasilis, Plemmons and Stewart [8] on the new concept of convergence of SOR in an *extended* sense are rigorously analyzed in §4, for the important case where J_2^2 (or J_q^q for some $q \le p$) has all real eigenvalues with the same sign.

2. The General p-Cyclic Case. For the determination of ω_{opt} in the p-cyclic singular case arising in Markov chains, we begin our analysis by giving the corresponding result for the nonsingular one. This is stated in [3] or, in a more compact form, in Theorem 2.2 of [4]. More specifically:

LEMMA 2.1. *Suppose we are given a nonsingular system of the form (1.1), and let ω_p and ρ_p denote the relaxation factor and the convergence factor, respectively, of the optimal p–cyclic SOR for which*

$$(2.1) \qquad \begin{array}{c} \sigma(J_p^p) \subset [-\alpha^p, \beta^p], \quad -\alpha^p, \ \beta^p \ \in \ \sigma(J_p^p), \\ 0 \le \alpha < \frac{p}{p-2}, \quad 0 \le \beta < 1. \end{array}$$

Then ω_p and ρ_p are determined from the equations

$$(2.2) \qquad \left(\frac{(\alpha_p + \beta_p)}{2} \, \omega \right)^p - \frac{(\alpha_p + \beta_p)}{(\beta_p - \alpha_p)} \, (\omega - 1) = 0$$

and

$$(2.3) \qquad \rho_p = \left(\frac{\alpha_p + \beta_p}{2} \right) (\omega_p - 1) = \left(\frac{(\alpha_p + \beta_p)}{2} \, \omega_p \right)^p$$

where ω_p is the unique positive root of (2.2) in

$$(2.4) \qquad \left(\min\{1, 1 + \frac{\beta_p - \alpha_p}{\alpha_p + \beta_p}\}, \quad \max\{1, 1 + \frac{\beta_p - \alpha_p}{\alpha_p + \beta_p}\} \right)$$

and where

$$(2.5) \qquad \begin{array}{llll} i) & \alpha_p = \frac{p-2}{p}\beta_p, & \beta_p = \beta & \text{iff } \frac{\alpha}{\beta} \le \frac{p-2}{p} \\ ii) & \alpha_p = \alpha, & \beta_p = \beta & \text{iff } \frac{p-2}{p} \le \frac{\alpha}{\beta} \le \frac{p}{p-2} \\ iii) & \alpha_p = \alpha, & \beta_p = \frac{p-2}{p}\alpha_p & \text{iff } \frac{p}{p-2} \le \frac{\alpha}{\beta}. \end{array}$$

Note: The limiting cases $\alpha = \beta (= 0$ or $\ne 0)$ lead to $\omega_p = 1$ and $\rho_p = \alpha^p = \beta^p$; while for $\alpha \ne 0$, $\beta = 0$ it is assumed that $\frac{\alpha}{\beta} = \infty$ and also for $p = 2$, $\frac{p}{p-2} = \infty$.

For the singular case we are studying here we recall a result from Theorem 3.1 of [7] which will be used in the sequel.

LEMMA 2.2. *If the block Jacobi matrix J_p in (1.7) satisfies the assumption* index $(I - J_p) = 1$, *then for all $\omega \in (0, 2) \setminus \{p/(p-1)\}$ it follows that* index$(I - \mathcal{L}_\omega) =$ index$(I - J_p) = 1$.

For the general singular case of interest in in this paper we assume that

$$(2.6) \qquad \sigma(J_p^p) \subset [-\alpha^p, \beta^p] \cup \{1\},$$

with α and β being defined as in (2.1). Thus, under the assumption that index$(I - J_p) = 1$ and with (2.6) replacing the first part of (2.1), the main result of this section is identically the same as that of Lemma 2.1.

Evidently, in (2.3) the optimal semiconvergence factor $\gamma_p = \gamma(\mathcal{L}_{w_p})$ must replace the optimal spectral radius $\rho_p = \rho(\mathcal{L}_{w_p})$. This results extends that obtained in Theorem 3.3 of [7], where the nonnegative case ($\alpha = 0 \leq \beta < 1$) was treated, to that of the general real case of $\sigma(J_p^p)$. The resulting theorem can thus be stated.

THEOREM 2.3.

Suppose we are given a possibly singular system of the form (1), and let ω_p and γ_p denote the relaxation factor and the convergence factor, respectively, of the optimal p–cyclic SOR for which

$$(2.7) \qquad \begin{aligned} \sigma(J_p^p) \subset [-\alpha^p, \beta^p] \cup \{1\}, \quad -\alpha^p, \ \beta^p \ \in \ \sigma(J_p^p), \\ 0 \leq \alpha < \tfrac{p}{p-2}, \quad 0 \leq \beta < 1, \end{aligned}$$

and, moreover, assume that index$(I - J_p) = 1$. *Then ω_p and γ_p are determined by equations (2.1) through (2.5), with γ_p replacing ρ_p.*

3. Best Cyclic Repartitioning. As was mentioned in the Introduction, Markhan, Neumann and Plemmons [9] were the first who considered the problem of repartitioning a block 3-cyclic consistently ordered matrix into a 2-cyclic form for optimal SOR convergence. The most recent result on the general problem of the best cyclic repartitioning seems to be that obtained by Galanis and Hadjidimos [4]. It covers the case where the spectrum $\sigma(J_p^p)$ is real, under the assumptions (2.1), where the conditions on α are relaxed to $0 \leq \alpha < \infty$. The result is given in Theorem 2.1 of [4]. In the following lemma we give the main part of Theorem 2.1 of [4] and provide its accompanying Table R at the end of the paper.

LEMMA 3.1. *Let J_p be the block Jacobi matrix (1.7) associated with the linear system (1.1), where A has the p–cyclic consistently ordered form (1.6), $p \geq 3$, and let $\sigma(J_p^p)$ satisfy (2.1), where the bound $\tfrac{p}{p-2}$ on α is replaced by ∞. Assume that A is repartitioned into a block q–cyclic consistently ordered form ($2 \leq q < p$) and denote by ω_q and ρ_q the relaxation factor and the spectral radius of the optimal q–cyclic SOR. Let r be the value of q that gives the best cyclic repartitioning; i.e., the smallest optimal spectral radius ρ_q. Then the value of r is given in Table R, where the quantities $\alpha_{\ell,\ell+1}$ and $\beta_{\ell,\ell+1}$ in the table are found from the expressions*

$$(3.1) \qquad \begin{aligned} \alpha_{\ell,\ell+1} &= \left(\frac{2\rho^{1/\ell} - (1+\rho)\beta^{p/\ell}}{1-\rho} \right)^{\ell/p}, \\ \beta_{\ell,\ell+1} &= \left(\frac{2\rho^{1/\ell} - (1-\rho)\alpha^{p/\ell}}{1+\rho} \right)^{\ell/p}. \end{aligned}$$

In (3.1), ρ is the unique root, in $(0,1)$, of the equation

$$\beta^p (\ell + \rho)^{\ell+1} - (\ell+1)^{\ell+1} \rho = 0,$$

for $\alpha_{\ell,\ell+1}$, and of the equation

$$\alpha^p (\ell - \rho)^{\ell+1} - (\ell+1)^{\ell+1} \rho = 0,$$

for $\beta_{\ell,\ell+1}$. *The values of* ω_r *and* ρ_r *are determined via* (2.2)–(2.5), *where in all these formulas,* r *replaces* p *and then* $\alpha^{p/r}$ *and* $\beta^{p/r}$ *replace* α *and* β, *respectively.*

Note: As was pointed out in Kontovasilis, Plemmons and Stewart [8], Lemma 3.1 is of a more general value since it also covers the case of complex eigenvalues $\mu \in \sigma(J_p)$, with $\mu \in C$, provided $|\mu| \leq \min\{\alpha, \beta\}$.

Suppose now that in the singular case of the p–cyclic consistently ordered matrix A index$(I - J_p) = 1$. Suppose also that $\sigma(J_p^p)$ is given by (2.7), in which $0 \leq \alpha < \frac{p}{p-2}$ has been replaced by $0 \leq \alpha < \infty$. Suppose also that A is repartitioned into a block q–cyclic consistently ordered form. From the analysis in §2 it is obvious that Lemma 2.2 and therefore Theorem 2.1 apply to the singular case for $q = p$ provided $\alpha < \frac{p}{p-2}$. For Lemma 2.2 to apply for any $2 \leq q \leq p - 1$ one must have $\omega \in (0, 2) \setminus \{\frac{q}{q-1}\}$. This, however, assumes that the relationship

$$(3.2) \qquad \text{index}(I - J_q) = 1, \quad q = 2, \ldots, p - 1,$$

is valid. There are certainly cases of vital practical importance where the implication (3.2) is a straightforward consequence of some further property of the matrix A. For example, (3.2) follows directly when A is a singular *irreducible* M-matrix (see, e.g, [1]), as in the case of the matrix coefficient in the Markov Chain problem with $A = I - P^T$, where P is the transition probability matrix and the chain is ergodic. However, (3.2) is always true under the assumption index$(I - J_p) = 1$. This is stated in the following theorem.

THEOREM 3.2. *Let* J_p *be the block Jacobi matrix (7) associated with the linear system (1), where* A *has the* p-*cyclic consistently ordered form (6),* $p \geq 3$, *and let* index$(I - J_p) = 1$. *Assume that* A *is repartitioned into a block* q-*cyclic consistently ordered form* $(2 \leq q < p)$ *and denote by* J_q *be the block Jacobi matrix corresponding to the new repartinioning. Then (3.2) holds.*

So, under no further assumption, Theorem 2.1 holds for any q, when $\alpha^{p/q} < \frac{q}{q-2}$. Hence, a statement let us call it Theorem 3.2 (which will not be formally stated here), completely analogous to Lemma 3.1 holds true. We note that since Theorem 3.2 refers to the singular case instead of the optimal spectral radius ρ_q, one must use the optimal semiconvergence factor $\gamma_q = \gamma(\mathcal{L}_{\omega_q})$ in its place. Thus we obtain formulas on how to repatition A for optimal SOR convergence for computing the stationary distribution vectors of Markov chains. (See Table R at the end of the paper.)

4. Optimal Extended SOR. In this section we provide a convergence analysis for extended SOR convergence for the important 2-cyclic consistently ordered case. Here we assume that J_2^2 has all real eigenvalues with the same sign. It is shown that small perturbations around the optimal ω in the extended SOR method affect the convergence factor much less than for the usual SOR method. This formally confirms the validity of observations about numerical tests showing this phenomenom reported in [8].

We first introduce the following notation from [8]:

$$(4.1) \quad \begin{aligned} \alpha(\omega) &:= \max\{|\lambda| := f_1(\lambda, \omega) = 0\} \geq 1, \\ \vartheta(\omega) &:= \max\{|\lambda| := f_\mu(\lambda, \omega) = 0, \ \mu \in \sigma(J_p), |\mu| < 1\}, \\ r(\omega) &:= \vartheta(\omega)/\alpha(\omega), \end{aligned}$$

where

$$(4.2) \qquad f_\mu(\lambda, \omega) := \lambda^p - \omega\mu\lambda^{p-1} + \omega - 1.$$

In [8] a detailed analysis led to the determination of the optimal parameter(s) ω_p and therefore of the optimal convergence factor $r^p(\omega_p)$ of the extended SOR in the case of nonnegative spectra $\sigma(J_p^p)$. In this section we study in more detail the behavior of the asymptotic convergence factor of the extended SOR in both cases of the nonnegative ($\alpha = 0$) and the nonpositive ($\beta = 0$) spectra $\sigma(J_2^2)$ (see (2.7)). Also, the detailed study of the behavior of $r^2(\omega)$ around ω_p will subsequently allow us to explain the phenomenon observed in [8]; namely, that small perturbations around ω_p affect the convergence factor in the extended SOR much less than small perturbations around the corresponding ω_p in the usual SOR. In both cases to be studied the interval for ω will be considered as being $(-\infty, \infty)\backslash(0, 2)$, i.e.,

$$(4.3) \qquad \omega \in (-\infty, 0] \cup [2, \infty)$$

while from (4.1)

$$(4.4) \qquad f_1(\lambda, \omega) = \lambda^2 - \omega\lambda + \omega - 1 = 0$$

and

$$(4.5) \qquad f_\mu(\lambda, \omega) = \lambda^2 - \omega\mu\lambda + \omega - 1 = 0,$$

with $\mu \in \sigma(J_2)$ and $|\mu| \neq 1$. From (4.1), (4.3) and (4.4) we immediately obtain that

$$(4.6) \qquad \alpha(\omega) = |\omega - 1| \geq 1$$

while from (4.1) and (4.5)

$$(4.7) \qquad \vartheta(\omega) := \frac{1}{2} \max |\omega\mu \pm (\mu^2\omega^2 - 4\omega + 4)^{1/2}|.$$

4.1. The Nonnegative Case. In this case it is assumed that $\sigma(J_2)$ satisfies the following conditions

$$(4.8) \qquad\qquad \sigma(J_2^2) \subset [0, \beta^2] \cup \{1\}, \quad 0 \le \beta < 1.$$

Let then $0 \le \mu^2 \le \beta^2$, $\mu \in \sigma(J_2)$, where without loss of generality, we may consider μ such that $0 \le \mu \le \beta$. Denoting the discriminant in (4.7) by

$$(4.9) \qquad\qquad D(\mu) := \mu^2 \omega^2 - 4\omega + 4$$

we readily have $D(\beta) \ge D(\mu) \ge D(0)$. So, we distinguish the three subcases a) $D(\beta) \le 0$, b) $D(0) \ge 0$, and c) $D(\beta) \ge 0 \ge D(0)$ which are studied separately.

In subcase (a), (4.5) has two complex conjugate roots of modulus $|\omega - 1|^{1/2}$ each and this is the case when

$$(4.10) \qquad\qquad \frac{2}{1 + (1 - \beta^2)^{1/2}} \le \omega \le \frac{2}{1 - (1 - \beta^2)^{1/2}},$$

whence

$$\vartheta(\omega) = |\omega - 1|^{1/2}.$$

Therefore, setting

$$(4.11) \qquad\qquad r^2(\omega) = \frac{\vartheta^2(\omega)}{\alpha^2(\omega)} = \frac{1}{|\omega - 1|} < 1,$$

we conclude that the extended SOR converges for all

$$\omega \in (2, \frac{2}{1 - (1 - \beta^2)^{1/2}}].$$

From (4.11) it is also readily seen that $r^2(\omega)$ is a strictly decreasing function of ω. Specifically as ω increases from 2 to

$$2/(1 - (1 - \beta^2)^{1/2})$$

$r^2(\omega)$ decreases from 1 to

$$(1 - (1 - \beta^2)^{1/2})/(1 + (1 - \beta^2)^{1/2}).$$

After the study of subcases (b) and (c) takes place, the main result for the nonnegative case, which is stated below, confirms the corresponding one obtained in [8].

TABLE 4.1

ω	$-\infty$		0	$-$	2		ω_2		∞
$r^2(\omega)$	β^2	$\nearrow$	1	$-$	1	$\searrow$	$r^2(\omega_2)$	$\nearrow$	β^2

THEOREM 4.1. *Let $\sigma(J_2)$ satisfy (4.8) and let $\text{index}(I - J_2) = 1$. Then, the extended SOR converges for all $\omega \in (-\infty, 0) \cup (2, \infty)$. The optimal relaxation factor is given by*

$$(4.12) \qquad \omega_2 = \frac{2}{1 - (1 - \beta^2)^{1/2}}$$

while for the (optimal) convergence factor there holds for $\omega_2 \neq \omega \in (-\infty, 0) \cup (2, \infty)$,

$$(4.13) \qquad 1 > r^2(\omega) > r^2(\omega_2) = \frac{1 - (1 - \beta^2)^{1/2}}{1 + (1 - \beta^2)^{1/2}}.$$

Furthermore, the behavior of $r^2(\omega)$ is illustrated in Table 1.

Note: As is seen from Table 1 when $\omega \to \pm\infty$, $r^2(\omega) \to \beta^2$, in other words, the extended SOR converges (in the limiting cases) as fast as the usual Gauss-Seidel method.

4.2. The Nonpositive Case. This time it is assumed that $\sigma(J_2)$ satisfies

$$(4.14) \qquad \sigma(J_2^2) \subset [-\alpha^2, 0] \cup \{1\}, \quad 0 \leq \alpha < 1,$$

so $\tilde{\sigma}(J_2) := \sigma(J_2) \setminus \{-1, 1\}$ is purely imaginary. Let then $i\mu \in \tilde{\sigma}(J_2)$ be any eigenvalue of J_2, where without loss of generality we may assume that $0 \leq \mu \leq \alpha$ whence $-\alpha^2 \leq -\mu^2 \leq 0$. Obviously (4.4) will remain the same, leading again to the expression (4.6) for $\alpha(\omega)$, while (4.5) will become

$$(4.15) \qquad f_{i\mu} = \lambda^2 - i\omega\mu\lambda + \omega - 1 = 0$$

and hence

$$(4.16) \qquad \vartheta(\omega) = \frac{1}{2} \max |i\omega\mu \pm (-\mu^2\omega^2 - 4\omega + 4)^{1/2}|.$$

The discriminant $D(\mu)$ in (4.16) is now

$$(4.17) \qquad D(\mu) := -\mu^2\omega^2 - 4\omega + 4$$

implying that $D(0) \geq D(\mu) \geq D(\alpha)$. Three subcases are considered again. Specifically, a) $D(\alpha) \geq 0$, b) $D(0) \leq 0$, and c) $D(0) \geq 0 \geq D(\alpha)$. As in the nonnegative case the simplest of the three subcases, that is (a), will be examined in the sequel. For this we have $D(\mu) \geq 0$ for all $\mu \in [0, \alpha]$, so $(D(\mu))^{1/2}$ is real. In view of the purely imaginary nature of $i\omega\mu$, (4.15) has two complex roots having the same imaginary parts and opposite in sign real parts. Hence, the two roots have equal moduli, consequently

$$\vartheta(\omega) = |\omega - 1|^{1/2}.$$

However, from $D(\alpha^2) \geq 0$ it is obtained that

$$\omega \in [\frac{2}{1 - (1 + \alpha^2)^{1/2}}, \frac{2}{1 + (1 + \alpha^2)^{1/2}}]$$

which together with

(4.18)
$$r^2(\omega) = \frac{\vartheta^2(\omega)}{\alpha^2(\omega)} = \frac{1}{|\omega - 1|} < 1,$$

which implies that $\omega \in (-\infty, 0) \cup (2, \infty)$, gives

$$\omega \in [\frac{2}{1 - (1 + \alpha^2)^{1/2}}, 0).$$

It can be readily found out that in the previous interval $r^2(\omega)$ strictly increases from the value

$$\frac{(1 + \alpha^2)^{1/2} - 1}{(1 + \alpha^2)^{1/2} + 1}$$

to the value 1.

After the examination of the subcases (b) and (c), the main result of the present section can be stated as follows:

THEOREM 4.2. *Let* $\sigma(J_2)$ *satisfy (4.14) and let* $\mathrm{index}(I - J_2) = 1$. *Then, the extended SOR converges for all* $\omega \in (-\infty, 0) \cup (\frac{2}{1-\alpha}, \infty)$ *The optimal relaxation factor is given by*

$$\omega_2 = \frac{2}{(1 + \alpha^2)^{1/2} - 1},$$

while for the (optimal) convergence factor there holds for $\omega_2 \neq \omega \in (-\infty, 0) \cup (2, \infty)$,

$$r^2(\omega) > r^2(\omega_2) = \frac{(1 + \alpha^2)^{1/2} - 1}{(1 + \alpha^2)^{1/2} + 1}.$$

Moreover, the behavior of $r^2(\omega)$ *is illustrated in Table 2.*

Note: As in the nonnegative case when $\omega \to \pm\infty$, $r^2(\omega) \to \alpha^2$ and the extended (SOR) converges as fast as the usual Gauss-Seidel method.

TABLE 4.2

ω	$-\infty$		ω_2		$0 \;-\; 2$		$\frac{2}{1-\alpha}$		∞
$r^2(\omega)$	α^2	$\searrow$	$r^2(\omega_2)$	$\nearrow$	$1 \quad -$	$\searrow$	1	$\searrow$	α^2

5. Final Comments. Our purpose in this paper has been to extend the p–cyclic SOR theory from nonsingular problems to consistent singular systems and to apply the results to the solution of large scale singular linear systems arising in Markov analysis. We have developed a convergence theory for p–cyclic SOR, and shown how to repartition for optimal convergence.

Results by Kontovasilis, Plemmons and Stewart [8] on the new concept of convergence of SOR in an *extended sense* have been analyzed for the case where the spectrum $\sigma(J_p^p)$ is either nonnegative or nonpositive. Further work in this area is needed before the concept of extended SOR convergence can be fully applied. We have analyzed in detail the phenomenon observed in [8]; that small perturbations around ω_p affect the convergence factor in the extended SOR much less than small perturbations around the corresponding ω_p in the usual SOR. The extended SOR method is thus preferable to the usual SOR method for solving singular linear systems of equations associated with Markov chains.

REFERENCES

[1] A. Berman and R. J. Plemmons. *Nonnegative Matrices in the Mathematical Sciences.* Academic Press, New York, 1979.

[2] F. Bonhoure, Y. Dallery, and W. Stewart. On the efficient use of periodicy properties for the efficient numerical solution of certain Markov chains. *MASI Tech. Rept. 91-40*, Université de Paris, France, 1991.

[3] M. Eiermann, W. Niethammer, and A. Ruttan. Optimal successive overrelaxation iterative methods for p-cyclic matrices. Numer. Math., 57:593–606, 1990.

[4] S. Galanis and A. Hadjidimos. How to repartition a block p-cyclic consistently ordered matrix for optimal SOR convergence. *SIAM J. Matrix Anal. Appl.*, 13:102–120, 1992.

[5] S. Galanis, A. Hadjidimos, and D. Noutsos. On the equivalence of the k-step iterative Euler methods and successive overrelaxation (SOR) methods for k-cyclic matrices. *Math. Comput. Simulation*, 30:213–230, 1988.

[6] G. H. Golub and J. de Pillis. Toward an effective two-parameter SOR method. in *Iterative Methods for Large Lin. Syst.*, D. Kincaid and L. Hayes, Eds., Academic Press, New York, 107–118, 1988.

[7] A. Hadjidimos. On the optimization of the classical iterative schemes for the solution of complex singular linear systems. *SIAM J. Alg. Disc. Meth.*, 6(4):555–566, 1985.

[8] K. Kontovasilis, R. J. Plemmons and W. J. Stewart. Block cyclic SOR for Markov chains with p-cyclic infintesimal generator. *Linear Algebra Appl.*, 154–156:145–223, 1991.

[9] T. L. Markham, M. Neumann, and R. J. Plemmons. Convergence of a direct-

iterative method for large-scale least squares problems *Linear Algebra Appl.*, 69:155–167, 1985.

[10] D. J. Pierce, A. Hadjidimos, and R. J. Plemmons. Optimality relationships for *p*-cyclic SOR. *Numer. Math.*, 56:635–643, 1990.

[11] R. S. Varga. *p*-cyclic matrices: A generalization of the Young-Frankel successive overrelaxation scheme. *Pacific J. Math.*, 9:617–628, 1959.

[12] R. S. Varga. *Matrix Iterative Analysis*. Prentice Hall, Englewood Cliffs, NJ, 1962.

[13] D. M. Young. Iterative methods for solving partial differential equations of elliptic type. *Trans. Amer. Math. Soc.*, 76:92–111, 1954.

[14] D. M. Young. *Iterative Solution of Large Linear Systems*. Academic Press, New York, 1971.

ITERATIVE METHODS FOR FINDING THE STATIONARY VECTOR FOR MARKOV CHAINS

DIANNE P. O'LEARY*

Abstract. This overview concerns methods for estimating the steady-state vector of an ergodic Markov chain. The problem can be cast as an ordinary eigenvalue problem, but since the eigenvalue is known, it can equally well be studied as a nullspace problem or as a linear system. We discuss iterative methods for each of these three formulations. Many of the applications, such as queuing modeling, have special structure that can be exploited computationally, and we give special emphasis to three ideas for exploiting this structure: decomposability, separability, and multilevel aggregation. Such ideas result in a large number of diverse algorithms, many of which are poorly understood.

Key words. Markov chains, stationary vectors, nullspace, decomposability, separability, multilevel iterations, multigrid, aggregation, small rank corrections, small norm corrections

1. Introduction. This overview concerns the numerical approximation of stationary vectors of discrete Markov chains. The rest of this introductory section contains a sketch of the problem, a classification of the numerical methods, and examples of exploitable structure in Markov chain problems. The three following sections concern methods for the eigenvector formulation, the nullspace formulation, and the linear system formulation. The fifth section focuses on more specialized techniques that can exploit information about nearby problems.

This paper has been structured as an *overview* rather than a *survey*. The goal is to outline major research directions that have proven fruitful, present a sample of successful algorithms and theoretical results, and delineate opportunities for future research. In order to do this within a reasonable amount of space, the completeness of a survey has been sacrificed and many important algorithms, theorems, and references have regretfully been omitted. Related articles in this volume include those of Plemmons and Schweitzer.

1.1. Definitions and Notation. Discrete time Markov chains can be represented by a matrix P, whose (i,j)-element p_{ij} is the probability of a transition from state i to state j. The elements of the rows of P necessarily sum to one; i.e.,

$$(1.1) \qquad Pe = e,$$

where e is the vector with each component equal to 1. If the n states of the Markov chain initially have a probability distribution z^T, where z_i

* Computer Science Department and Institute for Advanced Computer Studies, University of Maryland, College Park, Maryland 20742. This work was completed while the author was in residence at the Institute for Mathematics and Its Applications, University of Minnesota, supported by the General Research Board of the University of Maryland, College Park and by the IMA.

$(i = 1, ..., n)$ is interpreted as the probability that the ith state of the chain is occupied, then at the next time unit the probability distribution is $z^T P$.

Equation (1.1) implies that 1 is an eigenvalue of P corresponding to the right eigenvector e. If the matrix P is irreducible and acyclic, or, equivalently, if the Markov chain is ergodic (has no transient states and is aperiodic) (see [43] and [18] for definitions), then this eigenvalue is simple and greater in magnitude than the other eigenvalues. Moreover, Perron-Frobenius theory [43, p.30] tells us that, corresponding to the eigenvalue 1, P has a positive left eigenvector π^T that is unique under the normalization

(1.2) $$\pi^T e = 1.$$

The vector π^T is called the *steady state* or *stationary* vector of the chain, since, for almost every vector z of initial probabilities, it is the limit of the transient probabilities of the system:

$$\pi^T = \lim_{k \to \infty} z^T P^k, \qquad z^T \geq 0, \qquad z^T e = 1.$$

Iterative methods for estimating π^T are the chief concern of this overview.

The problem can be recast in the form

$$\pi^T Q = 0,$$

where $Q = I - P$. Although this form is a simple algebraic rearrangement of (1.2), it has important theoretical and computational consequences. In particular, the matrix Q is a *singular M-matrix*; that is, its diagonal elements are positive, its off-diagonal elements are nonpositive, and for any $\epsilon > 0$,

$$(\epsilon I + Q)^{-1} > 0.$$

This means that many of the techniques used for M-matrices can be brought to bear on the solution of Markov chains [43,23].

Two kinds of Markov chain problems have particularly simple solutions. A Markov chain is said to be *completely decomposable* or *uncoupled* if, after a suitable reordering of its states, P is a block diagonal matrix with $k \geq 2$ blocks. A problem in this form can be solved as k independent problems of smaller dimensions.

A Markov chain is *separable* if it can be written in the form

$$P = A \otimes I + I \otimes B,$$

where $\otimes$ denotes the Kronecker tensor product. Many Markov chains arising in queuing networks are separable, or nearly so. Separable chains have a *product form solution* that can be written in closed form (see [19,42] for details and references).

1.2. Classification of Numerical Methods. The computation of the steady state vector of a Markov chain is in principle a remarkably easy eigenvalue problem. The eigenvalue is known to be 1. It is the eigenvalue of largest magnitude, and its multiplicity is 1. We know the right eigenvector, e, and need only find the left eigenvector.

On the other hand, several factors make the problem quite challenging. There may be many other eigenvalues clustered around the value 1, making iterative methods quite slow. Problems arising in many applications (e.g., queuing theory) produce sparse matrices of enormous size, limiting practical methods to a small number of matrix-vector products. Most intriguing of all, many practical problems are *nearly* decomposable or *nearly* separable, and such structure must be exploited in order for algorithms to be practical.

We will discuss the general principle of exploiting information about nearby problems. We consider three meanings of "nearby":

1. *Nearly decomposable systems.* The matrix P for these problems is equal to a block diagonal matrix plus a correction that is small in norm. Nearly uncoupled Markov chains model systems in which groups of tightly coupled states are loosely coupled to one another. Such chains arise frequently in applications [10].
2. *Nearly separable systems.* The matrix P is equal to a separable matrix plus a small rank correction. Such problems arise in overflow queuing networks [5,6].
3. *Multilevel families of chains.* These models arise from analyzing a system at various levels of granularity. A coarse model may be able to provide approximate information for a more detailed model.

Because we know the eigenvalue, we have three choices for formulating the problem of finding the stationary vector:

1. *an eigenproblem* $\pi^T P = \pi^T$.
2. *a nullspace problem* $Q = I - P$, $\pi^T Q = 0$.
3. *a linear system* $\pi^T Q = 0$ with π_1 fixed to be 1 (or $\pi^T e = 1$, or some other normalization condition).

Various iterations have been proposed for each of these formulations, and we will focus on each in turn. Alternative algorithms that we will not discuss in this overview are direct methods for the second and third formulation (see the paper by G. W. Stewart in this volume) and methods of approximation based on fluid flow approximation and diffusion approximations to queuing networks.

2. Methods for the Eigenproblem Formulation. These methods are based on the fact that $\lim_{k \to \infty} P^k = e\pi^T$. There are many variants, all having the nice property that they require only the formation of vector-matrix products.

2.1. The Power Method and Subspace Iteration. The method of subspace iteration (also known as simultaneous iteration) extracts approximate solutions from the rows of the matrices $Z^T P^k$. When Z has only one row, the method is called the power method, and the iterates approach the solution by simulating the Markov chain until steady state is reached — often slowly. When Z has more than one row, a nonsymmetric analogue of Rayleigh-Ritz refinement is used to extract an approximation to the steady state vector. The theory and practice of the method are well understood [15,33,34].

Convergence of the power method (or simultaneous iteration) can be accelerated by matrix splitting techniques. A splitting of Q is a decomposition of the form

$$Q = M - N,$$

where M is nonsingular. Since

$$\pi^T M = \pi^T N,$$

a splitting gives rise to an iteration of the form

$$z_{k+1}^T = z_k^T N M^{-1}.$$

Splitting methods include Jacobi's method, successive over-relaxation, and their block variants.

Splittings have been subjected to intense study for solving inhomogeneous systems associated with discretizations of partial differential equations [43,44]. Because Q is a singular M-matrix, some of the convergence results carry over to the solution of Markov chains (e.g., see [22,23,17,21]). Although splitting methods can be used alone, they are most often used in conjunction with aggregation methods, discussed in §5.

2.2. Krylov Sequence Methods. In many respects these are some of the most promising and least well understood methods. They extract approximate solutions from the Krylov sequence $\{z^T, z^T P, z^T P^2, \ldots\}$. The basic algorithm is known as the method of Arnoldi, and some variants have been surveyed by Saad [28]. The nonsymmetric Lanczos method works with Krylov sequences in powers of P and P^T. The symmetric Lanczos method can be applied to PP^T. There are related variants, conjugate gradient methods, for solving linear systems. For brevity we will only discuss the method of Arnoldi here.

Arnoldi's method is based on the observation that if the vectors $\{z^T, z^T P, z^T P^2, \ldots\}$ are orthonormalized, say by the Gram-Schmidt algorithm, to produce a sequence $\{v_0^T, v_1^T, \ldots\}$, then the matrices $V_k = (v_0, v_1, \ldots, v_{k-1})$ satisfy

$$V_{k-1}^T P = H_k^T V_k^T,$$

where H_k is an $(k+1) \times k$ upper Hessenberg matrix. The eigenvalues of $\hat{H}_k$, the leading $k \times k$ submatrix of H_k, are approximations to eigenvalues of P, and the approximate eigenvectors are V_k times eigenvectors of $\hat{H}_k$. Under exact arithmetic, the matrix $\hat{H}_n^T$ is similar to P, and the approximate eigenvalues and eigenvectors are exact.

In practice, for the sake of time and storage, the iteration must be terminated long before $k = n$. This leads to the *Arnoldi algorithm with restarts*, in which the iteration is restarted every k steps with a refined z_0.

Even apart from its use with Markov chains, there are many unanswered questions about convergence properties of this algorithm.

Much study has been devoted to the symmetric version, the Lanczos algorithm [25]. The eigenvalues and eigenvectors of the (now tridiagonal) matrices H_k converge, the convergence being faster for the extreme eigenvalues. Since 1 is an extreme eigenvalue of the positive definite matrix PP^T with eigenvector π, the Lanczos method applied to PP^T is a good candidate for solving large Markov chains. However, some issues are not well resolved — in particular, those relating to to restarting the iteration when the reduced system grows too large.

The nonsymmetric iteration has received much less attention. The interlacing properties and monotone convergence are lost. Convergence is no longer assured; the restart strategy must be carefully designed. It is thus necessary to study the local behavior of the iterative Arnoldi algorithm and determine how accurate the initial approximation must be in order to guarantee convergence to the dominant eigenvalue. Other issues such as convergence bounds, computable error bounds, computational variants, the effects of inexact arithmetic, and the need for reorthogonalization of the basis vectors require serious study.

The question arises of how extra information about Markov chain problems might be used to guarantee and accelerate convergence of the Arnoldi algorithm.

One approach is to develop preconditioners appropriate to Markov models:

1. We can use similarity transformations to convert the problem $\pi^T P = \pi^T$ to $\hat{\pi}^T \hat{P} = \hat{\pi}^T$, where $\hat{\pi} = U^{-T}\pi$ and $\hat{P} = UPU^{-1}$ for some matrix U [27]. This does not change the distribution of the eigenvalues but can rotate the solution vector towards a known one.

2. We can choose a polynomial $\mathcal{P}(x)$ such that $\mathcal{P}(1) = 1$ and solve the problem $\pi^T \mathcal{P}(P) = \pi^T$. This changes the distribution of the eigenvalues. [29]

3. Matrix splitting techniques can be used to change the distribution of the eigenvalues. These matrix splittings can be developed from separable queuing models related to the given problem, by multigrid analogues, and by other aggregation ideas as in §5 [26].

Such ideas have been studied in the context of the power method, but their effectiveness in combination with Arnoldi's algorithm or the nonsymmetric Lanczos method is just beginning to be understood.

An alternate approach is to devise an algorithm which forms the Krylov subspace, as in the Arnoldi iteration, but rather than finding eigenvalues of the reduced operator, minimizes a function such as $||z^T A - z^T||$ over all vectors z^T of length 1 in the Krylov subspace [26]. Convergence properties and effective preconditioners should be studied.

3. Methods for the Null-Space Formulation. A variety of projection methods have proven effective for solving nonsymmetric systems of linear equations [2], but their behavior for homogeneous systems is less well understood. As an example of this class of methods, we consider the projection method of Kaczmarz [16], which for the solution of $\pi^T Q = 0$ has the following form. Let Q be partitioned by columns:

$$Q = (q_1 \; q_2 \; \cdots \; q_n).$$

Let

$$R_i = I - \frac{q_i q_i^T}{q_i^T q_i}$$

be the projection onto the orthogonal complement of the ith column of Q, and let

$$S = R_1 R_2 \cdots R_n.$$

Starting with some initial vector z_0, iterate according to the scheme

$$z_{k+1}^T = z_k^T S.$$

It can be shown that under mild restrictions on z_0^T, the sequence z_k^T converges to a multiple of π^T.

The appealing fact about Kaczmarz' method is its simplicity. It requires only one column of Q at a time, and it requires only one n-vector of storage. For this reason it is well suited to queuing models, where the number of states is very large.

Kaczmarz's method for inhomogeneous systems has been extensively analyzed [14,1,12]. In spite of this, very little is known about the rate of convergence of the method for Markov problems.

Bramley and Sameh [2] have used projection methods as a preconditioner for Krylov subspace methods for nonsymmetric linear systems. A similar idea should be quite effective for Markov chain problems.

4. Methods for the Linear System Formulation. Let P_{n-1} denote a submatrix of P formed by deleting any one column. Then the linear system

$$\pi^T P_{n-1} = 0, \quad \pi^T w = 1,$$

where w is any nonzero vector, is a nonsingular linear system whose solution determines the stationary vector π^T up to a normalization factor. Standard iterative techniques (Jacobi, successive overrelaxation, Krylov subspace methods) can be applied to this augmented system, and preconditioners can also be used to accelerate convergence.

Although such techniques have been studied, it is perhaps more natural to deal directly with the homogeneous system [17] or with the consistent system of $n + 1$ equations in the n unknowns π. Many of the iterative methods discussed above can be used on this overdetermined system.

5. Exploiting Nearby Problems. Since Markov chain problems tend to be quite enormous in dimension, it is important to minimize the amount of work required to compute the stationary vector. One unifying principle has been used quite successfully: exploit information about nearby, more easily analyzed Markov chains. A prime example of this is the development of successful algorithms for solving nearly decomposable systems. There is potential for progress on other classes of problems as well, in particular, for nearly separable systems and for systems that are naturally part of a multilevel family of problems.

5.1. Nearly Decomposable Systems: Aggregation methods These methods lump states together and solve the resulting reduced chain. The solution along with auxiliary information is then used to approximate the eigenvector. For nearly uncoupled chains, one step often results in a sufficiently accurate approximation. They can be combined with other iterative methods to yield hybrid methods.

When the Markov chain is nearly uncoupled, the Rayleigh-Ritz refinement is usually called aggregation. For these chains aggregation can be done with the single vector coming from the power method. The resulting algorithm is a method proposed and analyzed by McAllister, Stewart, and Stewart [20]. The surprising fact is that there are theoretical reasons for believing that this method is competitive with simultaneous iteration, though it requires less work.

There are many different aggregation methods, but all are based on a partitioning of π^T into subvectors. The states corresponding to each subvector are then *aggregated* into a single state, and the stationary vector for the aggregate problem is used to improve an approximation to π^T.

When one aggregation step is not enough, aggregation can be coupled with an iterative method (power method, [20], splitting methods [37,4], etc.). Xiaobai Sun [38] has developed a unified theory for these methods.

For general Markov chains, aggregation methods seem to have been first proposed by Takahashi [39,40]. There is an extensive literature on the methods (e.g., [7,8,9,13,31,30]). The performance of aggregation methods in general is spotty; they can be quite slow.

When a chain is nearly uncoupled, one aggregation step often gives a sufficiently accurate approximation. The properties of nearly uncoupled systems were first pointed out by Simon and Ando [32] and have been the subject of many investigations [11,35,36].

5.2. Nearly Separable Systems. Such problems arise in overflow queuing networks, where the matrix P is equal to a separable matrix plus a small rank correction.

Consider a very simple example involving two queues. Suppose that there are $n_k - 1$ spaces and s_k servers in the kth queue ($k = 1, 2$), and that customers arrive at the kth queue at rate λ_k and are served at rate μ_k. If we let state (k, l) correspond to having k customers in queue 1 and l customers in queue 2, then we can define the matrix P for the model.

If there is no interaction between the queues, then the matrix is

$$\hat{P} = P_{n_1}(\lambda_1, \mu_1, s_1) \otimes I + I \otimes P_{n_2}(\lambda_2, \mu_2, s_2),$$

where $P_m(\lambda, \mu, s)$ is the matrix of size m for a single queue with arrival rate λ, s servers, and service rate μ.

If overflow is allowed, e.g., if customers overflow into queue 2 whenever queue 1 is full, then the resulting matrix is

$$P = \hat{P} + E,$$

where E is a matrix with λ_1 in diagonal positions corresponding to queue 1 being full, and $-\lambda_1$ in the off-diagonal position of these rows corresponding to queue 1 being full and queue 2 having one more customer.

The relation between the matrices P and $\hat{P}$ can be exploited. The most obvious way is to use a splitting of P as $M = \hat{P}$ and $N = -E$. This splitting can be used to accelerate the iterative methods discussed above [5,6].

A less expensive way to use the relation between the separable and nonseparable problems is to initialize the iteration for the nonseparable one using the stationary vector for a related separable model. For our example, we model queue 2 as an independent queue with arrival rate $\hat{\lambda}_2 = \lambda_2 + \lambda_1 p_1^{full}$, where p_1^{full} is the probability that queue 1 is full. The resulting matrix is

$$\tilde{P} = P_{n_1}(\lambda_1, \mu_1, s_1) \otimes I + I \otimes P_{n_2}(\hat{\lambda}_2, \mu_2, s_2),$$

The stationary vector for $\tilde{P}$ is the outer product of the stationary vectors for P_{n_1} and P_{n_2}. Since the state of queue 1 is independent of the state of queue 2, the probability that queue 1 is full can be computed directly by finding the stationary vector for $P_{n_1}(\lambda_1, \mu_1, s_1)$, and then we need only find the stationary vector for $P_{n_2}(\lambda_2 + \lambda_1 p_1^{full}, \mu_2, s_2)$.

For more complicated problems, the probabilities that queues are full in the separable model are found by solving a system of nonlinear equations with one equation per queue:

$$p^{full} = f(p^{full}),$$

where f_k is the probability that queue k is full given the probabilities that
the other queues are full.

Given an arrival rate for a queue of size m, it is an $O(m)$ process
to compute the stationary vector, so the process of solving the nonlinear
system of equations costs only a fraction of the time required to compute
a single matrix-vector multiplication using the nonseparable model matrix
P.

Using the separable problem as an approximation to overflow queuing
networks is an old idea, but using it to initialize an iterative algorithm for
the overflow network seems to be new.

As a simple example, on a three queue problem with $n = (8, 8, 4)$,
$s = (5, 2, 1)$, $\lambda = (.9, .7, .5)$, and $\mu = (.1, .4, .5)$, allowing overflow from
queue 1 to queue 2, and from queue 2 to queue 3, the solution to the
separable problem is a factor of 4 closer to the stationary vector than is e^T,
and the power method converges in 57 iterations to a solution that satisfies
$\|\hat{z}^T P - \hat{z}\| \leq 10^{-4}$, a savings of 33 iterations over the iteration started from
e^T.

5.3. Multilevel Families of Problems. These models arise from
analyzing a system at various levels of granularity. As an example, a coarse
model of a queuing system may be able to provide approximate information
for a more detailed model.

Much research into such multilevel families has proceeded under the
title of *algebraic multigrid methods*. These were an outgrowth of the *geometric* multigrid methods for discretizations of partial differential equations, and correspond to semi-automatic aggregation techniques based on
the structure of the matrix rather than the connectivity of the variable dependencies in the underlying model. There is some overlap between these
investigations and the aggregation research.

It seems that there is progress to be made in a more geometric approach
to Markov chain problems and, in particular, to queuing theory problems.
The connectivity of the states imposes a mesh structure on the unknown
transition probabilities reminiscent of the mesh structure for discretizations
of differential equations. The proportions of the dimensions of the problems
are quite different, though. Differential equations typically are solved in 2
or 3 dimensions with a large number of mesh points per dimension. This
corresponds to 2 or 3 interacting queues with a large number of waiting
spaces in each queue. The Markov problems typically correspond to very
high dimensional problems with relatively few mesh points per dimension.

A multigrid approach to a partial differential equation takes advantage
of the convergence of discrete approximations as the number of mesh points
converges to infinity. This leads to a family of nested grids that can be used
in concert to accelerate convergence [3]. For queues, we can build a similar
family of nested approximations to states. Chan [5,6] cleverly exploited
this relation, and Park [24] made some further progress in understanding

the relation between the nested models, but more could be done.

6. Conclusions. In this overview we have considered algorithms for computing the stationary vector of a Markov chain using the eigenproblem formulation, the null-space formulation, and the linear system formulation. Although promising algorithms have been identified for each formulation, we lack data on the comparison of algorithms for different formulations, and much work remains to be done.

Although ideas such as aggregation for nearly decomposable systems have now been rather thoroughly studied, other ideas, such as exploiting the relation to nearby separable problems and to multilevel families, still lack a firm foundation.

These factors, coupled with the importance of Markov models in queuing theory and other applications, ensure that iterative methods for computing the stationary vector of a Markov chain will remain an exciting area for research in the next decade.

7. Acknowledgments. I am grateful to William J. Stewart and other workshop participants for their helpful comments on this overview.

REFERENCES

[1] Å. Björck and Tommy Elfving. Accelerated projection methods for computing pseudoinverse solutions of systems of linear equations. *BIT*, 19:145–163, 1979.

[2] Randy Bramley and Ahmed Sameh. Row projection methods for large nonsymmetric linear systems. Technical Report CSRD 957, Center for Supercomputing Research and Development, University of Illinois, Urbana, 1990.

[3] W. Briggs. *A Multigrid Tutorial.* SIAM, Philadelphia, Pennsylvania, 1987.

[4] W.-L. Cao and W. J. Stewart. Iterative aggregation/disaggregation techniques for nearly uncoupled Markov chains. *Journal of the ACM*, 32:702–719, 1985.

[5] R. H. Chan. Iterative methods for overflow queueing models I. *Numerische Mathematik*, 51:143–180, 1987.

[6] R. H. Chan. Iterative methods for overflow queueing models II. *Numerische Mathematik*, 54:57–78, 1988.

[7] F. Chatelin. Iterative aggregation/disaggregation methods. In G. Iazeolla, P. J. Courtois, and A. Hordijk, editors, *Mathematical Computer Performance and Reliability*, pages 199–207, New York, 1984. Elsevier Science Publishers.

[8] F. Chatelin and W. L. Miranker. Acceleration by aggregation of successive approximation methods. *Linear Algebra and Its Applications*, 43:17–47, 1982.

[9] F. Chatelin and W. L. Miranker. Aggregation/disaggregation for eigenvalue problems. *SIAM Journal on Numerical Analysis*, 21:567–582, 1984.

[10] P.-J. Courtois. Error analysis in nearly-completely decomposable stochastic systems. *Econometrica*, 43:691–709, 1975.

[11] P.-J. Courtois. *Decomposability.* Academic Press, New York, 1977.

[12] T. Elfving. Block-iterative methods for consistent and inconsistent linear equations. *Numerische Mathematik*, 35:1–12, 1980.

[13] M. Haviv. Aggregration/disaggregation methods for computing the stationary distribution of a Markov chain. *SIAM Journal on Numerical Analysis*, 24:952–966, 1987.

[14] A. S. Householder. *The Theory of Matrices in Numerical Analysis.* Dover Publishing, New York, 1964. Originally published by Ginn Blaisdell.

[15] A. Jennings and D. R. L. Orr. Applications of the simultaneous iteration method to undamped vibration problems. *International Journal for Numerical Methods in Engineering*, 3:13–24, 1971.

[16] S. Kaczmarz. Ängenäherte Auflösung von Systemen linearer Gleichungen. *Bull. Internat. Acad. Polon. Sci. Cl. A.*, pages 355–356, 1937. Cited in [14].

[17] L. Kaufman. Matrix methods for queuing problems. *SIAM J. on Scientific and Statistical Computing*, 4:525–552, 1983.

[18] John G. Kemeny and J. Laurie Snell. *Finite Markov Chains*. D. Van Nostrand Co. Inc., New York, 1960.

[19] L. Kleinrock. *Queueing Systems. Volume I: Theory*. John Wiley & Sons, New York, 1975.

[20] D. F. McAllister, G. W. Stewart, and W. J. Stewart. On a Rayleigh–Ritz refinement technique for nearly uncoupled stochastic matrices. *Linear Algebra and Its Applications*, 60:1–25, 1984.

[21] D. Mitra and P. Tsoucas. Convergence of relaxations for numerical solutions of stochastic problems. In G. Iazeolla, P.-J. Courtois, and O. J. Boxma, editors, *Second International Workshop on Applied Mathematics and Performance/Reliability Models of Computer/Communication Sytems*, pages 119–133, Amsterdam, 1987. North Holland.

[22] M. Neumann and R. J. Plemmons. Convergent nonnegative matrices and iterative methods for consistent linear systems. *Numerische Mathematik*, 31:265–279, 1978.

[23] M. Neumann and R. J. Plemmons. Generalized inverse-positivity and splittings of M-matrices. *Linear Algebra and Its Applications*, 23:21–35, 1979.

[24] Pil Seong Park. Iterative solution of sparse singular systems of equations arising from queuing networks. Technical Report UMIACS-TR-91-83, CS-TR-2690, University of Maryland, College Park, 1991.

[25] B. N. Parlett. *The Symmetric Eigenvalue Problem*. Prentice-Hall, Englewood Cliffs, New Jersey, 1980.

[26] Bernard Philippe, Youcef Saad, and William J. Stewart. Numerical methods in Markov chain modelling. *Operations Research*, to appear, 1992.

[27] Y. Saad. Chebyshev acceleration techniques for solving nonsymmetric eigenvalue problems. *Mathematics of Computation*, 42:567–588, 1984.

[28] Y. Saad. Projection methods for the numerical solution of Markov chains. In W. J. Stewart, editor, *Numerical Solution of Markov Chains*, pages 455–472, Amsterdam, 1991. North Holland.

[29] Youcef Saad. Chebyshev acceleration techniques for solving nonsymmetric eigenvalue problems. *Mathematics of Computation*, 42:567–588, 1984.

[30] P. J. Schweitzer. A survey of aggregation-disaggregation in large Markov chains. In W. J. Stewart, editor, *Numerical Methods for Markov Chains*, pages 63–88, Amsterdam, 1991. North Holland.

[31] P. J. Schweitzer and K. W. Kindle. An iterative aggregation-disaggregation algorithm for solving linear equations. *Applied Mathematics and Computation*, 18:313–353, 1986.

[32] H. A. Simon and A. Ando. Aggregation of variables in dynamic systems. *Econometrica*, 29:111–138, 1961.

[33] G. W. Stewart. Simultaneous iteration for computing invariant subspaces of non-Hermitian matrices. *Numerische Mathematik*, 25:123–136, 1976.

[34] G. W. Stewart. SRRIT — a FORTRAN subroutine to calculate the dominant invariant subspaces of a real matrix. Technical Report TRR-514, University of Maryland, Department of Computer Science, 1978.

[35] G. W. Stewart. Computable error bounds for aggregated Markov chains. *Journal of the ACM*, 30:271–285, 1983.

[36] G. W. Stewart. On the structure of nearly uncoupled Markov chains. In G. Iazeolla, P. J. Courtois, and A. Hordijk, editors, *Mathematical Computer Performance and Reliability*, pages 287–302, New York, 1984. Elsevier.

[37] G. W. Stewart, W. J. Stewart, and D. F. McAllister. A two-stage iteration for solv-
      ing nearly uncoupled Markov chains. Technical Report TR-1384, Department
      of Computer Science, University of Maryland, 1984.

[38] Xiaobai Sun. A unified analysis of numerical methods for solving nearly uncoupled
      markov chains. Technical report, Ph.D. thesis, Computer Science Department,
      University of Maryland, College Park, 1991.

[39] Y. Takahashi. *Some Problems for Applications of Markov Chains.* PhD thesis,
      Tokyo Institute of Technology, 1972. Cited in [41].

[40] Y. Takahashi. A sequencing model with an application to speed class sequencing
      in air traffic control. *J. Operations Research Society of Japan,* 17:1–28, 1974.
      Cited in [41].

[41] Y Takahashi. Weak D-Markov chain and its application to a queuing network. In
      G. Iazeolla, P. J. Courtois, and A. Hordijk, editors, *Mathematical Computer
      Performance and Reliability,* pages 153–166, New York, 1984. Elsevier.

[42] K. S. Trivedi. *Probability and Statistics with Reliability, Queuing, and Computer
      Science Applications.* Prentice–Hall, Englewook Cliffs, NJ, 1982.

[43] R. S. Varga. *Matrix Iterative Analysis.* Prentice-Hall, Englewood Cliffs, New
      Jersey, 1962.

[44] D. M. Young. *Iterative Solution of Large Linear Systems.* Academic Press, New
      York, 1971.

LOCAL CONVERGENCE OF (EXACT AND INEXACT) ITERATIVE AGGREGATION

DANIEL B. SZYLD*

Abstract. Iterative aggregation methods are studied in two cases: the solution of linear systems of equaitons and finding stationary distributions of Markov Chains. Local convergence proofs are outlined for the exact method, as well as for the method were the system in the smaller space is solved only approximately.

1. Introduction. The iterative aggregation method, sometimes called aggregation/disaggregation, has been an effective tool for the solution of linear systems and certain eigenvalue problems, and in particular for finding the stationary distribution of Markov chains; see Cao and Stewart [2], Chatelin and Miranker [4], [5], Haviv [8], Mandel and Sekerka [11], [10], Marek [12], [13], the surveys by Schweitzer [17], [18], and the references given therein. The idea of aggregation appeared naturally in input-output economic models, where goods and services are aggregated according to certain economic criteria; see the survey by Vakhutinsky, Dudkin and Ryvkin [22] and the extensive bibliography there.

In this paper we consider two problems: 1. The solution of linear systems of the form

$$(1.1) \qquad Ax = b,$$

where A is an $n \times n$ nonsingular matrix; and 2. Finding the stationary distribution of a Markov operator. In this context a Markov operator is an $n \times n$ nonnegative matrix B, with

$$(1.2) \qquad B^T e = e, \quad e = (1, 1, \cdots, 1)^T,$$

and the problem consists of finding $x \in \mathrm{IR}^n$ such that

$$(1.3) \qquad Bx = x \quad \text{and} \quad x^T e = 1.$$

It follows that $r(B) = 1$, where $r(B)$ denotes the spectral radius of B; see e.g. [1].

We consider the application of the iterative aggregation method for these two problems. Briefly, the iterative aggregation method consists of the following steps:

Start with an initial guess $x^{(0)}$, and for each step $k = 0, 1, \cdots$

(a) Aggregate (restrict) the current iterate to IR^m, $m < n$.

* Department of Mathematics, Temple University, Philadelphia, Pennsylvania 19122-2585, USA (szyld@euclid.math.temple.edu or na.szyld@na-net.ornl.gov). This work was supported by the National Science Foundation grants DMS-8807338 and INT-9196077.

(b) Solve an $m \times m$ linear system (or find the probability distribution of a Markov operator in $\mathbb{R}^m$).

(c) Disaggregate (lift) the solution (correction) to $\mathbb{R}^n$.

(d) Iterate (e.g. a relaxation step).

(e) Test for convergence.

Global convergence of the iterative aggregation method for Markov chains was shown only for the nearly uncoupled case by Cao and Stewart [2]; see also Stewart, Stewart and McCallister [19]. Local convergence for linear systems was proved both for finite [11] and infinite dimensions [12], and for Markov operators [13]. More recently, these local convergence results were extended to Banach spaces as well as to the inexact case [14], i.e., where the linear system in $\mathbb{R}^m$, in step (b), is in turn solved iteratively. This situation is similar to inner/outer iterations or two-stage methods, see e.g. [6], [7], [9], [16], or [20].

In this communication we review the local convergence results in [14] for the finite-dimensional case ($\mathbb{R}^n$) using an illustrative example and giving the general flavor of the proofs therein.

2. Iterative Aggregation. The aggregation step (a) can be viewed as a nonnegative aggregation operator $R : \mathbb{R}^n \to \mathbb{R}^m$, i.e., an $m \times n$ matrix with nonnegative entries. For example, partition the index set $\{1, \cdots, n\} = G_1 \cup \cdots \cup G_m$ (the sets G_i are nonempty and disjoint) and define $z = Rx$ as $z_i = \sum_{j \in G_i} x_j$. Thus R is a matrix whose columns have a single nonzero entry and this entry is a one, e.g. of the form

$$R = \begin{bmatrix} 1 & & & & 1 & \\ & 1 & & 1 & & 1 \\ & & 1 & & 1 & \end{bmatrix}.$$

Similarly, the disaggregation operator $S(x) : \mathbb{R}^m \to \mathbb{R}^n$ in step (c) is nonnegative, but depends on $x \in \mathcal{D}$, for some set $\mathcal{D} \subset \mathbb{R}^n_+$. To complete the example, we define $[S(x)z]_j = z_i x_j / (\sum_{\ell \in G_i} x_\ell)$, if $j \in G_i$. Here $\mathcal{D} = \{x \in \mathbb{R}^n : x \geq 0, Rx > 0\}$. In this example, $S(x)$ is a $n \times m$ nonnegative matrix with a single nonzero per row.

Let $P(x) = S(x)R$, let I_m denote the identity in $\mathbb{R}^m$, and let $\mathcal{D} \subset \mathbb{R}^n$ be a nonempty set. In general, the aggregation and disaggregation maps R and $S(x)$ have to satisfy the following two relations for $x \in \mathcal{D}$.

(2.1) $$RS(x) = I_m,$$
(2.2) $$P(x)x = S(x)Rx = x.$$

It follows from (2.1) that $[P(x)]^2 = P(x)$, i.e., that $P(x)$ is a projection on $\mathbb{R}^n$.

It is easy to see that the maps given in the example satisfy (2.1) and (2.2). This basic example appears in several references, e.g. [3], [8] and [17].

For the solution of (1.1), it is customary to consider a splitting $A = M - N$ with nonnegative matrices M^{-1} and $B = M^{-1}N$, i.e., a weak regular splitting [1]. We thus rewrite (1.1) as

$$(2.3) \qquad\qquad x = Bx + c$$

Let $B_m(x) = RBS(x)$. It follows from the nonnegativity of the three factors that $B_m(x)$ is nonnegative. Also, if $b \geq 0$, then $c = M^{-1}b \geq 0$.

We are able to formally define the

ALGORITHM 2.1 (ITERATIVE AGGREGATION). *Given a nonnegative operator B, a nonnegative vector c, an initial guess $x^{(0)}$, and a convergence parameter $\varepsilon > 0$, let $k = 0$.*

1. Solve the equation

$$(2.4) \qquad\qquad z - B_m(x)z = Rc.$$

      with $x = x^{(k)}$ and call the solution $z^{(k)}$.

2. Disaggregate and iterate according to the formula
      $x^{(k+1)} = BS(x^{(k)})z^{(k)} + c.$

3. Test if $\|x^{(k+1)} - x^{(k)}\| < \varepsilon$ (or other convergence test). If yes, STOP; otherwise $k := k+1$ and go to 1.

When $r(B_m(x)) < 1$, the iterative aggregation method can be expressed in terms of vectors in $\mathbb{R}^n$ as

$$(2.5) \qquad\qquad x^{(k+1)} = BU(x^{(k)}) + c,$$

where

$$(2.6) \qquad\qquad U(x) = S(x)[I_m - B_m(x)]^{-1}Rc.$$

It is immediate that the method is consistent, i.e. that $x^\star = BU(x^\star) + c$, where $x^\star$ is the solution of (2.3), and thus of (1.1). It follows also that $U(x^\star) = x^\star$.

The iterative aggregation can thus be viewed as transforming the linear system (2.3) into the nonlinear system

$$x = BU(x) + c$$

and use the standard fixed point method on it.

We want to show the local convergence of the iterative aggregation method, i.e. we want to show that $x^{(k)} \to x^\star$ whenever $x^{(0)} \in \mathcal{U}$, a neighborhood of $x^\star$. The proof in [14] follows by writing

$$BU(x) - BU(x^\star) = J(x)(x - x^\star),$$

where

$$(2.7) \qquad J(x) = J(B, x) = B[I - P(x)B]^{-1}[I - P(x)].$$

Then under the assumption that $r(J(x^\star)) \leq \beta < 1$, which is satisfied for the given example [11], an appropriate norm for which the operator $J(x)$ is contracting is found. Moreover, the speed of convergence is characterized by the estimate

$$\|x^{(k)} - x^\star\| \leq \alpha \rho^k$$

where $\rho = r(J(x^\star)) + \eta < 1$, with some $\eta > 0$ and α independent of k.

3. Inexact Correction. In practice, the system (2.4) is often solved iteratively. A splitting $I_m - B_m(x) = F(x) - G(x)$ is used, where $H(x) = F(x)^{-1}G(x) \geq 0$. A certain number, say p, of iterations is performed and (2.4) is replaced by

$$
\begin{aligned}
z_0^{(k)} &= Rx^{(k)} \\
\text{for } j &= 0, \cdots, p-1 \\
z_{j+1}^{(k)} &= H(x^{(k)})z_j^{(k)} + F(x^{(k)})^{-1}Rc
\end{aligned}
$$

Thus,

$$z_p^{(k)} = H(x^{(k)})^p Rx^{(k)} + \sum_{j=0}^{p-1} H(x^{(k)})^j F(x^{(k)})^{-1}Rc,$$

and the iterative aggregation method with inexact correction can be expressed in terms of vectors in $\mathbb{R}^n$ as

$$x^{(k+1)} = BU^{(p)}(x^{(k)}) + c,$$

cf. (2.5), where

$$U^{(p)}(x) = S(x)[H(x)^p Rx + \sum_{j=0}^{p-1} H(x)^j F(x)^{-1}Rc].$$

The local convergence properties of the method with inexact correction follows in a way similar to that of Algorithm 2.1, by defining an operator $J^{(p)}(x)$ such that

$$BU^{(p)}(x) - BU^{(p)}(x^\star) = J^{(p)}(x)(x - x^\star),$$

and by showing that $J^{(p)}(x) \to J(x)$ as $p \to \infty$ uniformly on x.

4. Iterative Aggregation for Markov Operators. In order to present and analize the iterative aggregation method for Markov operators there are some issues to resolve:
(i) What is the appropriate aggregated problem to solve? and
(ii) Since $r(B) = 1$ and $r(B_m) = 1$, we cannot write the operators (2.6) and (2.7).

For the first issue, we have the following Lemma, whose proof can be easily adapted from [14].

LEMMA 4.1. *Let B be a Markov operator satisfying (1.2). Let the nonnegative aggregation and disaggregation operators R and $S(u)$ satisfy (2.1) and (2.2) for $u \in \mathcal{D}$. Let $B_m(u) = RBS(u)$, $P(u) = S(u)R$, $\tilde{e} = S(u)^T e$. Then $B_m(u)^T \tilde{e} = \tilde{e}$, i.e., $B_m(u)$ is a Markov operator in $\mathbb{R}^m$ with respect to $\tilde{e}$. Furthermore assume that*

$$(4.1) \qquad [P(u)x]^T e = x^T e, \qquad u \in \mathcal{D}, \quad x \geq 0.$$

Then there exists a nonnegative solution (stationary distribution) to

$$B_m(u)w = w, \quad \text{and} \quad w^T \tilde{e} = [S(u)w]^T e = 1.$$

Condition (4.1) is natural since it guarantees that if $\tilde{e} = S(u)^T e$, then $(Rx)^T \tilde{e} = x^T e$.

We show that for the example considered in this paper, condition (4.1) holds. Let $u \in \mathcal{D}$, then

$$[S(u)Rx]_j = \frac{u_j \sum_{\ell \in G_i} x_\ell}{\sum_{\ell \in G_i} u_\ell}, \quad \text{if } j \in G_i$$

thus $\sum_{j \in G_i}[S(u)Rx]_j = \sum_{\ell \in G_i} x_\ell$, and therefore

$$[S(u)Rx]^T e = \sum_{j=1}^{n}[S(u)Rx]_j = \sum_{i=1}^{m}\sum_{j \in G_i}[S(u)Rx]_j = \sum_{j=1}^{n} x_j = x^T e.$$

Thus we have the following

ALGORITHM 4.2 (ITERATIVE AGGREGATION FOR MARKOV OPERATORS).
Given B, a Markov operator satisfying (1.2), an initial guess $x^{(0)}$, and a convergence parameter $\varepsilon > 0$, let $k = 0$.

    1. Solve the equation

$$(4.2) \qquad B_m(x)z = z, \qquad z^T S(x)^T e = 1$$

       with $x = x^{(k)}$ and call the solution $z^{(k)}$.

    2. Disaggregate and iterate according to the formula
      $x^{(k+1)} = BS(x^{(k)})z^{(k)}$.

    3. Test if $\|x^{(k+1)} - x^{(k)}\| < \varepsilon$ (or other convergence test). If yes, STOP; otherwise $k := k + 1$ and go to 1.

To show the local convergence of the algorithm we associate with B $(r(B) = 1)$ an operator V, $r(V) < 1$, $V \geq 0$, and nonnegative vectors $b^{(j)}$, e_j $j = 1, \cdots, \bar{k}$ such that

$$(4.3) \qquad B = V + \sum_{j=1}^{\bar{k}} b^{(j)} e_j^T, \quad e = \sum_{j=1}^{\bar{k}} e_j.$$

This is a theoretical development and even though this construction may be possible in practice, it is not used in Algorithm 4.2. The vectors $b^{(j)}$ have to be small enough so that $V \geq 0$. In the example, one can choose $\bar{k} = 1$ and $b = b^{(1)}$ such that $b_j = \min_i(Be_i)_j$, where $e_i = (0, \cdots, 1, \cdots, 0)^T$. The decomposition (4.3) allows us to work with the operator V, whose powers converge to zero and leave aside the other part of B, which is essentially a finite rank update. Thus in the proof of convergence one can use the contracting properties of $J(V) = J(V, x)$, cf. (2.7); see further details in [14].

Since we do not impose the assumption of irreducibility of the Markov operator B, the stationary probability distribution need no be unique. One can obtain all distributions following the ideas of Tanabe [21]; see also [15].

We end this communication by mentioning that in a way analogous to that described in Section 3, often in practice, the system (4.2) is not solved exactly. Instead, an iterative method is used, and the (inner) process is stopped after a certain number of (inner) iterations, or, equivalently, after certain (inner) convergence criteria is satisfied. The resulting Iterative Aggregation with Inexact Correction Algorithm for Markov Operators has also local convergence properties similar to those already described.

REFERENCES

[1] Abraham Berman and Robert J. Plemmons. *Nonnegative Matrices in the Mathematical Sciences*. Academic Press, New York, 1979.

[2] Wei-Lu Cao and William J. Stewart. Iterative aggregation/disaggregation techniques for nearly uncoupled Markov chains. *Journal of the ACM*, 32:702–719, 1985.

[3] Françoise Chatelin. Iterative aggregation / disaggregation methods. In G. Iazeolla, P. J. Courtois, and A. Hordijk, editors, *Mathematical Computer Performance and Reliability*, pages 199–207, Amsterdam – New York – Oxford, 1984. North-Holland.

[4] Françoise Chatelin and Willard L. Miranker. Acceleration of successive approximations. *Linear Algebra and its Applications*, 43:17–47, 1982.

[5] Françoise Chatelin and Willard L. Miranker. Aggregation/disaggregation for eigenvalue problems. *SIAM Journal on Numerical Analysis*, 21:567–582, 1984.

[6] Andreas Frommer and Daniel B. Szyld. *H*-splittings and two-stage iterative methods. Research Report 91-71, Department of Mathematics, Temple University, July 1991. *Numerische Mathematik*, to appear.

[7] Gene H. Golub and Michael L. Overton. The convergence of inexact Chebyshev and Richardson iterative methods for solving linear systems. *Numerische Mathematik*, 53:571–593, 1988.

[8] Moshe Haviv. Aggregation/disaggregation methods for computing the stationary distribution of a Markov chain. *SIAM Journal on Numerical Analysis*, 14:952–966, 1987.

[9] Paul J. Lanzkron, Donald J. Rose, and Daniel B. Szyld. Convergence of nested classical iterative methods for linear systems. *Numerische Mathematik*, 58:685–702, 1991.

[10] Jan Mandel. A convergence analysis of the iterative aggregation method with one parameter. *Linear Algebra and its Applications*, 59:159–169, 1984.

[11] Jan Mandel and Bohuslav Sekerka. A local convergence proof for the iterative

aggregation method. *Linear Algebra and its Applications*, 51:163–172, 1983.

[12] Ivo Marek. Aggregation and homogenization in reactor diffusion. In E. Adams, R. Ansoge, Ch. Grossman, and H. G. Ross, editors, *Discretization in Differential Equations and Enclosures,* Mathematical Research, Bd. 36, pages 145–154, Berlin, 1987. Akademie.

[13] Ivo Marek. Aggregation methods of computing stationary distribution of Markov processes. In W. Velte, editor, *Eigenvalue Problems and their Numerical Treatment,* Proceedings of the Conference held at the Mathematisches Forschunginstitut Oberwolfach, February 26 – March 2, 1990, International Series in Numerical Mathematics (ISNM) Vol. 96, pages 155–169, Basel, 1991. Birkhäuser.

[14] Ivo Marek and Daniel B. Szyld. Iterative aggregation with inexact correction. Research Report 91-52, Department of Mathematics, Temple University, May 1991.

[15] Ivo Marek and Daniel B. Szyld. Iterative and semiiterative methods for computing stationary probability vectors of Markov operators. Research Report 91-101, Department of Mathematics, Temple University, October 1991.

[16] Jorge J. Moré. Global convergence of Newton-Gauss-Seidel methods. *SIAM Journal on Numerical Analysis*, 8:325–336, 1971.

[17] Paul J. Schweitzer. Aggregation methods for large Markov chains. In G. Iazeolla, P. J. Courtois, and A. Hordijk, editors, *Mathematical Computer Performance and Reliability*, pages 275–286, Amsterdam – New York – Oxford, 1984. North-Holland.

[18] Paul J. Schweitzer. A survey of aggregation-disaggregation in large Markov chains. In William J. Stewart, editor, *Numerical Solution of Markov Chains*, pages 63–88, New Yrok - Basel - Hong Kong, 1991. Marcel Dekker.

[19] G. W. Stewart, W. J. Stewart, and D. F. McAllister. A two-staged iteration for solving nearly uncoupled Markov chains. Technical Report TR-1384, Department of Computer Science, University of Maryland, College Park, Maryland 20742, April 1984.

[20] Daniel B. Szyld and Mark T. Jones. Two-stage and multi-splitting methods for the parallel solution of linear systems. *SIAM Journal on Matrix Analysis and Applications*, 13:671–679, 1992.

[21] Kunio Tanabe. The conjugate gradient method for computing all extremal stationary probability vectors of a stochastic matrix. *Ann. Inst. Stat. Math., Part B*, 37:173–187, 1985.

[22] I. Y. Vakhutinsky, L. M. Dudkin, and A. A. Ryvkin. Iterative aggregation – a new approach to the solution of large-scale problems. *Econometrica*, 47:821–841, 1979.

AUTOMATED GENERATION AND ANALYSIS OF MARKOV REWARD MODELS USING STOCHASTIC REWARD NETS

GIANFRANCO CIARDO*, ALEX BLAKEMORE†,
PHILIP F. CHIMENTO, JR.‡, JOGESH K. MUPPALA* AND
KISHOR S. TRIVEDI§

Abstract. Markov and Markov reward models are widely used for the performance and reliability analysis of computer and communication systems. Models of real systems often contain thousands or even millions of states. We propose the use of Stochastic Reward Nets (SRNs) for the automatic generation of these large Markov reward models. SRNs do allow the concise specification of practical performance, reliability and performability models.

An added advantage of using SRNs lies in the possibility of analyzing the (time-independent) logical behavior of the modeled system. This helps both the validation of the system (is the right system being built?) and of the model (does the model correctly represent the system?).

We discuss the methods to convert SRNs into Markov reward processes automatically. We review the solution techniques for the steady state and transient analysis of SRNs and Markov reward processes. We also discuss methods for the sensitivity analysis of SRNs.

1. Introduction. Reliability block diagrams and fault trees are commonly used for system reliability and availability analysis [61]. These model types allow a concise description of the system under study and can be evaluated efficiently, but they cannot easily represent dependencies occurring in real systems [59]. Markov models, on the other hand, are capable of capturing various kinds of dependencies that occur in reliability/availability models [8,24,32].

Task precedence graphs [39,42,58] can be used for the performance analysis of concurrent programs with unlimited system resources. Product-form queueing networks [40,41] can be used to model contention for resources. The product-form assumptions are not satisfied, however, when behavior such as concurrency within a job, synchronization, and server failures is considered. Once again, Markov models do provide a framework to address all these concerns [21].

The common solution for modeling reliability/availability or performance would then appear to be the use of Markov models, but one major drawback of Markov models is the largeness of their state spaces. Stochas-

---

* Department of Computer Science, College of William and Mary, P. O. Box 8795, Williamsburg, VA 23815.

† Dept. of Computer Science, University of Maryland, College Park, MD 20742.

‡ IBM Corporation, Research Triangle Park, NC 27709.

§ Dept. of Electrical Engineering, Duke University, Durham, NC 27706. This work was supported in part by the National Science Foundation under Grant CCR-9108114 and by the Naval Surface Weapons Center under the ONR Grant N00014-91-5-4162.

tic Petri nets (SPNs) can be used to generate a large underlying Markov model automatically starting from a concise description.

Traditionally, performance analysis assumes a fault-free system. Separately, reliability and availability analysis is carried out to study system behavior in the presence of component faults, disregarding the performance levels of different configurations. Several types of interactions and corresponding tradeoffs have prompted researchers to consider combined evaluation of performance and reliability/availability [44,70]. Most work on the combined evaluation is based on the extension of Markov models to Markov reward models [30], where a reward is attached to each state of the Markov model.

Markov reward models have the potential to reflect concurrency, contention, fault-tolerance, and degradable performance [63]; they can be used to obtain not only program/system performance and system reliability/availability measures [14], but also combined measures of performance and reliability/availability [6,13,49,70]. Since the Markov chain is generated from a concise SPN model, it is necessary to express the reward structure in terms of SPN entities. In other words, the SPN becomes a "Stochastic Reward Net (SRN)" which can be automatically transformed into a Markov reward model.

Steady-state analysis of SRNs is often adequate to study the performance of a system, but time-dependent behavior is sometimes of greater interest: instantaneous availability, interval availability, and reliability (for a fault-tolerant system); response time distribution of a program (for performance evaluation of software); computational availability (for a degradable system) are examples where transient analysis is required. Except for a few instances [10,18,7,60], transient analysis of SPN models has not received much attention in the past.

Analytical models are often used during the design phase of a system, when exact values for all the input parameters may not be known yet. *Parametric sensitivity analysis* is performed so as to determine the parameters to which the model is sensitive and the degree of sensitivity. Frank [23] suggests the use of sensitivity functions as a method of estimating parametric sensitivities. Sensitivity functions are the derivatives of state probabilities with respect to the parameters. We will discuss this method of sensitivity analysis for SRNs.

The SRNs described in this paper, allow the concise specification of Markov reward models, the computation of instantaneous, cumulative, and time-averaged measures both in steady state or in the transient. The derivatives of these measures with respect to input parameters can be obtained as well. Efficient and numerically stable algorithms employing sparse matrix techniques are used to solve the underlying CTMC.

In Section 2, we introduce the SRN formalism. Logical and temporal analysis of SRNs are discussed in Sections 3 and 4, respectively, while Section 5 examines numerical issues connected to the solution of SRNs.

Section 6 presents some concluding remarks.

2. From Petri Nets to Stochastic Reward Nets. We introduce the SRN formalism incrementally. First, we define the (untimed) PNs in Section 2.1, which describes the basic PN formalism, and in Section 2.2, which contains a set of extensions. Then, we define the stochastic PN (SPN) formalism by adding timing and probabilistic information to the PNs, in Section 2.3. Finally, we define the SRN formalism by imposing a reward structure upon the stochastic process described by a SPN, in Section 2.4.

2.1. Petri nets. PNs were introduced by C. A. Petri in 1962 [52]. A PN is a bipartite directed graph with two classes of nodes, places and transitions. Arcs may only connect places to transitions or transitions to places. Graphically, places are represented by circles or ovals, while transitions are represented by rectangles or bars. An arc from place p to transition t is an *input arc* of t and an *output arc* of p. We also say that p is an *input place* of t and t is an *output transition* of p. Analogous definitions hold for an arc from t to p. Then, $\cdot t, t\cdot, \cdot p$, and $p\cdot$ denote the set of input and output places of transition t, and the set of input and output transitions of place p, respectively.

A PN graph may be marked by placing tokens, drawn as black circles, in places. If P is the set of places, a *marking* is a multiset, $\mu \in \mathbb{N}^{|P|}$, describing the number of tokens in each place. A marking represents the state of the model at a particular instant. This concept is central to PNs. The notation $\#(p, \mu)$ is used to indicate the number of tokens in place p in marking μ. If the marking is clear from the context, the notation $\#(p)$ can be used.

A transition is *enabled* when each of its input places contains at least one token. An enabled transition may *fire* by removing a token from each of its input places and depositing a token in each of its output places. Firing a transition is an atomic event. While timing becomes an issue only in the timed or stochastic PNs, sequencing is important even in the basic PNs. If two transitions are enabled in a PN, they cannot be fired "at the same time": a choice must be made concerning which one to fire first, the other can only fire after that, if it is still enabled. This convention greatly simplifies the analysis of PNs.

PNs are dynamic in that they describe how the modeled system evolves over time. A PN evolves by firing enabled transitions to change the marking. Figure 2.1 illustrates a simple PN before and after the firing of transition t_1, which is enabled because its sole input place, p_1, contains at least one token. Transition t_2 is disabled because, while p_1 contains a token, p_2 is empty and there is an input arc from p_2 to t_2. When firing transition t_1, a token is removed from place p_1 and, at the same time, a token is deposited into each of p_2 and p_3. In the new marking, both t_1 and t_2 are enabled.

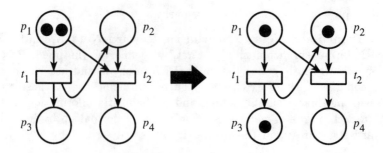

FIG. 2.1. *A PN before and after the firing of transition t_1.*

Let μ_i and μ_j be markings and t be a transition. We use the notation $\mu_i \xrightarrow{t}$ to denote that t is enabled in μ_i. If its firing in μ_i changes the marking to μ_j, we write $\mu_i \xrightarrow{t} \mu_j$. A sequence of transitions s is a *firing sequence* if the transitions may be legally fired in the order given by the sequence from some marking. We then use the notation $\mu_i \xrightarrow{s} \mu_j$ to mean that the firing sequence s, when started in marking μ_i, terminates in marking μ_j. The notation $\mu_i \xrightarrow{*} \mu_j$ indicates that it is possible to reach μ_j from μ_i with some firing sequence.

The system behavior depends not only upon the structure of the PN graph, but also upon the *initial marking*, μ_0. Changing the initial marking can completely change the behavior of the PN. We define a PN to include the initial marking in its specification, and will use the term PN graph to refer to a net with an unspecified initial marking.

DEFINITION 2.1. A PN is a 5-tuple $\mathbf{A} = \{P, T, D^-, D^+, \mu_0\}$ where:

- $P = \{p_1, ..., p_{|P|}\}$ is a finite set of places. Each place contains a non-negative number of tokens.
- $T = \{t_1, ..., t_{|T|}\}$ is a finite set of transitions ($P \cap T = \emptyset$).
- $D^- \in \{0,1\}^{|P \times T|}$, $D^+ \in \{0,1\}^{|P \times T|}$ describe the input and output arcs. There is an input arc from place p to transition t iff $D^-_{p,t} = 1$. Analogously, there is an output arc from transition t to place p iff $D^+_{p,t} = 1$.
- μ_0 is the initial marking.

A marking μ is *reachable* if $\mu_0 \xrightarrow{*} \mu$. The set of all reachable markings is termed the *reachability set*:

$$\mathcal{S} = \{\mu | \mu_0 \xrightarrow{*} \mu\}$$

The evolution of a PN can be completely described by its *reachability graph* $(\mathcal{S}, \mathcal{A})$. Each marking in the reachability set is a node in the reachability graph, while the arcs $\mathcal{A}$ describe the possible marking-to-marking transitions. If $\mu_i \in \mathcal{S}$ and $\mu_i \xrightarrow{t} \mu_j$, then $\mu_i \xrightarrow{t} \mu_j \in \mathcal{A}$. A single PN transition typically corresponds to many arcs in the reachability graph and a relatively simple PN may give rise to a large or even infinite reachability graph. Each arc in $\mathcal{A}$ defines not only the source and destination marking, but also the PN transition causing the change of marking. If more than one transition can cause this change, $\mathcal{A}$ contains multiple arcs with same source and destination (see Section 3).

Places often represent *local* conditions or states, while transitions represent possible events or activities. According to this interpretation, input arcs describe the conditions necessary for an event to occur, while input and output arcs together describe the effect that an event has upon the system, that is, the changes the event causes in the system state.

It is possible to determine the conditions required for a transition to fire and the subsequent effect solely from the arcs connected to that transition. Likewise, the events (transitions) which depend upon or may affect a condition (place) can be determined solely from the arcs connected to the place. Thus, a PN graph describes the system behavior in terms of local conditions and events, while the reachability graph describes the possible evolution of the *global* system state. This is why PNs have also been called condition/event nets.

PNs excel at modeling parallel systems, in spite of the apparently severe restriction that no two transitions fire simultaneously. When transitions are simultaneously enabled, the associated activities are proceeding in parallel. The first transition to fire simply represents the first activity to complete. Transition enabling corresponds to the *commencement* of an activity, while transition firing corresponds to the *completion* of an activity. It is quite possible that firing a transition will cause a previously enabled transition to become disabled. According to this interpretation, the interrupted activity was aborted before completing and had no lasting effect. Since PN semantics do not state which of two simultaneously enabled transitions must fire first, PN analysis must examine every possible ordering - thus reintroducing the notion of parallelism and non-determinism. The interpretation of the semantics of simultaneously enabled transitions arises again in the context of SPN models, where the choices for semantics and their implications are more complex.

The firing and enabling rules and the basic PN primitives can be combined to model many types of behavior. PNs are especially well suited for describing behavior that arises naturally in complex systems, such as sequencing, concurrency, synchronization, and conflict (see Figure 2.2).

2.2. PN extensions. Many extensions to PNs have been proposed both to increase the class of problems that can be represented (modeling power) and the practical ability to represent common behavior (modeling convenience). Unfortunately, increasing the modeling power decreases the decision power (the set of properties which may be analyzed). Consequently, proposed extensions to the basic PN formalism must be evaluated in terms of their effect upon modeling and decision power. Peterson gives an excellent overview of these tradeoffs [51].

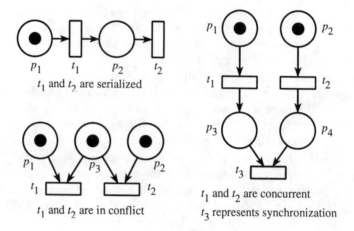

FIG. 2.2. *Types of behavior that can be represented by PNs.*

Extensions which affect only modeling convenience may be adopted without losing analytical ability. Such extensions can be removed by transforming an extended PN into an equivalent PN satisfying Definition 2.1. Theoretically, such extensions may be regarded as simply a convenient shorthand. In practice, they may drastically improve the ability to apply PNs to real problems.

One such extension has been so widely adopted that it is now considered part of the *standard PN* definition, *arc multiplicity*. In the presence of multiple arcs, the enabling and firing rules of Definition 2.1 must be revised. A transition is enabled iff there are at least as many tokens in each input place as the multiplicity of the corresponding input arc. Similarly, firing a transition removes and deposits as many tokens as the multiplicities of the input and output arcs, respectively. The set of input/output arcs is generalized to a multiset. The entries in matrices D^+ and D^- are non-negative integers. Multiple arcs are usually drawn as a single arc annotated with an expression denoting the multiplicity. The addition of multiple arcs does not change the modeling or decision power of the PN formalism.

Another frequently encountered extension is the *inhibitor arc*. Inhibitor arcs connect a place to a transition and are drawn with a small circle instead of the arrowhead. An inhibitor arc from place p to transition t disables t in any marking where p is not empty. Multiple inhibitor arcs can also be defined, in which case t is disabled whenever p contains at least as many tokens as the multiplicity of the inhibitor arc. Inhibitor arcs provide the ability to test whether a place is empty, which is not possible using standard PNs. This ability to perform *zero testing* extends the modeling power of PNs, and reduces their decision power, to equal that of Turing machines.

Inhibitor arcs are the most common "Turing extension", but several

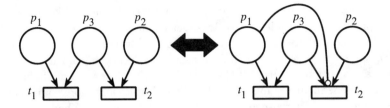

FIG. 2.3. *Equivalence of priorities and inhibitor arcs (t1 > t2).*

others have been introduced. In theory, once any Turing extension has been adopted, any other extension is redundant. In practice, each of the extensions described below is particularly well suited to representing some type of behavior.

Often, an activity must have precedence over another when they both require the same resource. Inhibitor arcs can be used to represent such constraints, but they may clutter the model. It is much more convenient to incorporate *transition priorities* directly into the formalism (see Figure 2.3). Traditionally, PN priorities have been defined by assigning an integer priority level to each transition, and adding the constraint that a transition may be enabled only if no higher priority transition is enabled. The extended PN definition that follows generalizes the notion of priorities by requiring only a partial order among transitions. This added flexibility provides a simple way to model the situation where $t_1 > t_2$, $t_3 > t_4$, but t_1 has no priority relation with respect to t_3 and t_4.

We insist that transition priorities be defined by a static relation – the relative priority of any two transitions cannot depend upon the marking of the PN. This restriction allows more efficient analysis techniques, as well as a simpler, less error prone model specification. However, there are situations where a transition needs to be disabled depending upon *marking dependent* conditions. It is always possible to define a marking dependent constraint to prevent transition enabling using the primitives discussed so far, but the construction may be cumbersome. A better alternative is to define a marking dependent predicate as an additional enabling criterion for each transition. We associate a *guard* with each transition; the transition is enabled only if the guard is satisfied. Guards are also known as enabling functions or, if the sense of the logic is reversed, as inhibiting functions.

The convenience of priorities and guards comes at a price. Standard PNs can capture the entire system behavior graphically, but, when priorities and guards are used, this ability is partially lost. Our choice is to include priorities and guards in the PN definition, letting the modeler determine the proper balance between graphical expressiveness and conciseness.

Finally, we allow the multiplicity of an arc to depend on the marking of the PN. *Marking-dependent arc multiplicities* belong to the same class of extensions as inhibitor arcs, priorities and guards, but unlike the others, they provide a convenient way to describe activities where the number of tokens needed to be transferred (or to enable the transition) depends upon the system state. A common use for marking dependent arcs is to allow a transition to flush all the tokens from a place with a single firing. Marking dependent arc multiplicities allow simpler and more compact PNs than would be otherwise possible in many situations. When exhaustive state space exploration techniques are employed, their use can dramatically reduce the state space. However, they preclude the structural analysis techniques described in Section 3.4.

Other extensions to PNs, which we do not include in our definition, are debit arcs and anti-tokens [65], colored tokens [35], and predicate-transition nets [56] (the last two fall in the category of *high-level nets*).

In the special case of PNs with finite reachability sets, all the Turing extensions provide only convenience and do not extend the class of systems that may be represented beyond that of finite state machines. This is the case of interest to us.

DEFINITION 2.2. An extended PN is a 8-tuple $\mathbf{A} = \{P, T, D^-, D^+, D^\circ, e, >, \mu_0\}$ where:

- P, T, and μ_0 are defined as in Definition 2.1.
- $\forall p \in P, \forall t \in T$, $D_{p,t}^- : \mathbb{N}^{|P|} \to \mathbb{N}$, $D_{p,t}^+ : \mathbb{N}^{|P|} \to \mathbb{N}$, and $D_{p,t}^\circ : \mathbb{N}^{|P|} \to \mathbb{N}$ are the marking-dependent multiplicities of the input arc from p to t, the output arc from t to p, and the inhibitor arc from p to t, respectively. If an arc multiplicity evaluates to zero in a marking, the arc is ignored (does not have any effect) in that marking.

We say that a transition $t \in T$ is arc-enabled in marking μ iff

$$\forall p \in P, D_{p,t}^-(\mu) \leq \#(p, \mu) \wedge \left(D_{p,t}^\circ(\mu) > \#(p, \mu) \vee D_{p,t}^\circ(\mu) = 0\right)$$

When transition t fires in marking μ_i, the new marking μ_j satisfies the following firing rule:

$$\forall p \in P, \#(p, \mu_j) = \#(p, \mu_i) - D_{p,t}^-(\mu_i) + D_{p,t}^+(\mu_i)$$

Note that both $D_{p,t}^-$ and $D_{p,t}^+$ are evaluated on μ_i *before* firing t.

- $\forall t \in T, g_t : \mathbb{N}^{|P|} \to \{true, false\}$ is the guard for transition t. If $g_t(\mu) = false$, t is disabled in μ.
- $>$ is a transitive and irreflexive relation imposing a priority among transitions. In a marking μ, t_1 is marking-enabled iff it is arc-enabled, $g_{t_1}(\mu) = true$, and no other transition t_2 exists such that $t_2 > t_1$, t_2 is arc-enabled, and $g_{t_2}(\mu) = true$.

2.3. Stochastic Petri nets. In the previous section, we adopted a particular PN formalism which can, in theory, describe any discrete-state behavior, being equivalent to Turing machines. By associating stochastic and timing information to it, we obtain the SPN formalism.

In general terms, we are interested in imposing a timing upon the flow of tokens. We do this by associating a random variable called *firing time* with transitions, representing the time that must elapse from the instant the transition is enabled to the instant it actually fires in isolation, that is, assuming that no other transition firing affects it. An alternative approach associates a waiting time with places [62], so that a token arriving into a place can enable a transition only after this waiting time is elapsed. The two approaches can model the same class of systems, but the latter has the disadvantage that the marking of the (untimed) PN does not represent the entire state of the model, since it is further necessary to distinguish whether each token in each place is "waiting" or "ready".

Much more fundamental are the choices about the *execution policy*, the *memory policy*, and the probabilistic distributions available to specify the timing (we follow the terminology used in [2], but see also [7]).

The execution policy specifies which transition will fire among those enabled in a marking. We adopt the *race policy*, where the transition whose firing time elapses first is the one that will fire. An alternative is the *preselection policy*, where the next transition to fire in a given marking is chosen among the enabled transitions using a probability distribution independent of their firing times.

If the race policy is adopted, a memory policy must be specified to determine the effect of the change of marking on the (remaining) firing time of the transitions that did not fire. With the *resampling policy*, each change of marking causes the other enabled transitions to resample a new firing time from their distribution. With the *enabling memory policy*, transitions which remain enabled after the change of marking retain the firing time initially sampled, reduced by the time they have been enabled without interruption; transitions which become disabled instead "lose the work done" and must resample a new firing time once they become enabled again. Finally, with the *age policy*, the time spent by a transition while enabled is never lost, even when it becomes subsequently disabled; a new sample is drawn from the firing time distribution only after the previous sample elapsed, causing the transition to fire.

These policies correspond to different types of behavior. In the same SPN, different policies might be needed for different transitions or even for the same transition in different markings.

According to the type of distribution allowed, the SPN formalism may correspond to a wide range of well-known stochastic processes. For example, assuming an enabling memory policy:

- If firing times are geometrically or exponentially distributed, a discrete or continuous-time Markov chain (DTMC or CTMC) is obtained, respectively [46].
- If firing times can have general distributions, a generalized semi-

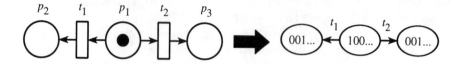

FIG. 2.4. *Influence of timing over the structural behavior of the SPN.*

Markov process (GSMP) is obtained [28].

- Under certain conditions restricting the type of distribution that concurrently enabled transitions can have, a semi-Markov process (SMP) is obtained [7,12].

Consider now a SPN with race policy where some or all of the transitions have a firing time distribution with discrete impulses, such as the constant distribution. In this case, the race policy is insufficient to specify the entire behavior of the SPN, since it does not specify which transition should be fired when several are enabled and their firing time elapses at the same instant (this event has probability zero if the distributions are continuous). The introduction of *firing probabilities*, usually specified as *weights* to be normalized, resolves these conflicts [3].

It is often desirable to ensure that the reachability graphs for the SPN and underlying PN models are equivalent. When this holds, all the PN analysis results of Section 3 can be applied without change. Unfortunately, timing may affect the logical behavior of the SPN as well. In this case, the reachability graph of the SPN has arcs or even nodes missing with respect to that of the underlying PN. For example, in Figure 2.4, transitions t_1 and t_2 are both enabled in the underlying PN. If their firing times are constants equal 1 and 2, respectively, t_2 cannot fire in that marking, since the token will be removed by t_1, which has a shorter firing time.

If all the distributions involved are continuous with support $[0, \infty)$, the reachability graphs of the PN and the SPN are indeed equivalent. This is one of the reasons why the firing times are often restricted to have an exponential distribution. Even more important in this case, though, is the fact that the SPN can be automatically transformed into a CTMC [46].

In the generalized stochastic Petri nets (GSPNs) [3], the transitions are allowed to either have an exponential distribution (*timed transitions*) or fire in zero time (*immediate transitions*). This destroys the equivalence between the SPN and the PN reachability graphs, since the firing time of immediate transitions is always smaller than that of timed transitions. The equivalence of the reachability graphs is easily restored, for example, by assigning a higher priority to the immediate transitions. Also the GSPNs can be automatically transformed into CTMCs, provided that the underlying process is regular, that is, the number of transition firings in a finite interval of time is finite with probability one. The presence of an infinite firing sequence containing only immediate transitions could violate this assumption. For finite reachability sets, this can only happen if a *vanishing loop* exists [12]. This case is of little practical interest and we do not consider it further.

Our definition of SPNs differs from that of GSPNs in several key aspects. From a structural point of view, both formalisms are equivalent to Turing machines. Our SPNs, though, provide guards, marking-dependent arc multiplicities, a more general approach to the specification of priorities, and the ability to decide in a marking-dependent fashion whether the firing time of a transition is exponentially distributed or zero, often resulting in more compact nets.

DEFINITION 2.3. A SPN is a 10-tuple $\mathbf{A} = \{P, T, D^-, D^+, D^\circ, e, >, \mu_0, \lambda, w\}$ where:

- P, T, D^-, D^+, D°, μ_0, e, and $>$, are defined as in Definition 2.2. They describe an extended PN.
- $\forall t \in T, \lambda_t : \mathbb{N}^{|P|} \to \mathbb{R}^+ \cup \{\infty\}$ is the rate of the exponential distribution for the firing time of transition t. If $\lambda_t(\mu) = \infty$, the firing time of t in μ is zero. This is a generalization of the GSPN definition, where transitions are *a priori* classified as timed or immediate. The definition of vanishing and tangible marking, though, is still applicable: a marking μ is said to be *vanishing* if there is a marking-enabled transition t in μ such that $\lambda_t(\mu) = \infty$; μ is said to be *tangible* otherwise.

  We indicate with $\mathcal{T}$ and $\mathcal{V}$ the tangible and vanishing portion of the reachability set, respectively: $\mathcal{T} \cup \mathcal{V} = \mathcal{S}$.

  We additionally impose the interpretation that, in a vanishing marking μ, all transitions t with $\lambda_t(\mu) < \infty$ are implicitly inhibited. Hence, a transition t in a marking μ is enabled in the usual sense and can fire iff it is marking-enabled and either μ is tangible or $\lambda_t(\mu) = \infty$.

- $\forall t \in T, w_t : \mathbb{N}^{|P|} \to \mathbb{R}^+$ describes the weight assigned to the firing of enabled transition t, whenever its rate λ_t evaluates to ∞. The probability of firing transition t enabled in a vanishing marking μ is:

$$\frac{w_t(\mu)}{\displaystyle\sum_{t_i : \mu \overset{t_i}{\longrightarrow}} w_{t_i}(\mu)}$$

If a marking-dependent weight specification is not needed, the definition of w can be reduced to

$$\forall t \in T, w_t \in \mathbb{R}^+$$

Both λ_t and w_t are partial functions. λ_t need only be defined in markings where transition t is marking-enabled, while w_t need only be defined when t is enabled and λ_t is infinite.

According to our definition, a SPN identifies a trivariate discrete-time stochastic process: $\{(t^{[n]}, \theta^{[n]}, \mu^{[n]}), n \in \mathbb{N}\}$. For $n = 0$, we use the convention $t^{[0]} = NULL$ and $\theta^{[0]} = 0$, that is, the SPN is in marking μ_0 with probability one at time zero. For $n > 0$, $t^{[n]} \in T$ is the n-th transition to fire; its firing leads to $\mu^{[n]}$, the n-th marking encountered, and $\theta^{[n]}$ is the time at which it fires. Hence, $\mu^{[n-1]} \xrightarrow{t^{[n]}} \mu^{[n]}$ and $\theta^{[n+1]} - \theta^{[n]} \geq 0$ is the sojourn time for the corresponding visit in marking $\mu^{[n]}$. This process is a SMP where the sojourn times are either exponentially distributed, for tangible markings, or zero, for vanishing markings. It is also possible to define a continuous-time process describing the marking at time θ, $\{\mu(\theta), \theta \geq 0\}$, which is completely determined given $\{(t^{[n]}, \theta^{[n]}, \mu^{[n]}), n \in \mathbb{N}\}$:

$$\mu(\theta) = \mu_{\max\{n : \theta^{[n]} \leq \theta\}}$$

This process considers only the evolution with respect to the tangible markings, that is, $Pr\{\mu(\theta) \in \mathcal{V}\} = 0$. $\{\mu(\theta), \theta \geq 0\}$ is a CTMC [3].

Marking dependent transition firing rates [3] can succinctly describe a complex behavior. For example, traditional queueing network models describe the effect of competition for a shared resource in terms of the queueing and service disciplines [5]. Common queueing disciplines include first-come-first-served (FCFS), processor sharing (PS), and last-come-first-served-preemptive-resume (LCFSPR). Common service disciplines include single (SS), multiple (MS), and infinite (IS) number of servers. In the SPN formalism, tokens (customers) are indistinguishable (single class), and the firing (service) times are exponentially distributed, hence the queueing discipline is not an issue. Any service discipline, though, can be represented in a SPN, using the appropriate firing rates.

Consider the SPN in Figure 2.5, where transition t and place p represent a service center and its associated queue, respectively. The IS discipline is obtained by defining the firing rate of t be defined as a constant $x \in \mathbb{R}^+$ times the number of tokens in the input place:

$$\lambda_t(\mu) = \#(p, \mu)x$$

The SS discipline corresponds to a constant firing rate:

$$\lambda_t(\mu) = x$$

A MS discipline with K servers is obtained by defining

$$\lambda_t(\mu) = \min\{\#(p, \mu), K\}x$$

In some models, several transitions may share a single resource. For example, transitions may represent the internal behavior of software processes. Logically, the processes may be independent, yet they do affect one another by competing for a shared resource such as the physical computer

FIG. 2.5. *Representing a single queue with a SPN.*

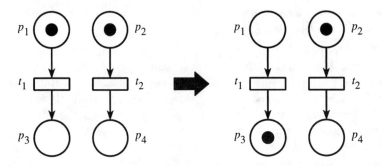

FIG. 2.6. *Representing shared resources with SPNs.*

system. The firing rates for the competing transitions can be modified to take this into account. Consider Figure 2.6, where t_1 and t_2 share a resource (this is is not apparent from the PN graph). In the first marking, both t_1 and t_2 are enabled and competing for the shared resource, so their rates should be half of what they would be if they each had exclusive use the resource. If t_1 fires, t_2 has exclusive use of the resource, so its firing rate is no longer degraded. Assuming that each token in p_1 and p_2 attempts to use the resource, the firing rates could be defined as

$$\lambda_{t_1}(\mu) = \frac{x_1 \#(p_1, \mu)}{\#(p_1, \mu) + \#(p_2, \mu)}$$

$$\lambda_{t_2}(\mu) = \frac{x_2 \#(p_2, \mu)}{\#(p_1, \mu) + \#(p_2, \mu)}$$

where x_1 and x_2 are the rates at which tokens would be processed by t_1 and t_2 in isolation, respectively.

Several extension are possible. There could be s transitions $t_i, \ldots t_s$ sharing a marking dependent number $K(\mu)$ of identical resources (this could be used to model failure and repair of servers). The number of requests issued to the resource by each transition t_i could be a more complex function than just the number of tokens in its input place, $f_i(\mu)$. The overhead of sharing resources could be non-negligible, requiring a marking-dependent function $0 < o(\mu) \leq 1$ to describe it. The firing rates could then be defined as:

$$\lambda_{t_i}(\mu) = x_i f_i(\mu) \frac{\min\left\{K(\mu), \sum_{i=1}^{s} f_i(\mu)\right\}}{\sum_{i=1}^{s} f_i(\mu)} o(\mu)$$

Finally, if the resource must be made unavailable in certain states, even if there are customers ready to use it, a guard can be associated with the appropriate transitions.

2.4. Proposed formalism. The SPN formalism underlying our SRN definition can model the same class of systems as the GSPN formalism: the power of a Turing machine is available to describe the structural behavior, while exponential or constant zero distributions are available to describe the timing of activities. The SPNs in Definition 2.3, though, are much more flexible because they have a richer set of constructs which often result in a more compact model for a given problem.

The treatment of the measures in the SRN is even more important: the SRN formalism considers the measure specification as an integral part of the model. Underlying an SRN is an independent semi-Markov reward process with reward rates associated to the markings and reward impulses associated to the transitions between markings. Our definition of SRN explicitly includes parameters (inputs) and the specification of multiple measures (outputs). A SRN with m inputs and n outputs defines a function from $\mathbb{R}^m$ to $\mathbb{R}^n$.

DEFINITION 2.4. A (non-parametric) SRN is an 11-tuple $\mathbf{A} = \{P, T, D^-, D^+, D^\circ, e, >, \mu_0, \lambda, w, M\}$ where:

- $P, T, D^-, D^+, D^\circ, \mu_0, e, >, \lambda$, and w are defined as in Definition 2.3. They describe a SPN.
- $M = \{(\rho_1, r_1, \psi_1), ..., (\rho_{|M|}, r_{|M|}, \psi_{|M|})\}$ is a finite set of measures, each specifying the computation of a single real value. A measure $(\rho, r, \psi) \in M$ has three components. The first and second components specify a reward structure over the underlying stochastic process $\{(t^{[n]}, \theta^{[n]}, \mu^{[n]}), n \in \mathbb{N}\}$.

$$\rho : \mathbb{N}^{|P|} \to \mathbb{R}$$

  is a reward rate: $\forall \mu \in \mathcal{S}, \rho(\mu)$ is the rate at which reward is accumulated when the marking is μ.

$$\forall t \in T, r_t : \mathbb{N}^{|P|} \to \mathbb{R}$$

  is a reward impulse: $\forall \mu \in \mathcal{S}, r_t(\mu)$ is the instantaneous reward gained when firing transition t in marking μ. Often, a marking-dependent reward impulse specification is not needed, so the definition of r can be reduced to

$$\forall t \in T, r_t \in \mathbb{R}$$

The reward structure specified by ρ and r over $\{(t^{[n]}, \theta^{[n]}, \mu^{[n]}), n \in \mathbb{N}\}$ completely determines the sequence $\{Y^{[n]}, n \in \mathbb{N}\}$, where

$$Y^{[n]} = \int_0^{\theta^{[n]}} \rho(\mu(u))du + \sum_{i=1}^n r_{t^{[i]}}(\mu^{[i-1]})$$

represents the reward accumulated until time $\theta^{[n]}$ and $Y^{[0]} = 0$. We can also define a continuous process $\{Y(\theta), \theta \geq 0\}$, describing the reward accumulated by the SRN up to an arbitrary time θ:

$$Y(\theta) = \int_0^{\theta} \rho(\mu(u))du + \sum_{n=1}^{\max\{n:\theta^{[n]}\leq\theta\}} r_{t^{[n]}}(\mu^{[n-1]})$$

The third component of a measure specification, ψ, is a function that computes a single real value from the stochastic process $\{(t^{[n]}, \theta^{[n]}, \mu^{[n]}, Y^{[n]}), n \in \mathbb{N}\}$ or, more simply, from $\{Y(\theta), \theta \geq 0\}$.

The generality of this definition is best illustrated by showing the wide range of measures the triplet (ρ, r, ψ) can capture (in a given SRN, some of these measures might be infinite):

- Expected number of transition firings up to time θ: this is simply $E[Y(\theta)]$ when all reward rates are zero and all reward impulses are one.
- Expected time-averaged reward up to time θ:

$$E\left[\frac{Y(\theta)}{\theta}\right]$$

- Expected instantaneous reward rate at time θ:

$$\lim_{\delta \to 0} E\left[\frac{Y(\theta + \delta) - Y(\theta)}{\delta}\right]$$

- Expected accumulated reward rate up to steady state:

$$E[Y] = \lim_{\theta \to \infty} E[Y(\theta)]$$

- Mean time to absorption: this is a particular case of the previous measure, obtained by setting the reward rate of transient and absorbing markings to one and zero, respectively, and all reward impulses to zero.
- Expected instantaneous reward rate in steady state:

$$\lim_{\theta \to \infty} \lim_{\delta \to 0} E\left[\frac{Y(\theta + \delta) - Y(\theta)}{\delta}\right]$$

which can also be expressed as the expected time-average reward up to steady state:

$$\lim_{\theta \to \infty} E\left[\frac{Y(\theta)}{\theta}\right]$$

- Supremum reward rate in any reachable marking:

$$\sup_{n \geq 0} \left\{ \rho(\mu) : Pr\{\mu^{[n]} = \mu\} > 0 \right\}$$

A parametric SRN is obtained allowing each component of an SRN to depend on a set of parameters $\beta = (\beta_1, ..., \beta_m) \in \mathbb{R}^m$: $\mathbf{A}(\beta) = \{P(\beta), T(\beta), D^-(\beta), D^+(\beta), D^\circ(\beta), e(\beta), >(\beta), \mu_0(\beta), \lambda(\beta), w(\beta), M(\beta)\}$. Once the parameters β are fixed, a simple (non-parametric) SRN is obtained.

A fundamental capability captured by our parametrization is that of specifying the initial marking not as a single marking, but as a probability vector defined over a set of markings. This is often required in transient analysis, if the initial state of the system is uncertain. If $\mu_0 = \mu_i, 1 \leq i \leq K$ with probability $\gamma_i(0)$, the following construction will suffice:

1. Add a place p_0.
2. Add transitions $t_0^i, 1 \leq i \leq K$, with rate $\lambda_{t_0^i} = \infty$ and weight $w_{t_0^i} = \gamma_i(0)$.
3. For each new transition t_0^i, set $D^-_{p_0,t_0^i} = 1$.
4. For each new transition t_0^i and for each original place p_j, set $D^+_{p_j,t_0^i} = \#(p_j, \mu_i)$.
5. Set μ_0 to have one token in p_0, no tokens elsewhere.

The modified SRN will initiate its activity by removing the token from p_0 and adding the appropriate tokens to create one of the possible initial markings, with the corresponding probability.

While this is not a practical method if the number of possible initial markings K is large, it shows that the parametrization of the initial marking probability falls into our framework.

In practice, more efficient *ad hoc* approaches for the specification of the initial probability vector might be employed by solution packages. For example, SPNP [17] uses the following approach:

1. Build the reachability set starting from a single initial marking μ_0 specified by the user.
2. Evaluate a user-specified marking-dependent non-negative expression f in each marking μ of the reachability set.
3. Assign initial probability $\gamma_i(0) = f(\mu_i)/\sum_{j \in \mathcal{S}} f(\mu_j)$ to marking μ_i.

3. Logical Analysis. A PN, a marking, or a reachability graph may exhibit several logical properties. It is important to determine whether these properties hold since they correspond to properties in the system being modeled and they may also affect the choice of analysis technique.

A place p is *k-bounded* iff the place contains at most k tokens in any reachable marking:

$$\forall \mu \in \mathcal{S}, \#(p, \mu) \leq k$$

In particular, a place is *safe* iff it is 1-bounded. A PN is k-bounded (safe) iff each of its places is k-bounded (safe).

A PN is *conservative* iff the total number of tokens remains constant in each reachable marking. A PN is *weighted conservative* iff there exists a set of positive integer weights such that the weighted sum of the number of tokens in each place remains constant in every reachable marking:

$$\exists c_1, \ldots, c_{|P|} \in \mathbb{N}^+, \forall \mu_i, \mu_j \in \mathcal{S}, \sum_{p \in P} c_p \#(p, \mu_i) = \sum_{p \in P} c_p \#(p, \mu_j)$$

A PN which exhibits any of the above properties (boundedness or conservation) has a finite reachability set.

A transition t is said to be *live* iff for each reachable marking, there exists some firing sequence that leads to a marking where the transition is enabled:

$$\forall \mu_i \in \mathcal{S}, \exists \mu_j : \mu_i \xrightarrow{*} \mu_j \wedge \mu_j \xrightarrow{t}$$

Actually, there are several degrees of liveness, the above definition being the most strict [51]. A PN is live iff each of its transitions is live. Liveness is an important property when studying potential deadlocks in a system.

A PN is *reversible* iff the initial marking can be reached from every reachable marking:

$$\forall \mu_i \in \mathcal{S}, \mu_i \xrightarrow{*} \mu_0$$

This definition may be relaxed to require only that there exist some marking μ_j that is reachable from every reachable marking [50]:

$$\exists \mu_j : \forall \mu_i \in \mathcal{S}, \mu_i \xrightarrow{*} \mu_j$$

The definition of reversibility commonly applied to PNs is unrelated to the concept of time-reversibility used to describe a class of Markov chains that admit a recursive solution [38].

Two transitions t_i and t_j are *mutually exclusive* iff they are never simultaneously enabled

$$\neg(\exists \mu \in \mathcal{S} : \mu \xrightarrow{t_i} \wedge \mu \xrightarrow{t_j})$$

A PN is *pure* if it does not contain a transition whose firing has no effect on some marking

$$\neg(\exists t \in T, \exists \mu \in \mathcal{S} : \mu \xrightarrow{t} \mu)$$

A PN is *simple* if it does not contain two distinct transitions t_i and t_j whose firing cause the same effect in some marking[1]

---

[1] For standard PNs with constant multiplicity arcs, simplicity is used to ensure that no redundant transitions or places exist. In our definition, instead, we are only concerned with redundant transitions.

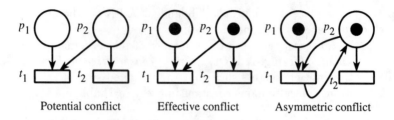

Potential conflict   Effective conflict   Asymmetric conflict

FIG. 3.1. *Different types of conflict.*

$$\neg(\exists t_i, t_j \in T, t_i \neq t_j, \exists \mu_1, \mu_2 \in \mathcal{S} : \mu_1 \overset{t_i}{\rightarrow} \mu_2 \wedge \mu_1 \overset{t_j}{\rightarrow} \mu_2)$$

Two transitions t_1 and t_2 are in *potential conflict* iff they share some input place: $\cdot t_1 \cap \cdot t_2 \neq \emptyset$. Potential conflict depends only on the structure of the PN graph. Two transitions t_1 and t_2 are instead in *effective conflict* in some marking μ iff they are both enabled in μ and firing one disables the other. Conflict is not necessarily symmetric. In the last PN in Figure 3.1, transition t_2 is in conflict with t_1, but t_1 is not in conflict with t_2.

Two transitions t_1 and t_2 are *concurrent* in some marking μ iff they are both enabled in μ and neither is in effective conflict with the other.

$$\mu \overset{t_1}{\rightarrow} \mu_1 \overset{t_2}{\rightarrow} \wedge \mu \overset{t_2}{\rightarrow} \mu_2 \overset{t_1}{\rightarrow}$$

If two transitions are concurrent and their input and output arcs have constant multiplicity, firing them in either order leads to the same marking. Let t_1 and t_2 be concurrent in marking μ as above, then

$$\mu \overset{t_1,t_2}{\rightarrow} \mu_3 \wedge \mu \overset{t_2,t_1}{\rightarrow} \mu_3$$

Combining concurrency and conflict may lead to a situation known as *confusion*. Confusion occurs when the manner and order in which separate conflicts are resolved affects the final outcome. In the presence of confusion, conflicts that involve distinct sets of transitions or even that occur at different times are not independent. For example, in Figure 3.2 there are two conflicts to resolve: the conflict between t_1 and t_2 and the conflict between t_3 and t_4. Assume that t_2 is chosen over t_1 and that t_3 is chosen over t_4. If t_3 fires first then t_5 becomes enabled and may fire removing the conflict between t_1 and t_2 altogether. If, on the other hand, t_2 fires before t_3 then the conflict between t_3 and t_4 is unaffected, and t_3 may fire. In this case, the final marking depends not only on how the conflicts are resolved, but in which order.

A more precise definition of confusion requires the notion of a *conflict set*. The conflict set of transition t enabled in marking μ is the maximal

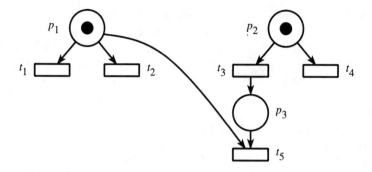

FIG. 3.2. *A confused PN.*

set of transitions which are in conflict with t.

$$cfl(t,\mu) = \{t_i \in T : \exists \mu_j \in S, \mu \xrightarrow{t_i} \mu_j \wedge \neg(\mu_j \xrightarrow{t})\}$$

Let t_1 and t_2 be simultaneously enabled in μ, then (μ, t_1, t_2) is a confusion iff $cfl(t_1, \mu) \neq cfl(t_1, \mu_j)$ where $\mu \xrightarrow{t_2} \mu_j$ and $t_2 \notin cfl(t_1, \mu)$. A PN is *confused* in marking μ iff there is a confusion in μ [57]. That is, a PN is confused when firing a transition changes the conflict set of a concurrent transition.

Confusion may accurately represent the behavior of the system being modeled or be artificially introduced by the interleaving semantics required for PN transition firings. Confusion becomes particularly critical to detect among the immediate transitions in a GSPN or SRN. It is disconcerting for model results to depend upon the interleaving of asynchronous *instantaneous* events that are not in conflict. In [4], a test for finding potential confusion among immediate transitions is presented.

3.1. Marking classification. The standard state classification schemes for Markov chains apply to SRN markings as well [69]. A marking is *recurrent* iff, upon exit from that marking, it will be reentered with probability one. Otherwise a marking is *transient*. We may further classify recurrent markings as *positive recurrent* and *null recurrent*, according to whether the mean time to return to the marking is finite or infinite, respectively.

We only consider the case of finite reachability sets, where no null recurrent markings exist and the determination whether a state is transient or recurrent depends solely upon the structure of the reachability graph, and not upon the transition firing rates:

$$\mu_i \text{ is transient } \Leftrightarrow \exists \mu_j \in \mathcal{S} : \mu_i \xrightarrow{*} \mu_j \wedge \neg(\mu_j \xrightarrow{*} \mu_i)$$

Every SRN with a finite reachability set will eventually enter a marking from where no transient state is reachable.

An SRN will instead return to a recurrent marking an arbitrary number of times after it has reached it at least once. However, a recurrent marking may not be reached at all in a particular evolution of the SRN, because there may be multiple sets of states that act as traps, such that once the SRN enters one state in the set, it may thereafter only enter states belonging to that same set. A set of mutually reachable recurrent states is called a *recurrent class*.

In general, a reachability graph contains a possibly empty set of transient markings and one or more sets of recurrent markings. If there is more than one recurrent class, then the initial marking must be transient, since there is at least one firing sequence that leads from the initial marking, possibly through some intermediate transient markings, to each recurrent class. Obviously, the recurrent markings are of special interest when studying the steady state behavior of the SRN, but transient markings can affect the probability of which recurrent class the SRN eventually enters. When there is a single recurrent class, transient markings are irrelevant to the study of the steady state behavior of the SRN.

For finite reachability graphs, the first step in determining the recurrent classes is to identify the strongly connected components of the reachability graph [1]. Each strongly connected component which cannot be exited constitutes a recurrent class. In particular, a single marking enabling no transition, or an *absorbing marking*, is a recurrent class by itself.

Finally, when discussing SRN markings, it is important to note whether they are vanishing or tangible as defined in Section 2.3. This depends solely upon the rates of the enabled transitions in that marking and is completely independent of any structural properties of the reachability graph. In particular a marking may be either tangible or vanishing and also be either transient or recurrent. Vanishing or tangible refers to the time an SRN remains in a marking per visit. Recurrent or transient refers to whether the SRN returns to a marking a finite or an infinite number of times.

3.2. Classes of PNs. It is natural to consider the effect of restricting or extending the definition of PNs upon the class of systems that may be modeled and the feasibility or efficiency of various types of analysis (see Figure 3.3).

A *marked graph* is a PN where every place has one input arc and one output arc. When drawing a marked graph, it is customary to draw only the transitions and to substitute each place and its input and output arcs with a single arc, so that the place is only implicitly represented and tokens reside on the arcs of the graph. Transitions in a marked graph are typically shown as circles.

Marked graphs can represent synchronization and concurrency but not conflict. In return for the restricted modeling power, marked graphs can be analyzed more efficiently than standard PNs. As a consequence of the definition of a marked graph, the steady-state throughput is the same for all

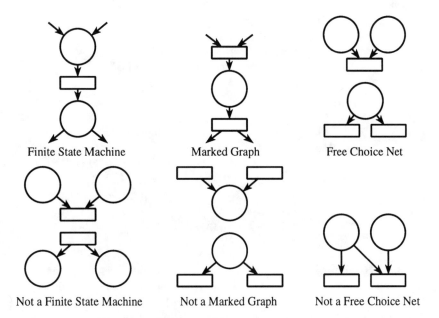

FIG. 3.3. *Classes of PNs.*

transitions (defined as the expected number of times a transition fires per unit time), independent of the firing time distributions. Thus it is possible to speak of the throughput of a marked graph. Though restrictive compared to standard PNs, marked graphs are an extension of PERT networks and task precedence graphs, which are simply safe acyclic marked graphs.

A finite state machine (FSM) is a PN where every transition has exactly one input arc and exactly one output arc. Similarly to marked graphs, it is customary to draw only the places since there is an implicit transition along every arc.

A PN is said to be *free choice* iff each transition that has more than one input place is the only output transition for each of its input places. The free choice restriction implies that each conflict may be resolved in isolation. Free choice PNs extend the class of systems that can be represented by marked graphs. They allow a limited type of conflict that can be resolved based solely on local information. Free choice PNs cannot represent confusion. Balbo *et al.* generalized the definition of free choice based on the concept of extended conflict sets [4].

Standard PNs can represent many forms of conflict, including those that are disallowed in free choice PNs, but cannot represent zero testing. As discussed in Section 2.2, adding inhibitor arcs increases the modeling power of PNs to equal that of Turing machines.

1) $\mathcal{S} = \{\mu_0\}$
2) $\mathcal{A} = \emptyset$
3) $\mathcal{S}^{new} = \{\mu_0\}$
4)   while $\mathcal{S}^{new} \neq \emptyset$ do
5)   choose a marking μ from $\mathcal{S}^{new}$
6) $\mathcal{S}^{new} = \mathcal{S}^{new} \setminus \{\mu\}$
7)   foreach $t \in T$ do
8)   if t enabled in μ
9) $\mu^{new} = \mu + D^{+}_{\bullet,t}(\mu) - D^{-}_{\bullet,t}(\mu)$
10)   if $\mu^{new} \notin \mathcal{S}$ then
11) $\mathcal{S}^{new} = \mathcal{S}^{new} \cup \{\mu^{new}\}$
12) $\mathcal{S} = \mathcal{S} \cup \{\mu^{new}\}$
13) $\mathcal{A} = \mathcal{A} \cup \{(\mu \xrightarrow{t} \mu^{new})\}$

FIG. 3.4. *Algorithm to generate the reachability graph* $(\mathcal{S}, \mathcal{A})$.

3.3. Reachability graph analysis. A standard technique for studying PNs is to construct and explore the reachability graph. Obviously, this approach requires the reachability graph to be finite.[2] It is possible to simultaneously generate the reachability set $\mathcal{S}$ and the reachability graph $(\mathcal{S}, \mathcal{A})$ using a simple generate-and-test algorithm. This is accomplished by visiting each reachable marking, starting from the initial marking. The algorithm is outlined in Figure 3.4.

If the SRN is not simple, $\mathcal{A}$ contains multiple arcs from μ_i to μ_j, labeled with different transitions. Furthermore, if the SRN is not pure, self-arcs $\mu_i \xrightarrow{t} \mu_i$ are also present.

When the reachability set is finite, all questions about a PN are decidable and can be answered by exploring the reachability graph. Unfortunately, the reachability set is extremely large in most practical problems, so tractability, rather than decidability, is often the concern when using reachability graph analysis.

When the size of the reachability graph prohibits exhaustive analysis, a subset of the reachability graph may be explored using simulation. Most properties cannot be determined with certainty using simulation (such as the absence of reachable markings satisfying a given condition), yet it is possible to gain some insight into the behavior of the system and to obtain timing information, using Monte Carlo simulation. Alternatively, it may be possible to avoid generating the reachability graph altogether, using the techniques illustrated in the next section.

---

[2] When the reachability graph is infinite, it is possible to construct a finite representation of it known as the *reachability tree*, [51,56]. Inspection of the reachability tree can answer only a subset of the questions that can be answered by inspecting the reachability graph.

3.4. Structural analysis. This section presents results applicable only to PNs with constant arc multiplicities. It is important to distinguish whether a property refers to the set of all reachable markings, as with mutual exclusion, or to a particular marking, as with conflict.

In some cases, it is possible to determine that a property is preserved in all reachable markings solely from information present in the PN graph. For example, in the standard PN formalism, two transitions that do not share any input places can never be in effective conflict. Such transitions are said to be *structurally* conflict-free. In [4], structural tests for logical properties such as conflict, mutual exclusion, and confusion are presented. Such tests are weaker than those based upon exploration of all reachable markings, but they are much more efficient.

For example, transitions t_1 and t_2 in the first PN in Figure 3.1 share an input place p_2, so they are in potential conflict. If the initial marking is such that p_1 and p_2 cannot both have a token at the same time, then the two transitions are never in effective conflict.

In contrast to reachability graph analysis, structural analysis techniques consider only the structure of the PN graph and examine the set of reachable markings only implicitly. Structural analysis is normally quite efficient, requiring a number of operations that is polynomial in the number of places and transitions.

Invariant analysis considers only the PN incidence matrix $(D^+ - D^-)$ which describes the "net marking change" when firing each transition, but not the conditions required for a transition to be enabled.

An s-invariant is a non-negative integer solution to the equation

$$(3.1) \qquad\qquad x(D^+ - D^-) = 0$$

The weighted sum $\sum_{p \in P} x_p \#(p)$ has the same value in any reachable marking. The s-invariant x *covers* a place if $x_p > 0$.

A t-invariant is a solution to the equation

$$(3.2) \qquad\qquad (D^+ - D^-)y^T = 0$$

(where the superscript T indicates transposition). If, $\forall t \in T$, a firing sequence s contains y_t occurrences of transition t, then s leads any marking back to itself. The existence of a t-invariant does not imply that any such *legal* firing sequence can actually occur.

If a and b are two s-invariants and $\alpha \in \mathbb{N}^+$, then so are αa and $a + b$. The analogous property holds for t-invariants. Hence, we are normally interested in a *minimal set of invariants*, from which all other invariants are obtained as linear combinations.

Invariant analysis has limitations. The set of invariants is independent of the initial marking, while the behavior of the PN can be substantially affected by it. Also, the incidence matrix does not capture the situation where a place p is both an input and an output for a transition t, since the entries $D_{p,t}^-$ and $D_{p,t}^+$ cancel each other in this case. This "control place", though, could have an important effect on the PN behavior. Nor can invariant analysis make use of the presence of inhibitor arcs, transition priorities, or guards, as they do not effect the incidence matrix. The invariants given by solutions to Equations 3.1 and 3.2 are still valid in these cases, but there are other conditions which might hold in the PN and are not captured by the invariants.

Marking invariants simply use the initial marking to fix the constant implicit in each s-invariant. While an s-invariant states that the weighted sum of the tokens in some set of places remains constant, the corresponding marking invariant determines the value of this constant from the initial marking.

Invariants may be used for several purposes. For example, if there exists an s-invariant that covers every place in a PN, then the PN is structurally bounded, it has a finite reachability graph for any initial marking. Consider a GSPN where every transition is defined *a priori* to be either timed or immediate for all markings, as in [3]. If no t-invariant contains only immediate transitions, then the state space does not contain any vanishing loops, regardless of the initial marking. This property may be exploited to warn the user of potential vanishing loops or to use a more general (but less efficient) solution method only when needed.

4. Temporal Analysis. This section shows how to perform the temporal analysis of a SRN, starting from its reachability graph $(\mathcal{S}, \mathcal{A})$. The steps we provide are those used by automatic tools such as SPNP [17]. For the type of numerical solution we discuss, we assume that the measures of interest are of the form given by the examples in Section 2.4. These measures are computed by solving the SMP $\{(t^{[n]}, \theta^{[n]}, \mu^{[n]}), n \in \mathbb{N}\}$ first, to compute the following quantities (see also Table 4.1):

- The expected amount of time spent in each tangible marking i during the interval $[0, \theta)$ or up to steady state:

$$\sigma(\theta) = [\sigma_i(\theta)] = \left[\int_0^\theta Pr\{\mu(u) = i\} du \right]$$

$$\sigma = [\sigma_i] = \left[\lim_{\theta \to \infty} \int_0^\theta Pr\{\mu(u) = i\} du \right]$$

  σ_i is of interest only if i is a transient marking, since $\sigma_i = \infty$ if i is recurrent. σ is often studied in connection with SRNs having absorbing markings, hence it is usually indicated as the expected time spent in each marking "up to absorption".

- The probability of being in each tangible marking i at a given time θ or in steady state:

$$\pi(\theta) = [\pi_i(\theta)] = [Pr\{\mu(\theta) = i\}]$$

TABLE 4.1
Eight quantities obtainable from the study of $\{(t^{[n]}, \theta^{[n]}, \mu^{[n]}), n \in \mathbb{N}\}$.

	Cumulative		Instantaneous	
	Trans.	**St. state**	**Trans.**	**St. state**
Marking i	$\sigma_i(\theta)$	σ_i	$\pi_i(\theta)$	π_i
Transition $i \rightarrow j$	$N_{i,j}(\theta)$	$N_{i,j}$	$\Phi_{i,j}(\theta)$	$\Phi_{i,j}$

$$\pi = [\pi_i] = \left[\lim_{\theta \to \infty} Pr\{\mu(\theta) = i\} \right]$$

- The expected number of marking-to-marking transitions from i to j, indicated with $i \rightarrow j$, up to time θ or up to steady state:

$$N(\theta) = [N_{i,j}(\theta)] = \left[E \left[\sum_{n=1}^{\max\{n : \theta^{[n]} \leq \theta\}} Pr\{\mu^{[n]} = i \wedge i \xrightarrow{t^{[n]}} j\} \right] \right]$$

$$N = [N_{i,j}] = \left[E \left[\sum_{n \geq 1} Pr\{\mu^{[n]} = i \wedge i \xrightarrow{t^{[n]}} j\} \right] \right]$$

$N_{i,j}$ is of interest only if i is transient.
- The expected frequency of marking-to-marking transitions $i \rightarrow j$ at time θ or in steady state:

$$\Phi(\theta) = [\Phi_{i,j}(\theta)] = \left[\lim_{\delta \to 0} E \left[\frac{N_{i,j}(\theta + \delta) - N_{i,j}(\theta)}{\delta} \right] \right]$$

$$\Phi = [\Phi_{i,j}] = \left[\lim_{\theta \to \infty} \lim_{\delta \to 0} E \left[\frac{N_{i,j}(\theta + \delta) - N_{i,j}(\theta)}{\delta} \right] \right]$$

The SRN measures are then obtained as a linear combination of the above quantities using the specified reward rates or impulses as the multiplying factors. $\sigma(\theta)$, σ, $\pi(\theta)$, and π are related to sojourns in the SRN markings, so they are multiplied by the corresponding reward rates: the multiplying factor associated to marking i is $\rho(i)$. $N(\theta)$, N, $\Phi(\theta)$, and Φ are instead related to transitions between markings of the SRN, so they are multiplied by the corresponding reward impulses: the multiplying factor associated to $i \rightarrow j$ is $r_t(i)$ if t is the only transition such that $i \xrightarrow{t} j$. In general, though, the SRN might not be simple, so i and j might not uniquely identify t. We must then define the reward impulse associated to the marking-to-marking transition $i{\to}j$, based on the reward impulses associated to transition firings:

$$r_{i,j} = \sum_{t\in T:i\overset{t}{\to}j} r_t(i)Pr\{t^{[n]} = t|\mu^{[n-1]} = i \wedge \mu^{[n]} = j\}$$

$$= \begin{cases} \dfrac{\displaystyle\sum_{t\in T:i\overset{t}{\to}j} r_t(i)\lambda_t(i)}{\displaystyle\sum_{t\in T:i\overset{t}{\to}} \lambda_t(i)} & \text{if } i \in T \\[2em] \dfrac{\displaystyle\sum_{t\in T:i\overset{t}{\to}j} r_t(i)w_t(i)}{\displaystyle\sum_{t\in T:i\overset{t}{\to}} w_t(i)} & \text{if } i \in V \end{cases}$$

The expected instantaneous reward at time θ is then obtained as

$$\lim_{\delta\to 0} E\left[\frac{Y(\theta + \delta) - Y(\theta)}{\delta}\right] = \sum_{i\in S}\rho(i)\pi_i(\theta) + \sum_{i,j\in S} r_{i,j}\Phi_{i,j}(\theta)$$

and the expected reward up to time θ as

$$E[Y(\theta)] = \sum_{i\in S}\rho(i)\sigma_i(\theta) + \sum_{i,j\in S} r_{i,j}N_{i,j}(\theta)$$

Time-averaged measures are obtained by dividing $E[Y(\theta)]$ by θ. The analogous steady state measures are obtained by using the steady state quantities π, σ, N, and Φ instead of their transient counterparts.

The reachability graph (S, A) implicitly describes $\{(t^{[n]}, \theta^{[n]}, \mu^{[n]}), n \in \mathbb{N}\}$ but, in practice, the following quantities are used.

- $h = [h_i] = [E[\theta^{[n+1]} - \theta^{[n]}|\mu^{[n]} = i]]$, a vector describing the expected holding time in each marking:

$$h_i = \begin{cases} \left(\displaystyle\sum_{t\in T:i\overset{t}{\to}} \lambda_t(i)\right)^{-1} & \text{if } i \in T \\[1.5em] 0 & \text{if } i \in V \end{cases}$$

- $\Pi = [\Pi_{i,j}] = [Pr\{\mu^{[n+1]} = j|\mu^{[n]} = i\}]$, a matrix describing the probability of marking-to-marking transitions. Π is stochastic, corresponding to the embedded DTMC process $\{\mu^{[n]}, n \in \mathbb{N}\}$.

$$
\Pi_{i,j} = \begin{cases} \displaystyle\sum_{t \in T : i \xrightarrow{t} j} \lambda_t(i) h_i & \text{if } i \in T \\[2em] \displaystyle\sum_{t \in T : i \xrightarrow{t} j} w_t(i) \left(\sum_{t \in T : i \xrightarrow{t}} w_t(i) \right)^{-1} & \text{if } i \in V \end{cases}
$$

- $\gamma(0) = [\gamma_i(0)] = [Pr\{\mu_0 = i\}]$, a vector describing the probability of being initially in each marking. The embedded DTMC process $\{\mu^{[n]}, n \in \mathbb{N}\}$ has the same initial probability vector. As discussed in Section 2.4, the initial distribution could be captured by the SRN, but it is more efficient and common to assume that $\gamma(0)$ is explicitly provided as input.

4.1. Preservation or elimination of the vanishing markings.
So far, we have considered the entire state space S of the SMP. In practice, the reward rates associated to the vanishing markings do not contribute in any way to the value of a measure, since $\sigma(\theta)$, σ_i, $\pi(\theta)$, and π_i are zero if i is a vanishing marking. A common approach to the solution of the SMP requires the elimination of the vanishing markings. If h, Π, and $\gamma(0)$ are partitioned according to the type of marking (vanishing or tangible) into:

$$
h = [0 \,|\, h_T] \qquad \Pi = \left[\begin{array}{c|c} \Pi_{V,V} & \Pi_{V,T} \\ \hline \Pi_{T,V} & \Pi_{T,T} \end{array} \right] \qquad \gamma(0) = [\gamma_V(0) \,|\, \gamma_T(0)]
$$

it is possible to define a new process that considers only the tangible-marking-to-tangible-marking transitions. This new process is described by h_T, Π^*, and $\pi(0)$, given by:

$$
\Pi^* = \Pi_{T,T} + \Pi_{T,V} \left(I - \Pi_{V,V} \right)^{-1} \Pi_{V,T}
$$

$$
\pi(0) = \gamma_T(0) + \gamma_V(0) \left(I - \Pi_{V,V} \right)^{-1} \Pi_{V,T}
$$

Π^* might have nonzero diagonal entries either because $\Pi_{T,T}$ does (if the PN is not pure) or because the reachability graph contains paths $i \xrightarrow{s} i$ where $i \in T$ but all the intermediate markings are vanishing. Figure 4.1 depicts these situations on a portion of a reachability graph before and after this elimination (tangible and vanishing markings are represented as ovals and rectangles, respectively).

Since the sojourn time in the tangible markings is exponentially distributed, the resulting process is the CTMC $\{\mu(\theta), \theta \geq 0\}$ already mentioned in Section 2.3 [3] and it is more commonly described by the infinites-

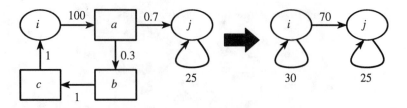

FIG. 4.1. *Origin of diagonal entries in the transition rate matrix.*

imal generator matrix Q plus the initial probability vector $\pi(0)$, where

$$Q_{i,j} = \begin{cases} \dfrac{\Pi^*_{i,j}(1 - \Pi^*_{i,i})}{h_i} & \text{if } i \neq j \\[2ex] -\dfrac{1 - \Pi^*_{i,i}}{h_i} & \text{if } i = j \end{cases}$$

The factor $(1 - \Pi^*_{i,i})$ is needed to account for the nonzero diagonal entries in Π. In other words, the expected sojourn time in tangible marking i is h_i for the SMP $\{(t^{[n]}, \theta^{[n]}, \mu^{[n]}), n \in \mathbb{N}\}$, but $h_i/(1 - \Pi^*_{i,i})$ for the CTMC $\{\mu(\theta), \theta \geq 0\}$, which does not consider transitions $i \rightarrow i$ as interruptions of the sojourn in marking i.

The elimination of the vanishing markings reduces the number of states in the process to be studied, but it also perturbs the structure of the reachability graph, which describes the marking-to-marking transitions. For example, if $r_{a,b} \neq 0$ in Figure 4.1, the arc $a \rightarrow b$ must be taken into account even if a and b have been eliminated.

We can then say that the temporal analysis of a SRN is concerned with computing $\sigma(\theta)$, σ, $\pi(\theta)$, π, $N(\theta)$, N, $\Phi(\theta)$, and Φ given h, Π, and $\gamma(0)$, or Q and $\pi(0)$. We call the first approach *preservation*, as opposed to the second one requiring the *elimination* of the vanishing markings.

4.2. Instantaneous steady state analysis. For steady state analysis, it is possible to use either preservation or elimination. As shown in [15], both approaches have advantages and disadvantages and the best choice among the two depends on the problem being solved. This discussion assumes that the state space contains a single recurrent class, see Section 4.4 for the general case.

4.2.1. Preservation. To compute the steady state instantaneous probabilities π and marking-to-marking transition frequencies Φ with preservation:

- Compute the steady state probability vector for the underlying DTMC:

$$(4.1) \qquad\qquad \gamma = \gamma \Pi \qquad \text{subject to} \quad \sum_{i \in \mathcal{S}} \gamma_i = 1$$

- Then:

$$\pi_i = \frac{\gamma_i h_i}{\displaystyle\sum_{k \in \mathcal{S}} \gamma_k h_k} = \frac{\gamma_i h_i}{\displaystyle\sum_{k \in \mathcal{T}} \gamma_k h_k}$$

- And:

$$\Phi_{i,j} = \Pi_{i,j} \phi_i = \Pi_{i,j} \frac{\gamma_i}{\displaystyle\sum_{k \in \mathcal{S}} \gamma_k h_k} = \frac{\gamma_i}{\displaystyle\sum_{k \in \mathcal{T}} \gamma_k h_k}$$

where $\phi = [\phi_i] = [\phi_{\mathcal{T}} | \phi_{\mathcal{V}}]$ represents the frequency with which each marking i is entered, or exited, in steady state.

4.2.2. Elimination. To compute the steady state instantaneous probabilities π and marking-to-marking transition frequencies Φ with elimination:

- Solve:

$$(4.2) \qquad \pi Q = 0 \qquad \text{subject to} \quad \sum_{i \in \mathcal{T}} \pi_i = 1$$

- If i is a tangible marking, $\phi_i = \pi_i / h_i$, hence:

$$\Phi_{i,j} = \Pi_{i,j} \phi_i = \Pi_{i,j} \frac{\pi_i}{h_i}$$

We compute ϕ_i as π_i / h_i and not as $-\pi_i / Q_{i,i} = \pi_i (1 - \Pi_{i,i}^*)/h_i$, so that ϕ_i represents the frequency at which tangible marking i is entered in steady state *before* the elimination of the vanishing markings. This ensures that ϕ_i has the same value independently of whether it is compute using preservation or elimination.

- For the vanishing markings, though, $\phi_{\mathcal{V}}$ must be obtained by solving the following equation:

$$(4.3) \qquad \phi_{\mathcal{V}}[I - \Pi_{\mathcal{V},\mathcal{V}}] = \phi_{\mathcal{T}} \Pi_{\mathcal{T},\mathcal{V}}$$

Once $\phi_{\mathcal{V}}$ is obtained, $\Phi_{i,j}$ for $i \in \mathcal{V}$ is computed as

$$\Phi_{i,j} = \Pi_{i,j} \phi_i$$

4.3. Cumulative steady state analysis. As with instantaneous steady state analysis, the study of the cumulative behavior up to steady state can be performed using either preservation or elimination.

4.3.1. Preservation. To compute the cumulative sojourn times (for the transient markings) σ and marking-to-marking transitions (from the transient markings) N up to steady state with preservation:

- Define $\Pi^{(0,0)}$ and $\gamma^{(0)}(0)$ to be the restrictions of Π and $\gamma(0)$ to the transient markings of the underlying DTMC, respectively.
- Define $n = [n_i] = [n_T | n_V]$ to be the expected number of visits to each transient marking up to steady state for the underlying DTMC. n is computed by solving

$$(4.4) \qquad n(I - \Pi^{(0,0)}) = \gamma^{(0)}(0)$$

- Then:

$$\sigma_i = n_i h_i$$

- And:

$$N_{i,j} = \Pi_{i,j} n_i$$

4.3.2. Elimination. To compute the cumulative sojourn times (for the transient markings) σ and marking-to-marking transitions (from the transient markings) N up to steady state with elimination:

- Define $Q^{(0,0)}$ and $\pi^{(0)}(0)$ to be the restrictions of Q and $\pi(0)$ to the transient markings, respectively.
- Solve:

$$(4.5) \qquad \sigma Q^{(0,0)} = -\pi^{(0)}(0)$$

- If i is a tangible marking, $n_i = \sigma_i / h_i$, hence:

$$N_{i,j} = \Pi_{i,j} n_i = \Pi_{i,j} \frac{\sigma_i}{h_i}$$

- For the vanishing markings, though, n_V must be obtained by solving the following equation:

$$(4.6) \qquad n_V[I - \Pi_{V,V}^{(0,0)}] = n_T \Pi_{T,V}^{(0,0)}$$

where $\Pi_{V,V}^{(0,0)}$ and $\Pi_{T,V}^{(0,0)}$ are the restrictions of $\Pi_{V,V}$ and $\Pi_{T,V}$ to the transient markings, respectively. Once n_V is obtained, $N_{i,j}$ is computed as

$$N_{i,j} = \Pi_{i,j} n_i$$

4.4. Multiple recurrent classes. Equation 4.1 admits a unique solution γ iff the state space $\mathcal{S}$ of the corresponding stochastic process contains a single recurrent class. This section describes how to approach the solution when multiple recurrent classes exist. Exactly the same discussion would apply to Equation 4.2 and π.

If m recurrent classes exist, the initial marking μ_0, and possibly other markings as well, are transient. We can partition $\mathcal{S}$ into $\mathcal{S}^{(0)}, \mathcal{S}^{(1)}, \ldots \mathcal{S}^{(m)}$, corresponding to the transient markings and to the m recurrent classes, respectively. Accordingly, Π and γ can be partitioned as

$$
\Pi = \left[
\begin{array}{c|c|c|c|c}
\Pi^{(0,0)} & \Pi^{(0,1)} & \Pi^{(0,2)} & \cdots & \Pi^{(0,m)} \\
\hline
0 & \Pi^{(1,1)} & 0 & \cdots & 0 \\
\hline
\cdots & \cdots & \Pi^{(2,2)} & \cdots & 0 \\
\hline
0 & 0 & \cdots & \cdots & \Pi^{(m,m)}
\end{array}
\right]
, \quad
\gamma = \left[\gamma^{(0)} \,\middle|\, \gamma^{(1)} \,\middle|\, \cdots \,\middle|\, \gamma^{(m)} \right]
$$

The solution of each equation

$$
\gamma^{(i)} \Pi^{(i,i)} = \gamma^{(i)} \qquad \text{subject to} \qquad \sum_{k \in \mathcal{S}^{(i)}} \gamma^{(i)}_k = 1
$$

for $i = 1, \ldots, m$, is unique, but

$$
\forall c_1, \ldots, c_m \in \mathbb{R}^+, \sum_{i=1}^{m} c_i = 1, \gamma = \left[0 \,\middle|\, c_1 \gamma^{(1)} \,\middle|\, \cdots \,\middle|\, c_m \gamma^{(m)} \right]
$$

is a solution to Equation 4.1.

To obtain the correct value for γ, the constants c_i must be set to the probability of entering recurrent class $\mathcal{S}^{(i)}$ given the initial probability vector $\gamma(0)$. This is obtained by first computing the vector n, as described in Section 4.3. Then,

$$
c_i = n \Pi^{(0,i)} \mathbf{1}^T
$$

where $\mathbf{1}$ is a vector of the appropriate dimension with all entries equal one.

This approach is particularly efficient since the identification of the recurrent classes has complexity $O(|\mathcal{A}|)$. Furthermore, it has the additional advantage of requiring the manipulation of smaller matrices.

4.5. Transient analysis. For transient analysis, elimination is the method of choice. To compute the transient instantaneous probabilities $\pi(\theta)$, solve the Kolmogorov ordinary differential equation

(4.7) $$\dot{\pi}(\theta) = \pi(\theta)Q \quad \text{with initial condition} \quad \pi(0)$$

where $\dot{\pi}(\theta)$ is the derivative of π with respect to θ.

To compute the cumulative sojourn times $\sigma(\theta)$, solve the Kolmogorov differential equation:

(4.8) $$\dot{\sigma}(\theta) = \sigma(\theta)Q + \pi(0), \quad \text{with initial condition} \quad \sigma(0) = 0,$$

which is obtained by integrating Equation 4.7 with respect to θ.

The computation of $\Phi(\theta)$ and $N(\theta)$ is exactly analogous to the steady state case and is omitted.

4.6. Sensitivity analysis. Analytical models are often used to evaluate alternatives during the design of a system, such as: (1) If the intercon-
nections among the components are changed, how is the system throughput
affected? (2) What effect does the increase in the number of buses have on
their utilization? (3) How is the response time affected if the speed of the
buses is increased?

Furthermore, exact values for the parameters used in the specification
of the system may be unknown. One way of overcoming this problem is
to solve the model for different values of the parameters and comparing
the solutions obtained. If the model is highly sensitive to variation in a
parameter, greater effort should be placed in estimating its exact value,
and, possibly, the system should be overengineered, to allow for errors in
the estimation process. The importance of performing a parametric or
sensitivity analysis has been stressed in [9,29,64].

Sensitivity analysis of SRN models can be performed at various levels:
(1) Changes in the structure of the SRN; (2) Changes in the initial number
of tokens in a place; (3) Changes in an independent parameter β which is
used in the definition of the rate or probability of one or more transitions.
Levels (1), (2) and (3) directly correspond to the three examples mentioned
above. For (1) and (2), the structure of the underlying reachability graph
is altered, hence, the entire analysis process must be performed for each
value of β. For (3), this approach is also possible, but it is not required,
since the variation in β only affects some of the entries in h, Π, and $\gamma(0)$,
with preservation, or Q and $\pi(0)$, with elimination.

4.6.1. Parametric sensitivity analysis. Assume that a parameter
β appears in the specification of one or more firing rates, firing probabilities,
or reward functions. Define x' to be derivative of x with respect to β. Given
an output measure, $\Psi(\beta)$, we can compute its derivative $\Psi'(\beta)$ and then
use the linear approximation

$$\Psi(\beta + \epsilon) \approx \Psi(\beta) + \Psi'(\beta)\epsilon$$

to obtain the value of Ψ for reasonably small variations ϵ in the parameter
β (for simplicity, we omit to write explicitly the dependency on β in the
remainder of this section).

If the measure is the expected instantaneous reward in steady state,

$$\Psi = \sum_{i \in S} \rho(i)\pi_i + \sum_{i,j \in S} r_{i,j}\Phi_{i,j}$$

and

$$\Psi' = \sum_{i \in S} (\rho'(i)\pi_i + \rho(i)\pi_i') + \sum_{i,j \in S} (r_{i,j}'\Phi_{i,j} + r_{i,j}\Phi_{i,j}')$$

To compute Ψ', $\rho(i)$, π_i, $r_{i,j}$, and $\Phi_{i,j}$ and the corresponding derivatives
$\rho'(i)$, π_i', $r_{i,j}'$, and $\Phi_{i,j}'$ are needed. If $\rho(i)$ and $r_{i,j}$ are given analytically,

the computation of their derivatives is trivial, so we focus on π_i' and $\Phi_{i,j}'$. To compute π' and Φ' with preservation:

- Compute γ first, then solve for γ' in

$$(4.9) \qquad \gamma'(I - \Pi) = \gamma\Pi' \qquad \text{subject to} \quad \sum_{i \in \mathcal{S}} \gamma_i' = 0$$

  This equation is obtained by taking the derivative of Equation 4.1 with respect to β.

- Then:

$$\pi_i' = \frac{\gamma_i' h_i + \gamma_i h_i'}{\sum_{k \in \mathcal{T}} \gamma_k h_k} - \frac{\gamma_i h_i \sum_{k \in \mathcal{T}} (\gamma_k' h_k + \gamma_k h_k')}{\left(\sum_{k \in \mathcal{T}} \gamma_k h_k\right)^2}$$

- And:

$$\Phi_{i,j}' = \frac{\gamma_i'}{\sum_{k \in \mathcal{T}} \gamma_k h_k} - \frac{\gamma_i \sum_{k \in \mathcal{T}} (\gamma_k' h_k + \gamma_k h_k')}{\left(\sum_{k \in \mathcal{T}} \gamma_k h_k\right)^2}$$

To compute π' and Φ' with elimination:

- Compute π first, then solve for π' in

$$(4.10) \qquad \pi'Q = -\pi Q' \qquad \text{subject to} \quad \sum_{i \in \mathcal{T}} \pi_i' = 0$$

  This equation is obtained by taking the derivative of Equation 4.2 with respect to β. While Π' is trivially obtained given that the functional dependency of the entries of Π on β is known, the computation of Q' is more complex. First, obtain $\Pi^{*'}$, the derivative of Π^* with respect to β:

$$\Pi^{*'} = \Pi_{\mathcal{T},\mathcal{T}}' + \Pi_{\mathcal{T},\mathcal{V}}' X + \Pi_{\mathcal{T},\mathcal{V}} X'$$

where X and X' are computed by solving, respectively,

$$[I - \Pi_{\mathcal{V},\mathcal{V}}]X = \Pi_{\mathcal{V},\mathcal{T}}$$

and

$$[I - \Pi_{\mathcal{V},\mathcal{V}}]X' = \Pi_{\mathcal{V},\mathcal{T}}' + \Pi_{\mathcal{V},\mathcal{V}}' X$$

Then, $Q'_{i,j}$ is given by,

$$
Q'_{i,j} = \begin{cases}
\dfrac{\Pi^{*'}_{i,j}(1 - \Pi^{*}_{i,i}) - \Pi^{*}_{i,j}\Pi^{*'}_{i,i}}{h_i} - \dfrac{\Pi^{*}_{i,j}(1 - \Pi^{*}_{i,i})h'_i}{h_i^2} & \text{if } i \neq j \\[3mm]
\dfrac{\Pi^{*'}_{i,i}}{h_i} + \dfrac{(1 - \Pi^{*}_{i,i})h'_i}{h_i^2} & \text{if } i = j
\end{cases}
$$

- If i is a tangible marking,

$$
\Phi_{i,j} = \Pi'_{i,j}\frac{\pi_i}{h_i} + \Pi_{i,j}\left(\frac{\pi'_i}{h_i} - \frac{\pi_i h'_i}{h_i^2}\right)
$$

- For the vanishing markings, $\phi'_\mathcal{V}$ must be obtained by solving the following equation:

$$
(4.11) \qquad \phi'_\mathcal{V}[I - \Pi_{\mathcal{V},\mathcal{V}}] = \phi'_\mathcal{T}\Pi_{\mathcal{T},\mathcal{V}} + \phi_\mathcal{T}\Pi'_{\mathcal{T},\mathcal{V}} + \Pi'_{\mathcal{V},\mathcal{V}}\phi_\mathcal{V}
$$

Once $\phi'_\mathcal{V}$ is obtained, $\Phi'_{i,j}$ is computed as

$$
\Phi'_{i,j} = \Pi'_{i,j}\phi_i + \Pi_{i,j}\phi'_i
$$

If Ψ is the expected accumulated reward up to steady state, σ'_i and $N'_{i,j}$ are needed. To compute σ'_i and $N'_{i,j}$ with preservation:

- Solve for n' using the equation:

$$
(4.12) \qquad n'\left(I - \Pi^{(0,0)}\right) = n\Pi^{(0,0)'}
$$

- Then:

$$
\sigma'_i = n'_i h_i + n_i h'_i
$$

- And:

$$
N_{i,j} = \Pi'_{i,j}n_i + \Pi_{i,j}n'_i
$$

To compute σ'_i and $N'_{i,j}$ with elimination:

- Solve:

$$
(4.13) \qquad \sigma'Q^{(0,0)} = -\sigma Q^{(0,0)'}
$$

where $Q^{(0,0)'}$ is the restriction of Q' to the transient markings.

- If i is a tangible marking,

$$
N_{i,j} = \Pi'_{i,j}\frac{\sigma_i}{h_i} + \Pi_{i,j}\left(\frac{\sigma'_i}{h_i} - \frac{\sigma_i h'_i}{h_i^2}\right)
$$

- For the vanishing markings, $n'_\mathcal{V}$ must be obtained by solving the following equation:

(4.14) $\qquad n'_\mathcal{V}[I - \Pi_{\mathcal{V},\mathcal{V}}] = n'_\mathcal{T}\Pi_{\mathcal{T},\mathcal{V}} + n_\mathcal{T}\Pi'_{\mathcal{T},\mathcal{V}} + n_\mathcal{V}\Pi'_{\mathcal{V},\mathcal{V}}$

Once $n'_\mathcal{V}$ is obtained, $N'_{i,j}$ is computed as

$$N_{i,j} = \Pi'_{i,j}n_i + \Pi_{i,j}n'_i$$

If Ψ is the instantaneous or accumulated reward at time θ, $\pi'(\theta)$ and $\Phi'(\theta)$, or $\sigma'(\theta)$ and $N(\theta)$ must be computed, respectively. To compute $\pi'(\theta)$ and $\sigma(\theta)$, solve the ordinary differential equations

$$\dot\pi'(\theta) = \pi'(\theta)Q + \pi(\theta)Q' \quad \text{with initial condition} \quad \pi'(0) = 0$$

and

$$\dot\sigma'(\theta) = \sigma'(\theta)Q + \sigma(\theta)Q' \qquad \sigma'(0) = 0,$$

obtained by taking the derivative of Equations 4.7 and 4.8 with respect to β [29,47].

The computation of $\Phi'(\theta)$ and $N'(\theta)$ is exactly analogous to the steady state case and is omitted.

5. Numerical Solution. This section describes numerical techniques that can be used to solve the equations encountered in Section 4 when the state space is finite.

5.1. Steady state analysis. From the perspective of linear algebra, Equations 4.1, 4.2, 4.3, 4.4, 4.5, 4.6, 4.9, 4.10, 4.11, 4.12, 4.13, and 4.14 can be expressed in the form

$$xA = b$$

where either $b = 0$ and A is a Q-matrix, that is, an ergodic matrix with non-negative off-diagonal entries and row-sum equal zero, or $b \neq 0$ and A is a submatrix of a Q-matrix, obtained by eliminating the rows and columns corresponding to a subset of the states. For example, Equation 4.1, $\gamma = \gamma\Pi$, can be rewritten as $\gamma(\Pi - I) = 0$, and $(\Pi - I)$ is a Q-matrix.

If the system is singular ($b = 0$), the problem can be stated as finding the left eigenvector $x \neq 0$ associated with the eigenvalue 0 of A. Since the CTMC is ergodic, there is exactly one redundant equation in the system, the Null Space of Q has dimension one, and we can uniquely determine x using one additional constraint, given by $x\mathbf{1}^T = 1$ in our case, since x is then a probability vector. Methods for computing eigenvectors, such as QR or non-symmetric Lanczos, can be quite expensive. In our particular case, though, we can use iterative methods to solve this linear equation, which can produce savings both in storage and in arithmetic operations. The same iterative methods can be used when $b \neq 0$, so we do not distinguish between these two possibilities.

Following Stewart and Goyal [64], we find that SOR (Successive Over-Relaxation) is a convenient and effective method of solving $xA = b$. One major advantage of an iterative method is that it computes the solution x in place, without modifying A: in theory, only A and x need to be stored. By storing A in compressed-row format [53], only $O(\eta)$ storage is required for A, where η is the number of non-zero entries in A. In practical problems, η is $O(n)$ where n is the number of states, rather than $O(n^2)$, the size of the full Q matrix.

SOR is a well-known method with good convergence characteristics [64,66]. Stewart and Goyal recommend estimating the optimal relaxation parameter ω from the differences between successive iterates. We use a variation of their method guiding the parameter ω directly by the magnitude of the difference vectors.

The iteration matrix for SOR is derived by splitting A into three components [19]:

$$A = (L + I + U)D$$

where L and U are strictly upper triangular and lower triangular, respectively.

The SOR iteration is [19]

$$x^{(k+1)} = x^{(k)}[(1-\omega)I - \omega D^{-1}L][I + \omega D^{-1}U]^{-1} + b\omega D^{-1}[I + \omega D^{-1}U]^{-1}$$

where $x^{(k)}$ is the k-th iterate for x and the SOR iteration matrix is:

$$M_\omega = [(1-\omega)I - \omega D^{-1}L][I + \omega D^{-1}U]^{-1}$$

If $b = 0$, the SOR iteration for the homogeneous system of equations is simply

$$x^{(k+1)} = x^{(k)}M_\omega$$

and $x^{(0)}$ should be chosen as recommended in [64], to ensure that the initial vector is not deficient in the direction of the solution vector. Also, as recommended in [64], we renormalize $x^{(k)}$ if the vector becomes too large or too small.

If $b \neq 0$, the choice of the initial vector is less critical, and no renormalization must be performed. In addition, convergence is guaranteed in this case, because in our applications, when $b \neq 0$ the matrix A is always strictly diagonally dominant. Convergence is not guaranteed for the homogeneous case.

Each iteration, we track the size of the difference vector $\delta^{(i)} = x^{(i)} - x^{(i-1)}, i \geq 1$. Specifically, we use the max vector norm to compute, for each iteration,

$$\frac{||\delta^{(i)}||_\infty}{||x^{(i)}||_\infty} = \frac{||x^{(i)} - x^{(i-1)}||_\infty}{||x^{(i)}||_\infty}$$

The solution vector can be of any length, so we normalize the norm of $\delta^{(i)}$ using the norm of $x^{(i)}$.

The algorithm starts with $\omega = 1$, and progresses by performing 10 iterations before changing to a new ω. When $\omega = 1$, $M_\omega = -D^{-1}L(I + D^{-1}U)^{-1}$, hence SOR reduces to Gauss-Seidel for the first ten iterations. Let k be the index of the ω's: that is

$$k = \left\lfloor \frac{i}{10} \right\rfloor$$

For every ω_k, compute Δ_k as

$$\Delta_k = \sum_{j=i}^{i+9} \frac{||\delta^{(j)}||_\infty}{||x^{(j)}||_\infty}$$

As we are accumulating the norms of the difference vectors, we normalize them by the norm of the current iterate. Every 30 iterations, the algorithm checks the progress of the computation in relation to the values of Δ_k. Let the index of this check be

$$\ell = \left\lfloor \frac{i}{30} \right\rfloor$$

At each check, the values of ω_k and ϵ_ℓ, used to increase or decrease ω as the computation proceeds, are updated ($\epsilon = 0.1$ initially). After every 10 iterations, unless we are checking the progress of the computation, we increase ω_k by the current ϵ_ℓ. That is,

$$\omega_{k+1} = \omega_k + \epsilon_\ell$$

Every 30th iteration, we adjust the value of ω_k and ϵ_ℓ depending upon the values of $\Delta_{k-2}, \Delta_{k-1}$, and Δ_k. The adjustments are described in Table 5.1.

Intuitively, we are causing the SOR routine to oscillate around the optimal value of ω as measured by the sum of difference norms between successive iterates. The key idea is that the ω producing the smallest sum of difference vector norms will be close to optimal. This heuristic appears to work well in practice. These quantities are related to an estimate of the subdominant eigenvalue of the iteration matrix [64, Equation (10)]. In our procedure, we attempt to smooth the difference vectors (which will oscillate) by summing their norms over 10 iterations. When ϵ_ℓ is small enough, say around 10^{-7}, we assume that the algorithm has found a value

TABLE 5.1
Adjustments to ϵ and ω for the SOR method.

Condition	Action	Explanation
$\Delta_{k-2} \leq \Delta_{k-1} \leq \Delta_k$	$\omega_{k+1} = \omega_k - 4\epsilon_\ell$ $\epsilon_{\ell+1} = 0.97\epsilon_\ell$	δ^i is increasing, we have overshot ω. Decrease ω so that it is smaller than ω_{k-2} and approach that point more slowly.
$\Delta_k \leq \Delta_{k-2} \leq \Delta_{k-1}$	$\omega_{k+1} = \omega_k - \epsilon_\ell$ $\epsilon_{\ell+1} = \epsilon_\ell$	The last set of 10 iterations produced the smallest accumulated difference norms. Back up to ω_{k-1} and perform another set of 30 iterations.
$\Delta_{k-2} \leq \Delta_k \leq \Delta_{k-1}$	$\omega_{k+1} = \omega_k - 3\epsilon_\ell$ $\epsilon_{\ell+1} = 0.89\epsilon_\ell$	The smallest sum of accumulated difference norms was at the beginning of the set. Back up to before ω_{k-2} and approach more slowly.
$\Delta_{k-1} \leq \Delta_k \leq \Delta_{k-2}$ or $\Delta_{k-1} \leq \Delta_{k-2} \leq \Delta_k$	$\epsilon_{\ell+1} = 0.5\epsilon_\ell$ $\omega_{k+1} = \omega_k - 3\epsilon_\ell$	In both of these cases, Δ_{k-1} is the smallest sum of difference norms. Decrease ϵ_ℓ by half, set $\omega_{k+1} = \omega_{k-1} - 1/2\epsilon_\ell$ and approach ω_{k-1} more slowly from below.
$\Delta_k \leq \Delta_{k-1} \leq \Delta_{k-2}$	$\omega_{k+1} = \omega_k - \epsilon_\ell$ $\epsilon_{\ell+1} = 0.97\epsilon_\ell$	The sum of difference norms is getting progressively smaller. Back up to ω_{k-1} and perform another set of 30 iterations.

close to the optimal value for ω and stop adjusting it. In the best case, this requires about 20 adjustments to ϵ_ℓ (600 iterations) and in the worst case about 500 adjustments to ϵ_ℓ (15,000 iterations). If, at any point, we exceed an upper bound on the number of iterations allowed, we terminate the adjustment of ϵ_ℓ and switch to Gauss-Seidel iteration. In practice, we have found this simple procedure to be very effective.

5.2. Instantaneous transient analysis. Several methods for the transient solution of a CTMC are available. Fully symbolic solution using Laplace transforms is possible only for CTMCs having a small number of states or a very regular structure [69]. Semi-symbolic solution of a CTMC in terms of time θ can be obtained via algebraic methods [67]. However, this algorithm has complexity $O(|T|^3)$, needs full storage for Q, and can be numerically unstable.

Thus, we resort to purely numerical solution techniques. We can write the general solution of Equation 4.7 as:

$$(5.1) \qquad \pi(\theta) = \pi(0)e^{Q\theta}$$

where the *matrix exponential* $e^{Q\theta}$ is given by the Taylor series [45]

$$e^{Q\theta} = \sum_{i=0}^{\infty} \frac{(Q\theta)^i}{i!}$$

Direct evaluation of the matrix exponential is subject to severe round-off problems since Q contains both positive and negative entries.

We can use solution methods for linear differential equations like Runge-Kutta to solve Equation 4.7 directly. Jensen's method [34] (also called Uniformization or Randomization by various authors [26,27,37,54]) is yet another numerical method based on infinite series summation. The above two methods have complexity $O(\eta q\theta)$ where

$$q = \max_{i \in T}\{|Q_{i,i}|\}$$

The computation requirements increase with q and θ, hence $q\theta$ has been identified as an index of the *stiffness* of a CTMC [54]. An implicit ODE method called *TR-BDF2* [54] is insensitive to the value of $q\theta$. An implicit Runge-Kutta method for stiff problems is described in [43].

5.2.1. Jensen's method. This section presents Jensen's method, which has many desirable properties, and a modification of it, which improves its performance for stiff problems. A significant advantage of this modification over implicit ODE methods is that it does not suffer from large overhead for non-stiff problems and, at the same time, can yield good accuracy.

The transient state probabilities of the CTMC are computed using Jensen's method as:

$$(5.2) \qquad \pi(\theta) = \sum_{i=0}^{\infty} \hat{\gamma}(i)e^{-q\theta}\frac{(q\theta)^i}{i!}$$

where $\hat{\gamma}(i)$ is the state probability vector of the underlying DTMC $\hat{Q} = Q/q + I$ at step i and is computed iteratively as

$$(5.3) \qquad \hat{\gamma}(i) = \hat{\gamma}(i-1)\hat{Q} \quad \text{starting from} \quad \hat{\gamma}(0) = \pi(0)$$

In practice, the summation in Equation 5.2 can and must be carried out only up to a finite number of terms k called the *right truncation point*. Furthermore, as $q\theta$ increases, the Poisson distribution thins on the left as well and the terms in the summation for small values of i become insignificant. Thus it is advisable to start the summation at a value $l > 0$, called the *left truncation point* [20,54]. Equation 5.2 reduces to

$$(5.4) \qquad \pi(\theta) \approx \sum_{i=l}^{k} \hat{\gamma}(i) e^{-q\theta} \frac{(q\theta)^i}{i!}$$

Given a truncation error tolerance requirement ϵ, we can precompute the number of terms of the series needed to satisfy this tolerance as

$$l = \max\left\{ j \in \mathbb{N} : \sum_{i=0}^{j-1} e^{-q\theta} \frac{(q\theta)^i}{i!} \leq \frac{\epsilon}{2} \right\}$$

$$k = \min\left\{ j \in \mathbb{N} : 1 - \sum_{i=0}^{j} e^{-q\theta} \frac{(q\theta)^i}{i!} \leq \frac{\epsilon}{2} \right\}$$

Since this method involves only additions and multiplications and no subtractions, it is not subject to severe roundoff errors. One of the main problems, though, is its $O(\eta q\theta)$ complexity [54]. The number of terms needed for Jensen's method between the left and the right truncation point is $O(\sqrt{q\theta})$. However, it is necessary to obtain the DTMC state probability vector at l, the left truncation point, and l is $O(q\theta)$. Thus we need to compute $O(\eta q\theta)$ matrix-vector multiplications. Instead of using successive matrix-vector multiplications to compute $\hat{\gamma}$, we could use the matrix squaring method and change the complexity from $O(\eta q\theta)$ to $O(|T|^3 \log(q\theta))$ [54]. However, this method results in fill-in (reducing sparsity) and is not feasible for CTMC with large state spaces. When $q\theta$ is large, computing the Poisson probabilities, especially near the tails of the distribution, may result in underflow problems [22]. This also causes round-off problems since the number of floating point operations needed is large.

We address some of the problems caused by large values of $q\theta$ from the practical point of view, by modifying Equation 5.2. This modification is based on recognizing when the underlying DTMC has reached steady state and rewriting the equations to avoid further computations. The total computation time becomes then proportional to the sub-dominant eigenvalue of the DTMC matrix rather than to $q\theta$. Thus, stiffness as seen by the modified algorithm is the same as that of using the *power method* [64] to compute the steady state solution. In our experience with a variety of problems, we have found significant improvement when using this modification.

We begin by observing that Equation 5.3, used to compute the probability vectors for the underlying DTMC, also represents the iteration equation of the power method used for computing the steady state solution of a CTMC. If the convergence of the power method is guaranteed, we can terminate the iteration in Equation 5.3 upon attaining steady state and obtain considerable computational savings. To ensure convergence in Equation 5.3, we require that

$$q > \max_{i \in T}\{|Q_{i,i}|\}$$

since this guarantees that the DTMC described by $\hat{Q}$ is aperiodic [25].

Assume that, observing the sequence $\hat{\gamma}(i)$, we establish that convergence has been achieved at the S-th iteration. Three cases arise: $S > k$, $l < S \leq k$, and $S \leq l$.

- $S > k$: steady state detection does not take place and $\pi(\theta)$ is computed using Equation 5.4.
- $l < S \leq k$: by substituting $\hat{\gamma}(i)$ with $\hat{\gamma}(S)$ for $i > S$, we can rewrite Equation 5.4 setting the right truncation point k to ∞:

$$
\begin{aligned}
\pi(\theta) &\approx \sum_{i=l}^{\infty} \hat{\gamma}(i) e^{-q\theta} \frac{(q\theta)^i}{i!} \\
&= \sum_{i=l}^{S} \hat{\gamma}(i) e^{-q\theta} \frac{(q\theta)^i}{i!} + \hat{\gamma}(S) \sum_{i=S+1}^{\infty} e^{-q\theta} \frac{(q\theta)^i}{i!} \\
&= \sum_{i=l}^{S} \hat{\gamma}(i) e^{-q\theta} \frac{(q\theta)^i}{i!} + \hat{\gamma}(S) \left(1 - \sum_{i=0}^{S} e^{-q\theta} \frac{(q\theta)^i}{i!} \right)
\end{aligned}
$$

- $S \leq l$: the DTMC reaches steady state before the left truncation point. In this case, no additional computation is necessary and $\pi(\theta)$ is set to $\hat{\gamma}(S)$.

For stiff problems, the number of terms needed to meet the truncation error tolerance requirements can be large, but substantial computational savings result if the DTMC steady state is detected. In our experience, this is often happens, especially for large values of θ.

The detection of steady state for the underlying DTMC requires extreme care. We have implemented the steady state detection based on the suggestions given in [64]. The usual test for convergence compares successive iterates, but if the method is converging slowly, the change between successive iterates might be smaller than the error tolerance specified. We might then incorrectly assume that the system has reached steady state when, instead, it is only experiencing slow convergence. To avoid this problem we compare iterates that are spaced m iterations apart, checking the difference between $\hat{\gamma}(i)$ and $\hat{\gamma}(i-m)$. Ideally, m should be varied according to the convergence rate, but this is difficult to implement in practice. For simplicity, we choose m based on the iteration number: $m = 5$ when number of iterations is less than 100, $m = 10$ when it is between 100 and 1000 and $m = 20$ when it is greater than 1000. We also test for steady state only every m iterations, saving computation.

5.3. Cumulative transient analysis. The computation of $\sigma(\theta)$ is similar to that of $\pi(\theta)$. A method similar to Jensen's method for solving Equation 4.8 is given in [55].

Integrating Equation 5.2 with respect to θ yields,

$$\sigma(\theta) = \frac{1}{q} \sum_{i=0}^{\infty} \hat{\gamma}(i) \sum_{j=i+1}^{\infty} e^{-q\theta} \frac{(q\theta)^j}{j!}$$

$$(5.5) \qquad = \frac{1}{q} \sum_{i=0}^{\infty} \hat{\gamma}(i) \left(1 - \sum_{j=0}^{i} e^{-q\theta} \frac{(q\theta)^j}{j!} \right)$$

This is again a summation of an infinite series which can be evaluated up to the first k significant terms [55] resulting in,

$$(5.6) \qquad \sigma(\theta) \approx \frac{1}{q} \sum_{i=0}^{k} \hat{\gamma}(i) \left(1 - \sum_{j=0}^{i} e^{-q\theta} \frac{(q\theta)^j}{j!} \right)$$

The error due to truncation can be bounded from above:

$$\frac{1}{q} \sum_{i=k+1}^{\infty} \sum_{j=i+1}^{\infty} e^{-q\theta} \frac{(q\theta)^j}{j!} \leq \frac{1}{q} \sum_{i=k+1}^{\infty} (i - (k+1)) e^{-q\theta} \frac{(q\theta)^i}{i!}$$

$$\leq \frac{1}{q} \sum_{i=k+1}^{\infty} i \, e^{-q\theta} \frac{(q\theta)^i}{i!} - \frac{1}{q} \sum_{i=k+1}^{\infty} (k+1) e^{-q\theta} \frac{(q\theta)^i}{i!}$$

$$\leq t \sum_{i=k}^{\infty} e^{-q\theta} \frac{(q\theta)^i}{i!} - \left(\frac{k+1}{q} \right) \sum_{i=k+1}^{\infty} e^{-q\theta} \frac{(q\theta)^i}{i!}$$

Given an error tolerance requirement ϵ, we can compute the number of terms k needed as:

$$k = \min \left\{ j \in \mathbb{N} : \theta \sum_{i=k}^{\infty} e^{-q\theta} \frac{(q\theta)^i}{i!} - \left(\frac{k+1}{q} \right) \sum_{i=k+1}^{\infty} e^{-q\theta} \frac{(q\theta)^i}{i!} \right\}$$

The detection of steady state for the underlying DTMC applies also to Equation 5.6. Two cases arise: $S > k$ and $S \leq k$.

- $S > k$: steady state detection does not take place and $\sigma(\theta)$ is computed using Equation 5.6.
- $S \leq k$: Equation 5.6 is modified as follows:

$$\hat{\sigma}(\theta) \approx \frac{1}{q} \sum_{i=0}^{\infty} \hat{\gamma}(i) \sum_{j=i+1}^{\infty} e^{-q\theta} \frac{(q\theta)^j}{j!}$$

$$= \frac{1}{q}\sum_{i=0}^{S}\hat{\gamma}(i)\sum_{j=i+1}^{\infty}e^{-q\theta}\frac{(q\theta)^{j}}{j!} + \frac{1}{q}\hat{\gamma}(S)\sum_{i=S+1}^{\infty}\sum_{j=i+1}^{\infty}e^{-q\theta}\frac{(q\theta)^{j}}{j!}$$

$$= \frac{1}{q}\sum_{i=0}^{S}\hat{\gamma}(i)\sum_{j=i+1}^{\infty}e^{-q\theta}\frac{(q\theta)^{j}}{j!}$$

$$+\frac{1}{q}\hat{\gamma}(S)\left(\sum_{i=0}^{\infty}\sum_{j=i+1}^{\infty}e^{-q\theta}\frac{(q\theta)^{j}}{j!} - \sum_{i=0}^{S}\sum_{j=i+1}^{\infty}e^{-q\theta}\frac{(q\theta)^{j}}{j!}\right)$$

$$= \frac{1}{q}\sum_{i=0}^{S}\hat{\gamma}(i)\sum_{j=i+1}^{\infty}e^{-q\theta}\frac{(q\theta)^{j}}{j!}$$

$$+\frac{1}{q}\hat{\gamma}(S)\left(qt - \sum_{i=0}^{S}\sum_{j=i+1}^{\infty}e^{-q\theta}\frac{(q\theta)^{j}}{j!}\right)$$

$$= \frac{1}{q}\sum_{i=0}^{S}\hat{\gamma}(i)\left(1 - \sum_{j=0}^{i}e^{-q\theta}\frac{(q\theta)^{j}}{j!}\right)$$

$$+\frac{1}{q}\hat{\gamma}(S)\left(qt - \sum_{i=0}^{S}\left(1 - \sum_{j=0}^{i}e^{-q\theta}\frac{(q\theta)^{j}}{j!}\right)\right)$$

6. Conclusion. Markov and Markov reward models are frequently used in performance and dependability analysis of discrete event systems. Models of real systems, however, tend to become extremely large. We have proposed the use of stochastic reward nets for the concise specification and automated generation of Markov reward models. We have presented a formal definition of stochastic reward nets and methods for their structural, temporal, and sensitivity analysis.

A number of papers have dealt with the application of the methods described in this paper to practical modeling problems [33,31,14,48,49]. Nevertheless, practical problems stretch the capabilities of current methods and tools. State truncation [36,48] and decomposition [16,68,11] methods are being investigated to solve the ubiquitous largeness problem.

Stochastic reward nets are extensions of stochastic Petri nets. The reader may consult [50,51,56] and the Springer-Verlag series, *Advances in Petri Nets* for further information on Petri nets. Stochastic extensions of PNs have been covered extensively in the proceedings of a series of IEEE workshops, *Petri Nets and Performance Models*.

REFERENCES

[1] A. V. Aho, J. E. Hopcroft, and J. D. Ullman. *The Design and Analysis of Computer Algorithms.* Addison-Wesley, Menlo Park, CA, USA, 1974.

[2] M. Ajmone Marsan, G. Balbo, A. Bobbio, G. Chiola, G. Conte, and A. Cumani. The effect of execution policies on the semantics and analyis of Stochastic Petri Nets. *IEEE Transactions on Software Engineering*, 15(7):832–846, July 1989.

[3] M. Ajmone Marsan, G. Balbo, and G. Conte. A class of Generalized Stochastic Petri Nets for the performance evaluation of multiprocessor systems. *ACM Transactions on Computer Systems*, 2(2):93–122, May 1984.

[4] G. Balbo, G. Chiola, G. Franceschinis, and G. Molinari Roet. On the efficient construction of the tangible reachability graph of generalized stochastic Petri nets. In *Proceedings of the IEEE International Workshop on Petri Nets and Performance Models*, Madison, WI, USA, Aug. 1987.

[5] F. Baskett, K. M. Chandy, R. R. Muntz, and F. Palacios-Gomez. Open, Closed, and Mixed networks of queues with different classes of customers. *Journal of the ACM*, 22(2):335–381, Apr. 1975.

[6] M. D. Beaudry. Performance-related reliability measures for computing systems. *IEEE Transactions on Computers*, C-27(6):540–547, June 1978.

[7] J. Bechta Dugan, K. S. Trivedi, R. M. Geist, and V. F. Nicola. Extended Stochastic Petri Nets: applications and analysis. In E. Gelenbe, editor, *Performance '84*, North-Holland, Amsterdam, 1985, pages 507-519.

[8] J. Bechta Dugan, K. S. Trivedi, M. K. Smotherman, and R. M. Geist. The Hybrid Automated Reliability Predictor. *AIAA Journal of Guidance, Control and Dynamics*, 9(3):319–331, May 1986.

[9] J. T. Blake, A. L. Reibman, and K. S. Trivedi. Sensitivity analysis of reliability and performance measures for multiprocessor systems. In *Proceedings of the 1988 ACM SIGMETRICS Conference on Measurement and Modeling of Computer Systems*, Santa Fe, NM, USA, pages 177-186, May 1988.

[10] G. Chiola. A software package for the analysis of Generalized Stochastic Petri Net models. In *Proceedings of the IEEE International Workshop on Timed Petri Nets*, Torino, Italy, July 1985.

[11] H. Choi and K. S. Trivedi. Approximate Performance Models of Polling Systems using Stochastic Petri Nets. In *Proceedings of the IEEE INFOCOM 92*, Florence, Italy, May 1992.

[12] G. Ciardo. *Analysis of large stochastic Petri net models*. PhD thesis, Duke University, Durham, NC, USA, 1989.

[13] G. Ciardo, R. A. Marie, B. Sericola, and K. S. Trivedi. Performability analysis using semi-Markov reward processes. *IEEE Transactions on Computers*, 39(10):1251–1264, Oct. 1990.

[14] G. Ciardo, J. Muppala, and K. S. Trivedi. Analyzing concurrent and fault-tolerant software using stochastic Petri nets. *Journal of Parallel and Distributed Computing*. To appear.

[15] G. Ciardo, J. Muppala, and K. S. Trivedi. On the solution of GSPN reward models. *Performance Evaluation*, 12(4):237–253, 1991.

[16] G. Ciardo and K. S. Trivedi. A decomposition approach for stochastic Petri net models. In *Proceedings of the Fourth IEEE International Workshop on Petri Nets and Performance Models (PNPM91)*, Melbourne, Australia, pages 74-83, Dec. 1991.

[17] G. Ciardo, K. S. Trivedi, and J. Muppala. SPNP: stochastic Petri net package. In *Proceedings of the Third IEEE International Workshop on Petri Nets and Performance Models (PNPM89)*, Kyoto, Japan, pages 142 - 151, Dec. 1989.

[18] A. Cumani. ESP - A package for the evaluation of stochastic Petri nets with phase-type distributed transitions times. In *Proceedings of the IEEE International Workshop on Timed Petri Nets*, Torino, Italy, July 1985.

[19] G. Dahlquist and A. Björck. *Numerical Methods*. Prentice-Hall, Englewood Cliffs, N.J., 1974.

[20] E. de Souza e Silva and H. R. Gail. Calculating availability and performability measures of repairable computer systems using randomization. *J. ACM*, 36(1):171–193, Jan. 1989.

[21] E. de Souza e Silva and R. R. Muntz. Queueing networks: solutions and applications. In H. Takagi, editor, *Stochastic Analysis of Computer and Communication Systems*. Elsevier Science Publishers B.V. (North-Holland), pages 319–400, 1990.

[22] B. L. Fox and P. W. Glynn. Computing poisson probabilities. *Commun. ACM.*, 31(4):440–445, Apr. 1988.

[23] P. M. Frank. *Introduction to System Sensitivity*. Academic Press, New York, NY, 1978.

[24] A. Goyal, W. C. Carter, E. de Souza e Silva, S. S. Lavenberg, and K. S. Trivedi. The System Availability Estimator. In *Proceedings of the Sixteenth International Symposium on Fault-Tolerant Computing*, pages 84–89, Vienna, Austria, July 1986.

[25] A. Goyal, S. Lavenberg, and K. S. Trivedi. Probabilistic modeling of computer system availability. *Annals of Operations Research*, 8:285–306, Mar. 1987.

[26] W. K. Grassmann. Means and variances of time averages in Markovian environments. *Eur. J. Oper. Res.*, 31(1):132–139, 1987.

[27] D. Gross and D. Miller. The randomization technique as a modeling tool and solution procedure for transient Markov processes. *Oper. Res.*, 32(2):926–944, Mar.-Apr. 1984.

[28] P. J. Haas and G. S. Shedler. Stochastic Petri net representation of discrete event simulations. *IEEE Transactions on Software Engineering*, 15(4):381–393, Apr. 1989.

[29] P. Heidelberger and A. Goyal. Sensitivity analysis of continuous time Markov chains using uniformization. In P. J. Courtois, G. Iazeolla, and O. J. Boxma, editors, *Computer Performance and Reliability* , North-Holland, Amsterdam, pages 93–104, 1988.

[30] R. A. Howard. *Dynamic Probabilistic Systems, Volume II: Semi-Markov and Decision Processes*. John Wiley and Sons, New York, NY, 1971.

[31] O. C. Ibe, H. Choi, and K. S. Trivedi. Performance Evaluation of Client-Server Systems. *IEEE Transactions on Parallel and Distributed Systems*. to appear.

[32] O. C. Ibe, R. C. Howe, and K. S. Trivedi. Approximate availability analysis of vaxcluster systems. *IEEE Trans. Reliability*, R-38(1):146–152, Apr. 1989.

[33] O. C. Ibe and K. S. Trivedi. Stochastic Petri net models of polling systems. *IEEE Journal on Selected Areas in Communications*, 8(9):1649–1657, Dec. 1990.

[34] A. Jensen. Markoff chains as an aid in the study of Markoff processes. *Skand. Aktuarietidskr.*, 36:87–91, 1953.

[35] K. Jensen. Coloured Petri nets and the invariant method. *Theoretical Computer Science*, 14:317–336, 1981.

[36] H. Kantz and K. S. Trivedi. Reliability Modeling of the MARS System: A Case Study in the Use of Different Tools and Techniques. In *Proceedings of the Fourth IEEE International Workshop on Petri Nets and Performance Models (PNPM91)*, Melbourne, Australia, Dec. 1991.

[37] J. Keilson. *Markov Chain Models — Rarity and Exponentiality*. Applied Mathematical Sciences Ser. Vol. 28. Springer-Verlag, 1979.

[38] F. P. Kelly. *Reversibility and Stochastic Networks*. Wiley, 1979.

[39] W. Kleinoder. Evaluation of task structures for hierarchical multiprocessor systems. In D. Potier, editor, *Modelling Techniques and Tools for Performance Analysis*. Elsevier Science Publishers B.V. (North Holland), 1985.

[40] S. S. Lavenberg, editor. *Computer Performance Modeling Handbook*. Academic Press, New York, 1983.

[41] E. D. Lazowska, J. Zahorjan, G. S. Graham, and K. C. Sevcick. *Quantitative System Performance*. Prentice-Hall, Englewood Cliffs, NJ, USA, 1984.

[42] V. W. Mak and S. F. Lundstrom. Predicting performance of parallel computations. *IEEE Transactions on Parallel and Distributed Systems*, 1(3):257–270, July 1990.

[43] M. Malhotra and K. S. Trivedi. Higher order methods for the transient analysis of Markov chains. In *Proc. Int. Conf. on the Performance of Distributed Systems and Integrated Communication Networks*, Kyoto, Japan, Sept. 1991.

[44] J. F. Meyer. Performability: a retrospective and some pointers to the future. *Performance Evaluation*, 14(3-4):139–156, 1992.

[45] C. Moler and C. F. V. Loan. Nineteen dubious ways to compute the exponential of a matrix. *SIAM Review*, 20(4):801–835, Oct. 1978.

[46] M. K. Molloy. *On the integration of delay and throughput measures in distributed processing models*. PhD thesis, UCLA, Los Angeles, CA, USA, 1981.

[47] J. K. Muppala. *Performance and Dependability Modeling Using Stochastic Reward Nets*. PhD thesis, Department of Electrical Engineering, Duke University, Durham, NC, Apr. 1991.

[48] J. K. Muppala, A. S. Sathaye, R. C. Howe, and K. S. Trivedi. *Dependability modeling of a heterogenous VAXcluster system using stochastic reward nets*. in: *Hardware and Software Fault Tolerance in Parallel Computing Systems*, D. Averesky (ed.), Ellis Horwood Ltd., 1992. to appear.

[49] J. K. Muppala, S. P. Woolet, and K. S. Trivedi. Real-time systems performance in the presence of failures. *IEEE Computer*, 24(5):37–47, May 1991.

[50] T. Murata. Petri Nets: properties, analysis and applications. *Proceedings of the IEEE*, 77(4):541–579, Apr. 1989.

[51] J. L. Peterson. *Petri Net Theory and the Modeling of Systems*. Prentice-Hall, Englewood Cliffs, NJ, USA, 1981.

[52] C. Petri. *Kommunikation mit Automaten*. PhD thesis, University of Bonn, Bonn, West Germany, 1962.

[53] S. Pissanetzky. *Sparse Matrix Technology*. Academic Press, Orlando, FL, USA, 1984.

[54] A. L. Reibman and K. S. Trivedi. Numerical transient analysis of Markov models. *Computers and Operations Research*, 15(1):19–36, 1988.

[55] A. L. Reibman and K. S. Trivedi. Transient analysis of cumulative measures of Markov model behavior. *Stochastic Models*, 5(4):683–710, 1989.

[56] W. Reisig. *Petri Nets*, volume 4 of *EATC Monographs on Theoretical Computer Science*. Springer-Verlag, New York, 1985.

[57] G. Rozenberg. Introduction to Petri nets, Dec. 1991. Tutorial Notes for the Fourth IEEE International Workshop on Petri Nets and Performance Models (PNPM91).

[58] R. A. Sahner and K. S. Trivedi. Performance and reliability analysis using directed acyclic graphs. *IEEE Transactions on Software Engineering*, SE-14(10):1105–1114, Oct. 1987.

[59] R. A. Sahner and K. S. Trivedi. Reliability modeling using SHARPE. *IEEE Transactions on Reliability*, R-36(2):186–193, June 1987.

[60] W. H. Sanders and J. F. Meyer. METASAN: a performability evaluation tool based on Stochastic Activity Networks. In *Proceedings of the ACM-IEEE Comp. Soc. Fall Joint Comp. Conf.*, Nov. 1986.

[61] M. L. Shooman. *Probabilistic Reliability: An Engineering Approach*. McGraw-Hill, New York, 1968.

[62] J. Sifakis. Use of Petri nets for performance evaluation. In H. Beilner and E. Gelenbe, editors, *Measuring, Modelling, and Evaluating Computer Systems*, pages 75–93. North Holland, 1977.

[63] R. M. Smith and K. S. Trivedi. The analysis of computer systems using Markov reward processes. In H. Takagi, editor, *Stochastic Analysis of Computer and Communication Systems*. Elsevier Science Publishers B.V. (North-Holland), 1990, pages 589-629.

[64] W. Stewart and A. Goyal. Matrix methods in large dependability models. Technical Report RC-11485, IBM T.J. Watson Res. Center, Yorktown Heights, NY, 10598, Nov. 1985.

[65] D. P. Stotts and P. Godfrey. Place/transition nets with debit arcs. *Information Processing Letters*, 41:25–33, Jan. 1992.

[66] G. Strang. *Introduction to Applied Mathematics*. Wellesley-Cambridge Press, Wellesley Massachusetts 02182, 1986.

[67] H. Tardif, K. S. Trivedi, and A. V. Ramesh. Closed-form transient analysis of Markov chains. Technical Report CS-1988, Dept. of Computer Science, Duke University, Durham, NC, 27706, June 1988.

[68] L. Tomek and K. S. Trivedi. Fixed-Point Iteration in Availability Modeling. In M. Dal Cin, editor, *Informatik-Fachberichte, Vol. 91: Fehlertolerierende Rechensysteme*. pages 229-240, Springer-Verlag, Berlin, 1991.

[69] K. S. Trivedi. *Probability & Statistics with Reliability, Queueing, and Computer Science Applications*. Prentice-Hall, Englewood Cliffs, NJ, USA, 1982.

[70] K. S. Trivedi, J. K. Muppala, S. P. Woolet, and B. R. Haverkort. Composite performance and dependability analysis. *Perf. Eval.*, 14(3-4):197–215, 1992.

MEANS AND VARIANCES IN MARKOV REWARD SYSTEMS*

WINFRIED K. GRASSMANN[†]

Abstract. In this paper, we study the total reward connected with a Markov reward process from time zero to time m. In particular, we determine the average reward within this time period, as well as its variance. Though the emphasis is on discrete time Markov processes, continuous-time reward processes will also be considered. For the discrete time reward process, the determination of the expected reward form 0 to m is of course trivial. It is of interest, however, that the deviation of this expectation from its steady state equivalent can be obtained from equations which are identical to the equations for the equilibrium probabilities, except that a vector of constants is added. We also consider the variance, both for transient systems and for systems in equilibrium. It is shown that the variance per time unit in equilibrium can also be obtained from equations which differ from the equations for the equilibrium probabilities only by a constant vector. Since there are three different sets of variables which satisfy similar equations, the LU factorization suggests itself. Instead of the LU factorization, we will use a UL factorization which reflects the probabilistic interpretation of the problem. This interpretation allows us to extend the factorization to systems with an infinite number of states as will be demonstrated, using the Wiener-Hopf factorization of the $GI/G/1$ queue as an example.

1. Introduction. There is an increasing interest in Markov reward process. Since all industrial decisions are based on costs and rewards, this is fully justified, and if anything, there is not enough rather than too much emphasis on reward processes. In the systems approach as used in engineering and management science, a state encapsules all information that is needed to predict the future, which makes any information about states in the past irrelevant. This implies that every stochastic system is automatically Markovian. Of course, to obtain a system in this sense, one often has to introduce supplementary variables.

We concentrate on discrete-time systems. There is a state space S, and at every time n, the system is in some state X_n, $X_n \in S$. Moreover, there is a function $g : S \to R$, the reward function. By changing the definition of g, this function can also be used to find other measures of interest, such as congestion figures, backlogs, and so on. If $Y_n = g(X_n)$, the rewards in the interval $[0, m)$ are given by $\sum_{n=0}^{m-1} Y_n$. We want to find the mean and the variance of this expression.

All decisions involving humans are are based on a finite planning horizons. To match this situation, one would need a transient analysis. However, many systems reach steady state fast enough that there is no appreciable difference between the transient and the steady state behavior,

---

* I gratefully acknowledge the support of the Natural Sciences and Engineering Research Council of Canada, Grant OPG0008112.

† Dept. of Computational Science, University of Saskatchewan, Saskatoon, Saskatchewan, Canada S7N 020

which means that a steady state analysis is sufficient. This motivates our first topic, which involves the determination of the difference between the reward in a transient and an equilibrium system. The second problem to be addressed is the variance of rewards within a finite time interval. We show that in the case of transient systems, this variance can be obtained from recursively calculated intermediate results. The computational effort required to do this is essentially the same as calculating transient probabilities. We also determine the cost variance per time unit over a large time interval.

It turns out that the equations for the steady state probabilities, the equation for finding the deviation from steady state, and the equations one has to solve to find the variance per time unit, are all identical, except for a constant vector. This suggests the use of a LU factorization. In fact, based on the GTH-algorithm [8], we give a UL-type factorization of the transition matrix which allows us to solve simultaneously all three equations in question.

We now give some definitions, and state some assumptions which will be used throughout the paper. As mentioned, the state at time n will be denoted by X_n, and we are looking at $Y_n = g(X_n)$, the reward at time n. The transition probabilities are denoted by p_{ij}, that is

$$p_{ij} = P\{X_{n+1} = j \mid X_n = i\} .$$

The transition matrix is denoted by P, that is, $P = [p_{ij}]$. The probability to be in state i at time n is π_i^n

$$\pi_i^n = P\{X_n = i\} .$$

Furthermore, we define

$$\pi^n = [\pi_i^n]^T$$

$$g = [g(i)]$$

$$\mu^n = \sum_{i \in S} g(i)\pi_i^n = \pi^n g$$

$$P_{ij}^k = \text{Entry of } P^k = P\{X_{n+k} = j \mid X_n = i\} .$$

We assume that the system is ergodic. This implies P_k is invertible, provided P_k is the matrix P with row k and column k deleted.

2. Deviation from Steady State. The question addressed in this section is how far $E(\sum_{n=0}^{m-1} Y_n)$ deviates from its steady state mean. To resolve this issue, consider first the Chapman-Kolmogorov equations, which lead to

$$(2.1) \qquad \pi^{n+1} = \pi^n P .$$

If Markov chain is ergodic, which means that it is non-separable and aperiodic, then there is a unique π with $\pi e = 1$ satisfying

$$(2.2) \qquad \pi = \pi P .$$

If $\mu^n = \pi^n g$ and $\mu = \pi g$, then the deviation of the reward in the period $[0, m)$ from its steady state can be expressed as

$$E\left(\sum_{n=0}^{m-1} Y_n\right) - m\mu = \sum_{n=0}^{m-1} (\mu_n - \mu) .$$

If we define $d^n = \pi^n - \pi$ and $R^m = \sum_{n=0}^{m-1} d^n$, we obtain

$$\sum_{n=0}^{m-1} (\mu_n - \mu) = \sum_{n=0}^{m-1} (\pi^n - \pi)g = \sum_{n=0}^{m-1} d^n g = R^m g .$$

To find an equation for d^n, we subtract (2.2) from (2.1) and obtain

$$(2.3) \qquad d^{n+1} = d^n P .$$

Since $d^n e = \pi^n e - \pi e = 0$

$$\sum_{n=0}^{m-1} d^n e = R^m e = 0 .$$

By taking the sum of (2.3) form 0 to $m - 1$, one obtains

$$\sum_{n=0}^{m-1} d^{n+1} = \sum_{n=0}^{m-1} d^n P$$

or

$$(2.4) \qquad R^m - d^0 + d^m = R^m P .$$

Since P_k is invertible, there is a unique solution R^m with $R^m e$ for each m. Since d^n goes to 0 as $n \to \infty$, the R^m converge toward a unique value R satisfying $Re = 0$ and

$$(2.5) \qquad R - d_0 = RP .$$

Obviously, R tends to be a pessimistic estimate for R^m.

3. Continuous Time Markov Reward Processes. Let $X(t)$ be the state in a continuous-time Markov chain. Since the system is continuous, $X(t)$ can change at any time t. The reward rate at time t is $Y(t) = g(X(t))$, and the reward from 0 to T is given as

$$\int_0^T Y(t)dt .$$

If $\pi_i(t) = P\{X(t) = i\}$, and if $\pi(t) = [\pi_i(t)]^T$, then the expected reward rate at time t is $\pi(t)g$. The expected reward from 0 to T becomes therefore

$$E\left(\int_0^T Y(t)dt \right) = \int_0^T \pi(t)dtg .$$

We propose to calculate $\pi(t)$ by Jensen's method [11], also known as randomization [4] [9]. Tis method uses the fact that there is a family of discrete-time Markov chains, parameterized by f, associated with every continuous-time Markov chain. If a_{ij}, $i \neq j$, are the transition rates of the continuous time Markov chain, one finds the corresponding p_{ij} of the discrete chain as a_{ij}/f. The probabilities p_{ii} are then calculated such that the sum across the row is one. Of course, f must be chosen in such a way that $p_{ii} \geq 0$ for all $i \in S$. This means that

$$f \geq \sum_{j \neq i} a_{ij} .$$

Given the π^n of any of the discrete time Markov chains, one can find the probability vector $\pi(t)$ representing the state in the associated continuous time Markov chain at time t as follows.

$$(3.1) \qquad \pi(t) = \sum_{n=0}^{\infty} \pi^n p(n; ft) .$$

Here, $p(n; ft)$ is the Poisson distribution with rate f. This equation suggests that one should choose as small a value for f as possible. To find the total reward from 0 to T, one integrates (3.1) from 0 to T which yields

$$\int_0^T \pi(t)dt = \sum_{n=0}^{\infty} \pi^n \int_0^T p(n; ft)dt = \sum_{n=0}^{\infty} \pi^n P(n; fT) .$$

Here, $P(n; fT)$ is defined as

$$\int_0^T p(n; ft)dt = \frac{1}{f} \sum_{r=n+1}^{\infty} p(r; fT) .$$

In some cases, the following alternative method is suggested to calculate the integral of $\pi(t)$.

$$\sum_{n=0}^{\infty}\sum_{r=0}^{m-1}\pi^r p(n; fT) .$$

However, as shown in Grassmann [1], this method always converges slower than the one derived earlier.

As in the discrete case, one can determine the values $d(t) = \pi(t) - \pi$, and $R = \int_0^\infty d(t)dt$. Since the techniques used to do this are identical to the ones in the discrete case, they will not be given here.

4. The Variance of the Reward in a Transient System. We now derive the variance of the rewards in $[0, m)$. When doing this, we concentrate on the discrete-time case, since the continuous-time case is discussed in [1] along similar lines. For further literature on this topic, see Whitt [13], and for the older literature, Reynolds [12]. One easily finds

$$\operatorname{Var}\left(\sum_{n=0}^{m-1} Y_n\right) = \sum_{n=0}^{m-1} \operatorname{Var}(Y_n) + 2\sum_{n=0}^{m-1}\sum_{r=0}^{n-1} \operatorname{Cov}(Y_r, Y_n) .$$

In principle, one can calculate this expression by calculating the covariances and variances first and plug them into the above formula. The covariances are of course given as

$$\operatorname{Cov}(Y_r, Y_n) = \sum\sum (g(i) - \mu_r)(g(j) - \mu_n)\pi_i^r P_{ij}^{n-r} , n > r .$$

However, this method is computationally very expensive for the large, but sparse matrices that typically arise when analyzing stochastic systems. We therefore propose the following solution. We define

$$S^m = \sum_{r=0}^{m-1} \operatorname{Cov}(Y_r, Y_m) = \sum_{j \in S} g(j) W_j^m$$

with

(4.1) $$W_j^m = \sum_{n=0}^{m-1}\sum_{i \in S} (g(i) - \mu_n)\pi_i^n P_{ij}^{m-n} .$$

The W_j^m can be calculated recursively as follows

(4.2) $$W_j^0 = 0$$

(4.3) $$W_j^m = \sum_{k \in S}(W_k^{m-1} + (g(k) - \mu_{m-1})\pi_k^{m-1})p_{kj} .$$

We prove by induction that (4.2) and (4.3) will give the correct W_j^m as defined by (4.1). For $m = 0$, the sum in question is empty, and (4.2) trivially implies (4.1). We now substitute W_k^{m-1} in (4.3) by (4.1) to obtain

$$W_j^m = \sum_{k\in S}\sum_{n=0}^{m-2}\sum_{i\in S}(g(i) - \mu_n)\pi_i^n P_{ik}^{m-n-1}p_{kj} + \sum_{k\in S}(g(k) - \mu_{m-1})\pi_k^{m-1}p_{kj}$$

$$= \sum_{n=0}^{m-2}\sum_{i\in S}(g(i) - \mu_n)\pi_i^n P_{ij}^{m-n} + \sum_{i\in S}(g(i) - \mu_{m-1})\pi_i^{m-1}p_{ij}$$

$$= \sum_{n=0}^{m-1}\sum_{i\in S}(g(i) - \mu_n)\pi_i^n P_{ik}^{m-n} .$$

Since this expression is equal to (4.1), the result follows by complete induction. Note that the number of operations needed to find W_j^n according to (4.3) is essentially the same as finding the π_j^n, a result which is quite remarkable.

5. The Variance in Steady State. In this section, we assume that X_0 has the invariant distribution. Moreover, we consider V, which is given by

$$V = \lim_{m\to\infty}\frac{1}{m}\text{Var}\left(\sum_{n=0}^{m-1}Y_n\right) .$$

V can be interpreted as the average variance over a long time period. In the case of independent random variables, V is equal to the variance of an individual obseration. Our analysis starts with the following result [10]

(5.1) $$V = \text{Var}(Y) + 2\sum_{n=1}^{\infty}\text{Cov}(Y_0, Y_n) .$$

V can be obtained from a vector A, and this vector is determined from an equation having the same structure as (2.5). This is now discussed. We define

$$\alpha_j^n = \sum_{i\in S}\pi_i^0(g(i) - \mu_0)P_{ij}^n$$

$$\alpha^n = [\alpha_j^n]^T$$

Once α^n is known, the covariance between Y_0 and Y_n becomes

$$\text{Cov}(Y_0, Y_n) = \sum_{i,j\in S}(g(i) - \mu_0)g(j)\pi_i^0 P_{ij}^n = \alpha^n g .$$

The α^n, $n \geq 0$, can be calculated recursively as follows.

$$\alpha^0 = [\pi_i^0 (g(i) - \mu_0)]$$

(5.2)
$$\alpha^{n+1} = \alpha^n P$$

In analogy with R^m and R we define

$$A^m = \sum_{n=1}^{m-1} \alpha^n .$$

$$A = \sum_{n=1}^{\infty} \alpha^n .$$

Summing (5.2) over n yields

$$\sum_{n=1}^{m-1} \alpha^{n+1} = \sum_{n=1}^{m-1} \alpha^n P$$

or

$$A^m - \alpha^1 + \alpha^m = A^m P .$$

This equation has the same structure as equation (2.5). Hence, if we can show that $A^m e = 0$ and that $\lim \alpha^m = 0$, we have proven that A^m reaches a limit. In this case, we have

(5.3)
$$A - \alpha^1 = AP .$$

First, $A^m e = 0$ if $\alpha^n e = 0$ for $n \geq 0$. We have

$$\alpha^0 e = \sum_{i \in S} \alpha_j^0 = \sum_{i \in S} \pi_i^0 (g(i) - \mu_0) = \mu_0 - \mu_0 = 0 .$$

It is easy to prove that $\alpha^n e = 0$ implies $\alpha^{n+1} = 0$. Hence, $\alpha^n e = 0$ for all $n > 0$, which implies $A^m e = 0$. Moreover, except for the unit eigenvalue, all eigenvalues of P are strictly less than 1. This means that α^n must converge, and since $\alpha^n e = 0$, it cannot converge to a vector proportional to the equilibrium probabilities. By default, it must converge to a vector of zeros. Hence, $\alpha^n \to 0$, and there is a unique solution A of (5.3) satisfying $Ae = 0$. As soon as A is found, V is given as

(5.4)
$$V = \sum_{i \in S} (g(i) - \mu_n)^2 \pi_i^n + 2Ag .$$

In the case of continuous-time Markov chains, a similar result can be obtained [2].

6. A UL Factorization. In the previous sections, we showed that the deviations of the rewards in $[0, m)$ from steady state are found as Rg, where R is found from (2.5). Similarly, V, the variance per time unit, is given by (5.4), where A can be found from (5.3). The three systems (2.1), (2.5), and (5.3) can be solved simultaneously by doing an LU factorization. Here, we use a modification of the LU factorization which has a stochastic interpretation, and which can be generalized to solve do the factorization in case the state space is infinite. In particular, we do the elimination starting from state N, and working our way backward to state zero, which is the reverse order normally used in linear algebra. This means that we really obtain a UL rather than a LU factorization. Moreover, U will be defined to be strictly upper diagonal, which yields a nicer probabilistic interpretation.

For the moment, we assume that the set of states is given by the integers between 0 and N. We also define values p_{ij}^n recursively, with $p_{ij}^N = p_{ij}$, and

$$(6.1) \qquad p_{ij}^n = p_{ij} + \sum_{m=n+1}^{N} p_{ij}^m Q_m^{-1} p_{mj}^m$$

$$(6.2) \qquad Q_m = 1 - p_{mm}^m = \sum_{j=0}^{m-1} p_{mj}^m .$$

The p_{ij}^n as defined here are distinct from the P_{ij}^n used earlier. The p_{ij}^n are the values one obtains by the GTH algorithm [8], which is a modification of Gaussian elimination. Moreover, the p_{ij}^n have a probabilistic meaning. To see this, we define a *path* to be a sequence of states, e.g. $s_1, s_2, \ldots s_n$. Furthermore, an n-terminated path is defined to be a path terminating as soon as it enters a state $j \le n$. p_{ij}^n is now the probability that an n-terminated path ends in j, given it starts in i [5]. We now define a strictly lower triangular matrix $L = [L_{ij}]$, and a strictly upper triangular matrix $U = [U_{ij}]$ as follows.

$$L_{ij} = p_{ij}^i, \ j < i \qquad U_{ij} = p_{ij}^j, \ j > i .$$

From (6.1), one obtains

$$p_{nj}^n = p_{nj} + \sum_{m=n+1}^{\infty} p_{nm}^m Q_m^{-1} p_{mj}^m$$

$$p_{in}^n = p_{in} + \sum_{m=n+1}^{\infty} p_{im}^m Q_m^{-1} p_{mn}^m$$

or

$$(6.3) \qquad L_{nj} = p_{nj} + \sum_{m=n+1}^{\infty} U_{nm} Q_m^{-1} L_{mj}$$

(6.4)
$$U_{in} = p_{in} + \sum_{m=n+1}^{\infty} U_{im} Q_m^{-1} L_{mn}$$

$$Q_m = \sum_{j=0}^{m-1} L_{mj} .$$

If Q is defined to be the diagonal matrix with the the elements of the diagonal given by the Q_m, then (6.3) and (6.4), together with the fact that $1 - Q_m = p_{mm}^m$ lead to the following result

(6.5)
$$U + (I - Q) + L = P + U Q^{-1} L .$$

This approach does not quite work for the upper left hand corner because it results in a division by zero. To fix that, we define $U_{00} = Q_{00} = 1$. To make the connection with the more traditional UL factorization, we write (6.5) as

$$I - P = (U Q^{-1} - I)(L - Q) .$$

Once $H = U Q^{-1}$ is obtained, one can use the normal LU theory to find the solutions of (2.1), (2.5) and (5.3). To find π from (2.1), we first calculate a measure π^*, which is proportional to π, and which has $\pi_0^* = 1$. As shown in [8], π_i^* is equal to the expected number of visits to state i between two visits to state 0. Starting with $\pi_0^* = 1$, one can apply normal back-substitution

(6.6)
$$\pi_j^* = \sum_{i=0}^{j-1} \pi_i^* H_{ij} .$$

Now

$$\pi_j = \pi_j^* / \sum_{i \in S} \pi_i^* .$$

To find R_j, we first must use the L matrix to calculate the constants. From the standard arguments of UL factorization, one obtains

$$\beta_j = d_j^0 + \sum_{m=j+1}^{\infty} \beta_m Q_m^{-1} L_{mj} .$$

The β_j are calculated, starting with $\beta_k = 0$, $k > N$, and working backward to β_0. At this stage, we have a problem because R_0 is unknown. We set

$$R_j = r_j + r_{0j} R_0$$

where r_j and r_{0j} are to be determined recursively. From $R_0 = r_0 + r_{00}R_0$, one finds $r_0 = 0$ and $r_{00} = 1$. Now

$$R_j = \beta_j + \sum_{i=0}^{j-1} R_i H_{ij} = \beta_j + \sum_{i=0}^{j-1} (r_i + r_{0i} R_0) H_{ij} \ .$$

This yields in turn

(6.7)
$$r_{0j} = \sum_{i=0}^{j-1} r_{0i} H_{ij}$$

and

$$r_j = \beta_j + \sum_{i=0}^{j-1} r_i H_{ij} \ .$$

By comparing (6.6) and (6.7) one finds

$$r_{0j} = \pi_j^*, \quad j \geq 0 \ .$$

Hence, we need only evaluate r_j, using equation (6.7). Once this is done, we use

$$0 = \sum_{j \in S} R_j = \sum (r_j + \pi_j^* R_0) = \sum_{j \in S} r_j + R_0 \sum_{j \in S} \pi_j^* \ .$$

Hence

$$R_0 = -\sum_{j \in S} r_j / \sum_{j \in S} \pi_j^*$$

and

$$R_j = r_j - \pi_j \sum_{i \in S} r_i \ .$$

The vector A can be found in exactly the same manner.

7. The $GI/G/1$ Paradigm. In this section, we show, by means of an example discussed in [7] and [3], how the theory presented can be generalised to the infinite state case. In principle, one is allowed to truncate the transition matrix, and the UL factorization of the truncated chain converges to the non-truncated one as the truncation point converges to infinity [6]. This fact can be used as a basis of factorization algorithms. Note that the convergence of the factorization does not imply the convergence of the probabilities calculated, and vice versa. Moreover, the main value of the idea of truncation is theoretical. The actual algorithms used for calculating the factorization may or may not use truncation. For details, see [3].

The example to be discussed arises in connection with the discrete $GI/G/1$ queue [7], but also in certain bulk queues. Specifically, we consider problems in which for all $j > 0$

$$p_{ij} = p_{i+k,j+k} = p_{j-i} \ .$$

In this case, one has [3]

$$L_{ij} = l_{i-j}, \quad j < i$$

$$U_{ij} = u_{j-i}, \quad j > i$$

$$q = \sum_{i=1}^{\infty} l_i \ .$$

Equations (6.3) and (6.4) now yield

$$l_j = p_{-j} + \sum u_m q^{-1} l_{m+j}$$

$$u_j = p_j + \sum_{m=1}^{\infty} u_{m+j} q^{-1} l_j \ .$$

This looks very much like the Wiener Hopf factorization of the $GI/G/1$ queue. It is in fact easy to show that l_j is a descending ladder height. According to the probabilistic interpretation of the p_{ij}^n, the $l_j = p_{n,n-j}^{n-j}$ are equal to the probabilities of starting at n and ending at $n - j$, and this is exactly the definition of a descending ladder height. Since the Wiener-Hopf factorization is unique, this proves that the factorization given in (6.5) is leads in fact to a Wiener-Hopf factorization.

The Wiener-Hopf factorization is also applicable for the continuous case. In fact, it is usually formulated for the continuous case. One has

$$l(t) = p(-t) + \int_{0+}^{\infty} u(x) q^{-1} l(t + x) dx$$

$$u(t) = p(t) + \int_{0+}^{\infty} u(t + x) q^{-1} l(t) dx$$

$$q = \int_{0+}^{\infty} l(x) dx \ .$$

In both the discrete state and the continuous state case, the functions l_i, u_i and q, respectively, $l(t)$, $u(t)$ and q contains all information needed to find R and A, and, with it, the variance of a time average. This shows that our theory can be extended to the infinite state case.

REFERENCES

[1] Grassmann, W. K. 1987. Means and variances of time averages in Markovian environments. EJOR 31, 132-139.

[2] Grassmann, W. K. 1987. The Asymptotic Variance of a Time Average in a Birth-Death Process Annals of Opertions Research 8, 165-174.

[3] Grassmann, W. K. 1985. The factorization of queueing equations and their interpretation. J. Opl. Res. Soc. 36, 1041-1050.

[4] Grassmann, W. K. 1977. Transient solutions in Markovian queueing systems. Comput. Oper. Res. 4, 47-53.

[5] Grassmann, W. K. and Heyman, D. P. 1990. Equilibrium distribution of block-structured Markov chains with repeating rows. J. Appl. Prob. 27 557-576.

[6] Grassmann, W. K. and Heyman, D. P. Computation of steady-state probabilities for infinite-state Markov chains with repeating rows. Preprint.

[7] Grassmann, W. K. and Jain, J.L. 1989. Numerical solutions of the waiting time distribution and the idle time distribution of the arithmetic $GI/G/1$ queue Operations Research 37, 141-150.

[8] Grassmann, W. K., Taksar, M. I. and Heyman, D. P. 1985. Regenerative analysis and steady state distributions for Markov chains. Operations Research 33, 1107-1116.

[9] Gross, D. and Miller, D. R. 1984. The randomization technique as a modeling tool and solution procedure for transient Markov processes. Operations Research 32, 343-361

[10] Jenkins, J.H. 1972. The relative efficiency of direct and maximum likelihood estimates of mean waiting time in the simple queue, $M/M/1$.

[11] Jensen, A. 1953. Markoff chains as an aid in the study of Markoff processes. Scand. Actuar. 36, 87-91.

[12] Reynolds, J. F. 1975. The covariance structure of queues and related processes-A survey of recent work. Adv. Appl. Prob. 7, 383-417.

[13] Whitt, W. 1989. Asymptotic formulas for Markov processes with applications to simulation. Preprint

A DIRECT ALGORITHM FOR COMPUTING THE STATIONARY DISTRIBUTION OF A P-CYCLIC MARKOV CHAIN

DANIEL P. HEYMAN*

Abstract. We consider the problem of computing the stationary distribution of a Markov chain when the block form of the transition matrix is cyclic. We show how to adapt Gaussain elimination to exploit this structure. When there are p blocks of approximately equal size, the savings in computational effort from exploiting this structure is a factor of about $3/p^2$.

1. Introduction. A finite, irreducible stochastic matrix P is p-*cyclic* if it can be put into the block-partioned form

$$
(1.1) \qquad P = \begin{pmatrix}
0 & A_1 & 0 & \cdots & 0 & 0 \\
0 & 0 & A_2 & \cdots & 0 & 0 \\
0 & 0 & 0 & \cdots & 0 & 0 \\
\cdots & \cdots & \cdots & \cdots & \cdots & \cdots \\
0 & 0 & 0 & \cdots & 0 & A_{p-1} \\
A_p & 0 & 0 & \cdots & 0 & 0
\end{pmatrix}.
$$

Let r_i be the number of rows, and c_i be the number of columns, in A_i. Each of the diagonal blocks must be square, so $r_1 = c_p$ and $r_i = c_{i-1}$, $2 \leq i \leq p$. The fact that P is stochastic implies that each A_i is also stochastic. We make no assumptions about the structure of the blocks; i.e., we treat them as dense.

The generator matrix of a continuous-time Markov chain is p-cyclic if it has the form in (1.1) with diagonal matrices on the principle diagonal, with the diagonal entries making each row sum equal to zero.

Matrices with these properties were shown to arise in a variety of discrete-event systems in Bonhoure (1990). Kontovasilis, Plemmons and Stewart (1991) show that the block SOR method converges in a sense more general than the usual for these matrices. Bonhoure, Dallery and Stewart (1991) compare various algorithms for computing the stationary distribution of a Markov chain, including LU decomposition and several iterative methods. When the off-diagonal blocks are approximately square, and p is near four, they found empirically that the LU factorization takes about $1/p$ of the time and space required when the p-cyclic structure is not exploited. Our analysis will make this observation manifest. Two other papers in these Proceedings concern p-cyclic Markov chains; Plemmons (1992) and W. Stewart (1992).

In this paper, we will apply the GTH algorithm presented in Grassmann, Taksar and Heyman (1985) to matrices with the form (1.1). This

---

* Bellcore, 331 Newman Springs Road, Red Bank, NJ 07701-7020.

will provide a stable way of computing the stationary distribution, and several aspects of the computations become transparent. We will also offer some brief remarks about choosing the best way to order the blocks.

1.1. The Algorithm. Let $P = (p_{ij})$ be the transition matrix of a Markov chain with states $1, 2, \cdots, N$, and let π be a stationary distribution of P, i.e.,

$$(1.2) \qquad \pi = \pi P \quad \text{and} \quad \sum_{i=1}^{N} \pi_i = 1.$$

When P is irreducible, the solution to (1.2) is unique.

1.2. Review of the GTH Algorithm. The GTH algorithm for obtaining π is as follows:

> 1. For $n = N, N - 1, \cdots, 2$, do the following:
>   - (a)   Let $S = \sum_{j=1}^{n-1} p_{nj}$.
>   - (b)   Let $p_{in} = p_{in}/S, \quad i < n$.
>   - (c)   Let $p_{ij} = p_{ij} + p_{in}p_{nj}, \quad i, j < n$.
> 2. Let $TOT = 1$ and $\pi_1 = 1$.
> 3. For $j = 2, 3, \cdots, N$ do the following :
>   - (a)   Let $\pi_j = p_{1j} + \sum_{k=2}^{j-1} \pi_k p_{kj}$.
>   - (b)   Let $TOT = TOT + \pi_j$.
> 4. Let $\pi_j = \pi_j/TOT, \ j = 1, 2, \cdots, N$.

This algorithm is Gaussian elimination that starts from the last equation and works towards the first equation. Part (a) of step 1 keeps the numerical errors in the computation of the pivot element small; see the analysis of a related algorithm in Stewart and Zhang (1991) and G. Stewart (1992) in this volume. Since the diagonal elements are not used and steps 1, 2, and 3 produce a positive solution that need not sum to one, the algorithm can be applied directly to the generator of a continuous-time Markov chain with the diagonal elements set to zero. Thus, the discrete- and continuous-time Markov chains both can be considered to have the form in (1.1).

1.3. Applying the GTH Algorithm to Cyclic Matrices. When the GTH algorithm is applied to matrices with the form in (1.1), parts (a) and (b) of step 1 need not be done when $n > r_1$ because $S = 1$. Part (c) can be ignored when either p_{in} or p_{nj} is zero, so in the first r_p iterations, part (c) is executed only for the elements in the (initially empty) block in position $(p - 1, 1)$ in the block-partitioned form. Each iteration affects c_1 columns and r_{p-1} rows, and it takes r_p iterations to eliminate block p. Thus, eliminating block p requires $r_1 r_p r_{p-1}$ flops, where a flop is the

computation in part (c); see Golub and Van Loan (1983). Thus, after r_p iterations, P has been transformed into P_p, where

$$
P_p = \begin{pmatrix}
0 & A_1 & 0 & \cdots & 0 & 0 \\
0 & 0 & A_2 & \cdots & 0 & 0 \\
\cdots & \cdots & \cdots & \cdots & & \cdots \\
0 & 0 & 0 & \cdots & A_{p-2} & 0 \\
B_{p-1} & 0 & 0 & \cdots & 0 & A_{p-1} \\
A_p & 0 & 0 & \cdots & 0 & 0
\end{pmatrix}.
$$

The computations in step 1 show that after r_p iterations, the submatrix of P_p that excludes the last r_p rows and columns is stochastic. Therefore it has the same form as (1.1) with $p - 1$ replacing p. Consequently, we can apply the same procedure on the rows corresponding to blocks $p - 1$, $p - 2, \cdots, 2$. When these computations are done, the total effort expended is $r_1 \sum_{k=2}^{p} r_k r_{k-1}$. At this stage in the computations, P has been transformed into P_2, where

$$
(1.3) \qquad P_2 = \begin{pmatrix}
B_1 & A_1 & 0 & \cdots & 0 & 0 \\
B_2 & 0 & A_2 & \cdots & 0 & 0 \\
B_3 & 0 & 0 & \cdots & 0 & 0 \\
\cdots & \cdots & \cdots & \cdots & \cdots & \cdots \\
B_{p-1} & 0 & 0 & \cdots & 0 & A_{p-1} \\
A_p & 0 & 0 & \cdots & 0 & 0
\end{pmatrix}.
$$

The matrix B_1 has no apparent special structure, so all three parts of step 1 need to be performed. Since this essentially Gaussian elimination, it requires $r_1^3/3$ flops, plus small multiples of r_1^2 and r_1.

The flops required to do part (a) of step 3 is less than the number of elements in B_1 for states $1, 2, \cdots, r_1$, and is the number of elements in A_{k-1} for the states that correspond to columns of these blocks. Both part (b) of step 3 and step 4 require N flops.

Thus, the time complexity of this algorithm, T say, is given by

$$
(1.4) \qquad T = \Theta \left[r_1 \left(\sum_{k=2}^{p} r_k r_{k-1} + \frac{r_1^2}{3} \right) \right].
$$

The fill-in, F say, is the space taken by the blocks B_j in (1.3), so

$$
(1.5) \qquad F = c_p(N - r_p) = r_1(N - r_p).
$$

When the blocks are all nearly the same size, $k = N/p$ say, then (1.4) shows that the time savings over Gaussian elimination that does not exploit the cyclic structure is about $3/p^2$, and the savings in fill-in is about $(p - 1)/p$, which is about $1/p$. When p is in the vicinity of four, the former is close to $1/p$, which is consistent with the empirical results in Bonhoure et al. (1991).

1.4. Matrix Interpretation of the Algorithm. The detailed calculations above can be compactly expressed in matrix terms. Part (c) of step 1 is $B_{k-1} = A_k A_{k-1}$ for $k > 2$, so

$$(1.6) \qquad\qquad B_1 = \prod_{k=1}^{p} A_k.$$

Partitioning π conformally with P, so $\pi = (\pi_1, \pi_2, \cdots, \pi_p)$, step 3 is

$$(1.7) \qquad\qquad \pi_1 = \pi_1 B_1,$$

and

$$(1.8) \qquad\qquad \pi_k = \pi_{k-1} A_{k-1}, \quad k = 2, 3, \cdots, p.$$

These equations show that the solution of (1.2) is attained when a positive solution of (1.7) is at hand. Equation (1.8) is just (1.2) applied to a matrix with the form in (1.1).

The iterations in (1.8) require an accurate solution of (1.7), which can be obtained with the GTH algorithm. The matrix-matrix products in (1.6) and the vector-matrix products in (1.8) are intrinsically suitable for vector-processing and SIMD parallel computers. Unpublished experiments by the author shows the same to be valid for the GTH algorithm, so it would not be a bottleneck.

Alternatively, Sheskin (1985) and Meyer (1989) show that B_1 is the Shur complement of 0 in the Northwest corner of P. That is, if we partition P into

$$P = \begin{pmatrix} 0 & u \\ v & Q \end{pmatrix}.$$

then

$$(1.9) \qquad\qquad B_1 = u(I - Q)^{-1}v.$$

It is easy to exploit the structure in (1.1) to obtain $(I - Q)^{-1}$, which we illustrate for the case $p = 5$. Here, $u = (A_1, 0, 0, 0)$, $v = (0, 0, 0, A_5)'$, and

$$Q = \begin{pmatrix} 0 & A_2 & 0 & 0 \\ 0 & 0 & A_3 & 0 \\ 0 & 0 & 0 & A_4 \\ 0 & 0 & 0 & 0 \end{pmatrix}.$$

Then

$$(I - Q)^{-1} = \begin{pmatrix} I & A_2 & A_2 A_3 & A_2 A_3 A_4 \\ 0 & I & A_3 & A_3 A_4 \\ 0 & 0 & I & A_4 \\ 0 & 0 & 0 & I \end{pmatrix},$$

and (1.9) yields (1.6).

2. Optimal Ordering of the States. We usually have a choice in numbering the states, so we can choose which states comprise block one. The cyclic ordering will then determine the blocks for the remaining states. Equations (1.4) and (1.5) indicate that it is beneficial to have r_1 as small as possible. Equation (1.4) suggests that we want the r_k's to alternate between large and small, and equation (1.5) shows that we want r_p to be as large as possible. It might not be surprising that the ordering that minimizes the computational complexity need not minimize the fill-in, as we show in the following example.

Example. Take $p = 4$ and let the blocks be named a, b, c, and d. The cyclic order is a before b before c before d before a. Set $r_a = 2$, $r_b = 20$, $r_c = 5$, and $r_d = 3$. A subscript on T or F will indicate which block is block 1. Only the dominant term of T will be considered. From (1.4) and (1.5) we obtain $F_a = 44 > 40 = F_c$ and $T_a < 153 < 310 < F_c$.

Since there are only p choices, and (1.4) only requires $p + 3$ multiplications and p additions, brute force is a reasonable way to calculate the minimum time-complexity ordering. Calculating the minimum fill-in ordering is even easier.

REFERENCES

[1] F. Bonhoure. *Solution numerique de chaines de Markov particulieres pour l'etude de systemes a evenements discrets.* Ph.D. thesis, Universite Pierre et Marie Curie, December, 1990.

[2] F. Bonhoure, Y. Dallery, and W. J. Stewart. *On the use of periodicity properties for the efficient numerical solution of certain Markov chains.* Technical Report MASI No. 91-40, Universite de Paris VI, June, 1991.

[3] W. K. Grassmann, M. I. Taksar, and D. P. Heyman. *Regenerative analysis and steady state distributions for Markov chains.* Oper. Res. 33 (1985), pp. 1107–1116.

[4] G. H. Golub and C. F. Van Loan. *Matrix Computations: First Edition.* The Johns Hopkins Press, Baltimore, 1983.

[5] K. Kontovasilis, R. J. Plemmons, and W. J. Stewart. *Block SOR for the computation of the steady state distribution of finite Markov chains with p-cyclic infinitesimal generators.* Linear Algebra Appl., 154-156 (1991), pp. 145–223.

[6] C. D. Meyer. *Stochastic complementation, uncoupling Markov chains, and the theory of nearly reducible systems.* SIAM Review 31 (1989), pp. 240–272.

[7] R. J. Plemmons. this volume 1992.

[8] T. J. Sheskin. *A Markov partioning algorithm for computing steady-state probabilities.* Oper. Res. 33 (1985), pp. 228–235.

[9] G. W. Stewart and G. Zhang. *On a direct method for nearly uncoupled Markov chains.* Numerische Mathematik 38 (1991), pp. 179–192.

[10] G. W. Stewart. this volume 1992.

[11] W. J. Stewart this volume 1992.

APPROXIMATE ANALYSIS OF A DISCRETE-TIME QUEUEING MODEL OF THE SHARED BUFFER ATM SWITCH

S. HONG, H.G. PERROS,* AND H. YAMASHITA[†]

Abstract. We model the shared buffer ATM switch as a discrete-time queueing system. The arrival process to each port of the ATM switch is assumed to be bursty and it is modelled by an Interrupted Bernoulli Process. The discrete-time queueing system is analyzed approximately. It is first decomposed into subsystems, and then each subsystem is analyzed separately. The results from the subsystems are combined together through an iterative scheme. The analysis of each subsystem involves the construction of the superposition of all the arrival processes to the switch. Comparisons against simulation data showed that the approximate results have a good accuracy.

Key words. ATM, shared buffer switch, bursty arrivals, interrupted Bernoulli process, discrete-time queueing, superposition, aggregation, approximations.

1. Introduction. Many ATM switch architectures have been proposed so far. These architectures can be classified into the following three categories: space-division, medium sharing, and buffer sharing (see Tobagi [19]). In this paper, we are concerned with modelling the shared buffer switch architecture. In this type of switch architecture, there is a single memory which is shared by all input and output ports. All incoming and outgoing cells are kept in the same memory. The size of the memory is fixed so that it corresponds to a specific cell loss. An example of this type of switch architecture is the Prelude architecture (see Devault, Cochennec, and Servel [4]). Also see Kuwahara et al [10], and Lee et al [11].

Analytic performance models of the shared buffer switch architecture have been proposed under the assumption of Bernoulli arrivals. These models are based on the analysis of a single Geo/D/1 infinite capacity queue. The results obtained from this queue are used to analyze the whole switch which is represented by N identical Geo/D/1 queues. In particular, Bruneel and Steyaert [3] obtained the N-fold convolution of the queue-length distribution of a Geo/D/1 queue. Using this convolution, they derived the queue-length distribution of N Geo/D/1 queues. Eckberg and Hou [5] proposed a heuristic method which involved the following two steps. First, derive the mean and variance of the number of cells in N Geo/D/1 queues. Then, choose a well known distribution, characterized by its first two moments, which closely approximates the real queue-length distribution in a shared buffer switch. For further details see Petit and Desmet [16]. See also Boyer et al [2], and Petit et al [15].

---

* Computer Science Department and, Center for Communication and Signal Processing, North Carolina State University, Raleigh, NC 27695-8206, U.S.A.

† Work done while on a sabbatical leave of absence at the Computer Science Dept. of N. C. state Univ., Mechanical Engineering Dept., Sophia University, Kioi-cho 7-1, Chiyoda-ku, Tokyo 102, Japan

In this paper, we model the shared buffer switch by a discrete-time queueing system which explicitly represents the input and output ports and the queueing of cells in the shared buffer. The arrival process to each input port is assumed to be an *Interrupted Bernoulli Process* (IBP). We analyze the queueing system approximately by decomposing it into subsystems. Each subsystem is analyzed separately. The results from the subsystems are combined together through the means of an iterative scheme.

The analysis of a subsystem is quite complex due to the large number of arrival processes. One way of analyzing a subsystem is to first obtain the superposition of all the arrival processes, and then analyze it assuming a single arrival process. This approach reduces the dimensionality of the problem. However, it requires the construction of the superposition process which is a fairly complex problem in itself. This problem has received a lot of attention in the open literature. One approach for obtaining the superposition process focuses on approximating it by a renewal process, see Albin and also Perros and Onvural [14]. Heffes and Lucantoni [7] proposed a method for superposing approximately voice sources using a *Markov Modulated Poisson Process* (MMPP). An alternative method for constructing an MMPP approximation to the superposition of identical bursty sources was proposed by Baiocchi et al [1]. Finally, Heffes [6] obtained an MMPP approximation to the superposition of different MMPP arrival processes using a set of simple expressions.

An alternative way to solving the superpostion problems is to analyze approximately all the arrival processes together with the queue where the arrivals wait. Such a system, typically, represents an ATM multiplexer. Various solutions have been proposed for the analysis of an ATM multiplexer, see for instance, Norros et al [13], Sengupta [18], and Nagarajan, Kurose, and Towsley [12]. For further reference, see Pujolle and Perros [17]. We note that most of the methodologies reported in the literature for the analysis of an ATM multiplexer have been obtained assuming identical arrival processes. However, this assumption is unrealistic since in a real ATM network it is highly unlikely that all patterns of arriving cells will be identical.

The remaing of the paper is organized as follows. In the following section, we describe in detail the queueing model of the shared buffer switch. In section 3, we present the approximation algorithm, and in section 4, we validate the approximation algorithm by comparing it against simulation data. Finally, the conclusions are given in section 5.

2. Model Description. An ATM switch routes incoming cells from input ports to their destination output ports. The mechanism to route incoming cells to their requested output ports can be implemented in various ways. In the shared buffer switch architecture, all incoming cells are stored in a shared buffer and their memory addresses are stored in address queues. There is one address queue per output port. All memory addresses in an address queue point to cells in the shared buffer which are destined for the same output port. The address queue can be implemented in different ways . For instance, an address queue can be implemented as a linked list (see Kuwahara et al [10]) or it can be stored in a FIFO buffer which is dedicated to a specific output port (see Lee et al [11]). The shared memory has a finite capacity. Also, limitations can be imposed on the size of each address queue. A cell will be lost if it arrives to find the shared buffer full. It will also be lost if the shared buffer is not full, but its address queue is full. The switch architecture is synchronized. Between two synchronization points any incoming cells that are in process of arriving at the input ports are written to the memory, and each output port transmits a cell (if there is one in its address queue).

An important factor that affects the performance of an ATM switch is the burstiness of the arrival process to each port. In this paper, each arrival process is assumed to be bursty and it is modelled by an *Interrupted Bernoulli Process* (IBP). The burstiness of an arrival process is measured by the squared coefficient of variation of the interarrival time. An IBP process may find itself either in the busy state or in the idle state. Arrivals occur in a Bernoulli fashion only when the process is in the busy state. No arrivals occur if the process is in the idle state. Let us assume that at the end of slot t the process is in the idle (or busy) state. Then, in the next slot $t + 1$ it will remain in the idle (or busy) state with probability q (or p), or it will change to busy (or idle) with probability $1 - q$ (or $1 - p$). The transitions between the busy and idle states are shown in figure 2.1. If during a slot the process is in the busy state, then with probability λ a cell will arrive during the slot, and with probability $1 - \lambda$ no arrival will occur. In this paper, we assume that $\lambda = 1$. That is, at every busy slot there is an arrival.

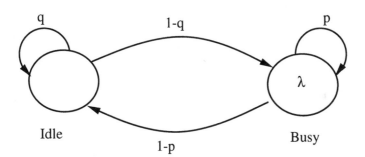

FIG. 2.1. *Transition between busy and idle states*

Let $\tilde{t}$ be the interarrival time of a cell. Then, it can be shown that the probability ρ that any slot containes a cell and the squared coefficient of variation of the interarrival time C^2 are as follows:

$$\rho = \frac{\lambda(1-q)}{2-p-q},$$

and

$$
\begin{aligned}
C^2 &= \frac{Var(\hat{t})}{E[\hat{t}]^2} \\
&= 1 + \lambda(\frac{(1-p)(p+q)}{(2-p-q)^2} - 1).
\end{aligned}
$$

For $\lambda = 1$, we have

(2.1) $$\rho = \frac{1-q}{2-p-q},$$

and

(2.2) $$C^2 = \frac{(p+q)(1-p)}{(2-(p+q))^2}.$$

We model the shared buffer switch by the queueing system shown in figure 2.2. The queueing model consists of N single server queues. Each of these queues represents an address queue in the shared buffer switch. The server of each queue represents an output port. There are N arrival streams, one per input port. Each arrival stream is assumed to be bursty and it is modelled as an IBP. We assume that the N arrival processes are heterogeneous. A cell upon arrival at the ith input port joins the jth queue with probability b_{ij}, where $\sum_{j=1}^{N} b_{ij} = 1$, $i = 1, 2, \cdots, N$. The total number of cells in all queues cannot exceed M, the buffer size of the swtich. Also, it is possible that the total number of cells in each queue i may be limited to M_i, where $M_i < M$. In this paper, we will assume that $M_i = M$. The case where $M_i < M$ can be easily incorporated.

All servers are synchronized so that they start and end service at the same time. The service time is assumed to be equal to one slot. At the end of each slot, a cell from each queue (provided that the queue is not empty) departs. Obviously, at most N cells can depart. Each arrival process is also slotted, with a slot size equal to the service time. (That is, the speed of an incoming link is assumed to be equal to the speed of an outgoing link.) We assume that the slot boundaries of the arrival processes lie somewhere between the service time slot boundaries as shown in figure 2.3. An arriving cell to an empty queue cannot start its service until the begining of the next service slot.

A continuous-time version of this queueing model was analyzed by Yamashita, Perros, and Hong [20]. The analysis presented here is similar in spirit to the analysis presented in [20]. Also, we note that under the assumption of exponential service times and exponential interarrival times, this queueing system is a truncated process of a reversible queueing system and, therefore, it has a product-form solution (see Kelly [8]).

Incoming links   Output ports

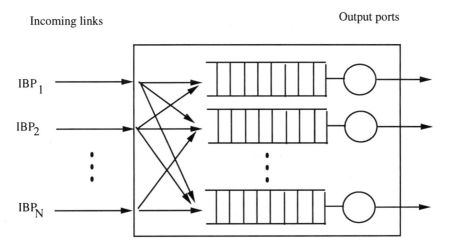

FIG. 2.2. *The queueing model of the shared buffer switch*

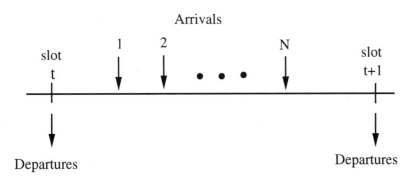

FIG. 2.3. *Relation between arrivals and departures*

3. The Approximation Algorithm. The state of our model is described immediately after the beginning of each slot by the vector $(\underline{w}; \underline{n}) = (w_1, ..., w_N; n_1, ..., n_N)$, where w_i is the state of the ith IBP arrival process, $w_i = 0, 1$, and n_i is the number of cells in the ith queue, $n_i = 0, 1, 2, ..., M$, such that $\sum_{i=1}^{N} n_i \leq M$. If $w_i = 0$, then the IBP arrival process is in the idle state, and if $w_i = 1$, then the process is in the busy state. Based on this state description, we can compute the transition probability P_{ij} from state i to state j, and subsequently the matrix of transition probabilities $P = [P_{ij}]$. Finally, solving the steady-state equation $\underline{\pi} P = \underline{\pi}$, where $\underline{\pi}$ is the steady-state probability vector, we can obtain the exact probabilities of all states numerically.

This numerical method can only be used to solve small systems. It becomes intractable as the state space increases. In this paper, we analyze this queueing system using the notion of decomposition. The queueing system is decomposed into subsystems, and then each subsystem is analyzed separately. The results from the subsystems are combined together through the means of an iterative scheme. Furthermore, in order ot reduce the complexity of each subsystem, we aggregate the N arrival processes to a single superposition. The main steps of the approximation algorithm are as follows.

We first consider in isolation the transition matrix of the N heterogeneous arrival processes. The states of this matrix are described by the vector $\underline{w} = (w_1, ..., w_N)$, where $w_i = 0, 1$ and $i = 1, 2, ..., N$. This transition matrix may be very large, seeing that it consists of 2^N states. We aggregate this transition matrix to a considerably smaller transition matrix involving $N + 1$ states, where each aggregate state k gives the number of the arrival processes which are in the busy state, $k = 0, 1, ..., N$. This aggregate matrix describes the superposition of the N arrival processes. If the superposition process is in state k, then there are k active arrival processes and consequent k cells will arrive (since each arrival is an on-off process).

We now consider the queueing system assuming a single arrival stream of cells, described by the above superposition process. The queueing system is analyzed by descomposing it into N subsystems. Each subsystem comprises one queue of the queueing system. The remaining queues are represented in the subsystem indirectly by simply keeping count of the total number of cells. As it will be shown below, the state space of a subsystem is not very big. Consequently, each subsystem is analyzed numerically. The N subsystem are combined together using an iterative procedure. We now proceed to describe the approximation algorithm in detail.

3.1. Aggregation of the N Arrival Processes. Let us consider the N heterogeneous arrival processes in isolation. The state of this system is described by the vector $\underline{w} = (w_1, w_2, ..., w_N)$. Let $r(\underline{w} \rightarrow \underline{w}')$ be the transition probability from state $\underline{w}$ to state $\underline{w}'$. Define S_i to be the set of all states $\underline{w}$ in which there are exactly i arrival processes in the busy state. Using this definition we can lump the state space into $N + 1$ sets of states S_i, $i = 0, 1, 2, ..., N$. The transition matrix with the lumped states is shown in figure 3.1.

Let R_{ij} be the transition probability from lump i to lump j. Then, we have

$$(3.1) \qquad R_{ij} = \sum_{\underline{w} \in S_i} Pr[\underline{w}|S_i][\sum_{\underline{w}' \in S_j} r(\underline{w} \rightarrow \underline{w}')],$$

where

$$Pr[\underline{w}|S_i] = \frac{Pr[\underline{w}]}{Pr[S_i]}$$

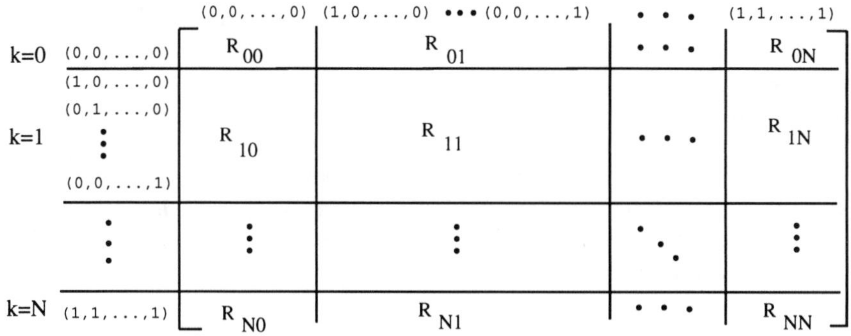

		$(0,0,\ldots,0)$	$(1,0,\ldots,0)$	$\bullet\bullet\bullet$	$(0,0,\ldots,1)$	$\bullet\quad\bullet$		$(1,1,\ldots,1)$
k=0	$(0,0,\ldots,0)$	R_{00}	R_{01}			$\bullet\quad\bullet\quad\bullet$		R_{0N}
k=1	$(1,0,\ldots,0)$ $(0,1,\ldots,0)$ $\vdots$ $(0,0,\ldots,1)$	R_{10}	R_{11}			$\bullet\quad\bullet\quad\bullet$		R_{1N}
		$\vdots$	$\vdots$		$\vdots$		$\ddots$	$\vdots$
k=N	$(1,1,\ldots,1)$	R_{N0}	R_{N1}			$\bullet\quad\bullet\quad\bullet$		R_{NN}

FIG. 3.1. *The transition matrix with the lumped states*

and

$$Pr[S_i] = \sum_{\underline{w} \in S_i} Pr[\underline{w}].$$

Seeing that the N arrival processes are independent from each other, we have that $Pr[\underline{w}] = Pr[w_1] \cdot Pr[w_2] \cdots Pr[w_N]$, where

$$Pr[w_i = 0] = \frac{1 - p_i}{2 - (p_i + q_i)},$$

and

$$Pr[w_i = 1] = \frac{1 - q_i}{2 - (p_i + q_i)}.$$

Thus, we have

$$Pr[\underline{w}] = [\prod_{i \in W_0} \frac{1 - p_i}{2 - (p_i + q_i)}][\prod_{i \in W_1} \frac{1 - q_i}{2 - (p_i + q_i)}],$$

where for a given state $\underline{w}$, W_0 and W_1 are the sets of the arrival processes which are in the idle state and the busy state respectively.

The generation of the aggregate transition matrix (hereafter denoted as A) is a time consuming operation because we first have to generate the transition matrix associated with the N arrival processes and subsequently calculate the transition probabilities R_{ij}, $i, j = 0, 1, 2, \ldots, N$.

The computational effort for the generation of the aggregate matrix A can be significantly reduced by observing that the parameters p and q of a bursty IBP process are very likely to be close to 1. For instance, in the case of voice, the squared coefficient of variation of the inter-arrival time C^2 is 18.1(see Heffes and Lucantoni [7]). Assuming a link utilization of 0.1, and using equation(2.1) and (2.2), we find that $p = 0.922$ and $q = 0.991$. The higher the value of C^2, the closer to 1 p and q get. It is highly unlikely, therefore, that a bursty source will change its current state at the next slot, seeing that $1 - p \simeq 0$ and $1 - q \simeq 0$. Based on this argument, we neglect all the transitions where more than two arrival processes change their states in the next slot. This results in great savings when it comes to generate the transition matrix of the N arrival processes. As it will be seen in the validation section, neglecting these transtions does not affect the accuracy of the results. The structure of the non-zero elements of the resulting aggregate matrix A is shown in figure 3.2.

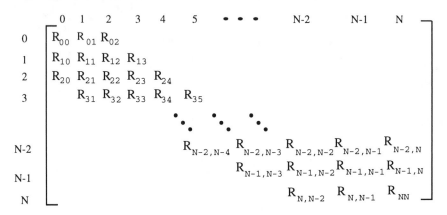

FIG. 3.2. *Aggregate transition matrix A*

3.2. Analysis of a Subsystem. As it was mentioned earlier on, we analyze the queueing system under study by decomposing it into N sub-systems, one per queue. In this section, we describe how a subsystem is analyzed. The subsystems are combined together using the iterative procedure described in section 3.3.

Let us consider the ith queue. The remaining queues will be denoted by the symbol $N - \{i\}$. In order to study the ith queue in isolation, we need to know how many cells are in $N - \{i\}$. Consequently, we analyze queue i by studying a subsystem of the original queueing system which consists of the following: (a) the arrival stream characterized by the superposition process, (b) queue i, and (c) the number of cells in $N - \{i\}$. The queueing structure of a subsystem i is shown in figure 3.3. The state of a subsystem i is given by the vector (k, n_i, n), where k is the state of the superposition process, n_i is the number of cells in the ith queue, and n is the number of cells in the entire system, i.e., $n - n_i$ is the number of cells in the remaining queues $N - \{i\}$. The total number of states in the subsystem is $\frac{(N+1)(M+1)(M+2)}{2}$. The synchronization of the departures and arrivals in a subsystem is the same as in the original queueing system. This is shown in figure 2.3. For presentation purposes, we shall refer to time t as being immediately after the end of slot t.

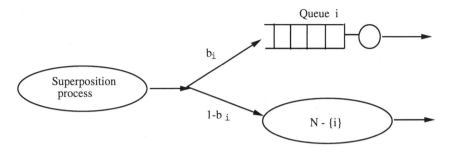

FIG. 3.3. *Queueing structure of subsystem i*

Each subsystem is analyzed numerically by first generating its undelying transition matrix and then solving the resulting system of linear equations to obtain its stationary probability vector. The automatic generation of the states of a subsystem is not discussed here since it is a straightforward procedure. Below, we describe how the one-step transition probabilities are calculated. A one-step transition from state (k, n_i, n) to state (k', n'_i, n') takes place in subsystem i if one or more of the following events occur:

1. state change of the arrival process
2. cell arrivals to queue i
3. cell arrivals to $N - \{i\}$
4. cell departures from queue i
5. cell departures from $N - \{i\}$

In case (1), the state of the arrival process is described by the number k of the arrival processes which are in busy states. The transition probability from k to k' is $R_{kk'}$ and it is given by expression (3.1). Let us consider now cases (2) and (3). Since there are k arrival processes which are in busy state at time t, k cells will arrive to queue i or to the remaining queues $N - \{i\}$ between time t and time $t + 1$. If there is not enough space in the shared buffer to accomadate k cells, some of them will be lost. The probability that a cell joins queue i or $N - \{i\}$ is computed using the branching probabilities b_{ji} $j, i = 1, ..., N$, i.e., the probability that a cell arriving at input port j is routed to output port i. We assume that every cell has the same branching probabilities regardless of the input port at which it arrived. Thus, with probability b_i, ,where $b_i = \sum_{j=1}^{N} b_{ji}$, an arriving cell joins the ith queue. In case (4), a cell always depart at the end of a slot if queue i is not empty with probability 1.

In case (5), the number of cells that depart from the remaining queues $N - \{i\}$ depends on how many of these queues are busy. From the state description (k, n_i, n), we only know that there are $n - n_i$ cells in $N - \{i\}$. We do not know how these cells are distributed over the $N - \{i\}$ queues. For instance, all the cells could be in the same queue. In this case, only one cell will depart at the end of the next slot. On the other hand, all $N-1$ queues may contain at least one cell (provided that $n-n_i > N-1$). In this case, there will be $N-1$ departures. In order to calculate the transition probabilities out of state (k, n_i, n) due to cell departures from $N - \{i\}$, we use the probability distribution $P_{\alpha_i}(l)$ that there are l busy queues in $N - \{i\}$, given that there are $\alpha_i = n - n_i$ cells in $N - \{i\}$. This probability is obtained approximately as follows.

As it will be described below in section 3.3, at each iteration we will have an estimate of the mean number of non-empty queues, $f_i(\alpha_i)$, in $N - \{i\}$ for subsystem i, given that there are α_i cells in $N - \{i\}$. For each α_i, $\alpha_i = 1, 2, \cdots, M$, the unknown probability distribution $P_{\alpha_i}(l)$ satisfies the following constraints:

$$(3.2) \qquad \sum_{l=1}^{N-1} P_{\alpha_i}(l) = 1$$

$$(3.3) \qquad \sum_{l=1}^{N-1} l P_{\alpha_i}(l) = f_i(\alpha_i),$$

where we assume that $P_{\alpha_i}(l) = 0$ if $l > \alpha_i$.

Since only the first moment of the probability distribution is given, there is an infinite number of probability distributions that satisfy constraint (3.3). We choose a distribution whose first moment matches $f_i(\alpha_i)$ using the notion of maximum entropy (see Kouvatsos and Xenios [9] and the references within). The maximum entropy distribution is the distribution that maximizes the entropy function $\sum_{l=1}^{N-1} P_{\alpha_i}(l) ln(P_{\alpha_i}(l))$ subject to constraints (3.2) and (3.3). Using the method of the Lagrange's undetermined multipliers we obtain:

$$(3.4) \qquad P_{\alpha_i}(l) = \frac{1}{Z} e^{-l\beta},$$

where Z is the normalizing constant given by

$$(3.5) \qquad Z = e^{-\beta_0} = e^{-\beta} + e^{-2\beta} + \cdots + e^{-N\beta},$$

and β_0 is the Lagrange multiplier associated with the normalizing constraint (3.2). It can be easily shown that the Lagrangian multiplyer β satisfies the following relation:

$$(3.6) \qquad -\frac{\partial \beta_0}{\partial \beta} = f_i(\alpha_i).$$

β is determined from equation (3.6) numerically. Therefore, the probability distribution, $P_{\alpha_i}(l)$, can be obtained using eqations (3.4), (3.5) and (3.6).

The algorithm that computes the transition probability from (k, n_i, n) to (k', n_i', n'), $Pr\{(k, n_i, n) \to (k', n_i', n')\}$, is summarized below. We use the following notation: a_i and a_R are the number of cells that arrive at queue i and $N - \{i\}$ respectively, a_N is the number of cells that enter the whole system without blocking, n_R is the number of cells in $N - \{i\}$, d_i and d_R are the number of cells to depart from queue i and $N - \{i\}$ respectively. All random variables are considered at time t except n_R', and a_R' which are considered at the time $t + 1$.

Algorithm

step 1: $a_i = n_i' + d_i - n_i$, where d_i is 1 if queue is non-empty, 0 otherwise.

step 2: *For* $d_R = 1$ *to* $\mathrm{Min}(N\text{-}1, n_R)$

$\quad$ *begin*

$\qquad a_R' = n_R' + d_R - n_R$

$\qquad$ *if* $a_i + a_R = k$ (if cells are not blocked) *then*

$\qquad Pr\{(k, n_i, n) \to (k', n_i', n')\} = R_{kk'} Pr\{d_R\}$

$$\times \left[\binom{k}{a_i} b_i^{a_i}(1 - b_i)^{k - a_i} \right].$$

$\qquad$ *if* $a_i + a_R < k$ (if cells are blocked) *then*

$\qquad a_N = a_i + a_R$

$\qquad Pr\{(k, n_i, n) \to (k', n_i', n')\} = R_{kk'} Pr\{d_R\}$

$$\times \left[\binom{k}{a_i} b_i^{a_i}(1 - b_i)^{k - a_i} \frac{\binom{a_i}{a_i}\binom{k - a_i}{a_R}}{\binom{k}{a_N}} \right.$$

$$+ \binom{k}{a_i + 1} b_i^{a_i + 1}(1 - b_i)^{k - a_i - 1} \frac{\binom{a_i + 1}{a_i}\binom{k - a_i - 1}{a_R}}{\binom{k}{a_N}}$$

$$\vdots$$

$$\left. + \binom{k}{k - a_R} b_i^{k - a_R}(1 - b_i)^{a_R} \frac{\binom{k - a_R}{a_i}\binom{a_R}{a_R}}{\binom{k}{a_N}} \right]$$

$\quad$ *end*;

Once the transition probability matrix is set up, the stationary probability vector is obtained numerically.

3.3. The Iterative Procedure. Each subsystem i is analyzed in isolation if we know the mean number of non-empty queues $f_i(\alpha_i)$ in $N - \{i\}$, given that there are α_i cells in $N - \{i\}$. The results obtained from all the subsystems are combined through an iterative procedure. Each iteration involves the solution of the N subsystems in isolation. At the end of each iteration a convergence test is carried out. If the procedure has not converged, the values of $f_i(\alpha_i)$, $\alpha_i = 1, 2, \cdots, M$, $i = 1, 2, \cdots, N$ are adjusted and another iteration is carried out. For each subsystem i, $f_i(\alpha_i)$ satisfies the following conditions:

1. $f_i(1) = 1$
2. $f_i(\alpha_i + 1) > f_i(\alpha_i)$, $\alpha_i = 1, 2, ..., M$
3. $f_i(\alpha_i + 1) - f_i(\alpha_i) > f_i(\alpha_i + 2) - f_i(\alpha_i + 1)$
4. $f_i(\alpha_i) < min[n - 1, \alpha_i]$

$f_i(\alpha_i)$ is calculated as follows. Let $g_i(\alpha_i)$ be the mean number of non-empty queues in queue $N - \{i\}$ when α_i cells arrive at the entire queueing system assuming that the service time at each queue is infinite. $g_i(\alpha_i)$ is an upper bound of $f_i(\alpha_i)$, and it is given by the following expression.

$$(3.7) \qquad g_i(\alpha_i) = \sum_{l=1, l \neq i}^{n} [1 - (\frac{\sum_{j=1, j \neq i, l}^{n} \rho_j}{\sum_{j=1, j \neq i}^{n} \rho_j})^{\alpha_i}].$$

where ρ_j is the average arrival rate of the jth arrival process (given by expression (2.1)). Clearly $1 \leq f_i(\alpha_i) \leq g_i(\alpha_i)$. $f_i(\alpha_i)$ is approximated as follows:

$$(3.8) \qquad f_i(\alpha_i, \beta) = 1 + \beta[g_i(\alpha_i) - 1],$$

where $0 \leq \beta \leq 1$. Note that $g_i(\alpha_i)$ satisfies the above four conditions, and so does $f_i(\alpha_i, \beta)$. We estimate $f_i(\alpha_i)$ iteratively by adjusting β up and down accordingly in order to meet a convergence criterion.

In this study, we use the mean number of cells in the entire queueing system as a convergence criterion because it is computationally simple. Since we obtain the stationary probabilities $P_i(k, n_i, n)$ for each subsystem i, the mean number of cells in the entire queueing system can be computed in the following two different ways:

$$(3.9) \qquad E_1 = \sum_{i=1}^{N} \sum_{n_i=1}^{M} n_i P_i(n_i)$$

$$(3.10) \qquad E_2 = \frac{1}{N} \sum_{i=1}^{N} \sum_{n=1}^{M} n P_i(n)$$

where $P_i(n_i)$ is the marginal probability that there are n_i cells in the ith queue, and $P_i(n)$ is the marginal probability that there are n cells in the entire queueing system as calculated from subsystem i. These two probabilities are obtained as follows:

$$P_i(n_i) = \sum_{k=0}^{N} \sum_{n=n_i}^{M} P_i(k, n_i, n),$$

and

$$P_i(n) = \sum_{k=0}^{N} \sum_{n_i=0}^{n} P_i(k, n_i, n).$$

Convergence occurs if $|E_1 - E_2| \leq \epsilon$. If convergence has not occured, we adjust β and do another iteration. In particular, we note that $E_1 - E_2$ is a monotonically increasing function of $f_i(\alpha_i)$. Thus, we can decide whether the $f_i(\alpha_i)$ is overestimated or underestimated depending on which mean value is larger, and accordingly change $f_i(\alpha_i)$ in order to reduce the difference E_1 and E_2. Below, we summarize the approximation algorithm. The superscript s denotes the iteration number.

Algorithm

step 0: Set $\beta_L^0 = 0$ and $\beta_H^0 = 1$ and $s = 0$
step 1: $s = s + 1$, $\beta_M^S = (\beta_H^{S-1} + \beta_L^{S-1})/2$ and calculate $f_i^S(\alpha, \beta_M^S)$ for $\alpha=1,...,$M and i=1,2,...,N using (3.7) and (3.8).
step 2: Calculate the probability distribution of nonempty queues, $P_{\alpha_i}(l)$ using (3.4), (3.5), and (3.6) for $l = 1, 2, \cdots, N - 1$.
step 3: Set-up the transition probability matrix for the each subsystem using the algorithm described in section 3.2, and calculate the stationary vector of $P_i(k, n_i, n)$ using the SOR method for $i = 1, 2, ..., N$.
step 4: Calculate E_1^S and E_2^S using (3.9) and (3.10) respectively.
step 5: If $|E_1 - E_2| \leq \epsilon$ then stop.
     if $E_1 - E_2 > \epsilon$ then set $\beta_H^S = \beta_M^S$, $\beta_L^S = \beta_L^{S-1}$ and go to step 1.
     if $E_1 - E_2 < -\epsilon$ then $\beta_H^S = \beta_H^{S-1}$, $\beta_L^S = \beta_M^S$ and go to step 1.

Due to the nature of this iterative procedure it is possible that the quantities $f_i(\alpha_i)$ may converge before $|E_1 - E_2| < \epsilon$. In this case, no further improvement on the approximation results can be achieved. Consequently, for convergence it is important to also check if $f_i(\alpha_i)$ have converged. In order to simplify the convergence criteria, it was found emprically that it suffices to check if the absolute value of the relative error between E_1 and E_2 is less than ε, i.e. $|\frac{E_1 - E_2}{E_1}| \leq \varepsilon$.

4. Validation. In this section, we discuss the validation of the approximation algorithm described in section 3 for the analysis of the shared buffer switch. The approximation algorithm was used to analyze an 8×8 and a 16×16 shared buffer switch. The approximation results were compared against simulation data. Representative results are summarized in tables 5.2 to 5.5. Each table gives approximate and simulation results for the global queue-length distribution $P(n)$, $n = 0, 1, 2, \cdots, M$, and for the queue-length distribution $P_1(n_1)$, $n_1 = 0, 1, 2, \cdots, M$, of queue 1 which is the most heavily utilized queue (hot spot). Approximate and simulation results for the mean number of cells, m.q.l., in the switch is also given. The absolute errors for $P(n)$ and $P_1(n_1)$ given in tables 5.2 to 5.5 have been plotted in figures 5.1 to 5.4 respectively. The parameters of the arrival processes are given in table 5.1.

In general, the approximation algorithm gives good results. The approximation results for the queue-length distribution $P_i(n_i)$ of each queue i seem to be slightly more accurate than the global queue-length distribution $P(n)$. CPU complexity problem were encounter when $N > 20$. The case of $N = 8$ can be done efficiently on a workstation. The case $N = 16$ required a mainframe system.

5. Conclusions. In this paper, we presented an approximation algorithm for analyzing a discrete-time queueing model of the shared buffer ATM switch. The arrival process to each port of the switch was assumed to be bursty, and it was modelled by an IBP. The discrete-time queueing model was analyzed approximately using the notion of decomposition. The queueing model was decomposed into subsystems, and each subsystem was analyzed separately. The results from the subsystems were combined together through the means of an iterative scheme. In order to reduce the complexity of each subsystem, the arrival processes to the switch were aggregated to a single superposition process. Comparisions against simulation data showed that the approximation algorithm has a good accuracy.

TABLE 5.1

Parameters of the arrival processes

table 5.2	$\rho_i = 0.2, i = 1, 2, \cdots, 8;$ $C_1^2 = C_2^2 = 500, C_3^2 = C_4^2 = 200, C_5^2 = C_6^2 = 100, C_7^2 = C_8^2 = 50;$ $b_{i1} = 0.5, b_{i2} = \cdots = b_{i5} = 0.1, b_{i6} = 0.05, b_{i7} = b_{i8} = 0.025$
table 5.3	same as in table 5.2
table 5.4	$\rho_i = 0.1, i = 1, 2, \cdots, 16;$ $C_1^2 = \cdots = C_4^2 = 500, C_5^2 = \cdots = C_8^2 = 200, C_9^2 = \cdots = C_{12}^2 = 100,$ $C_{13}^2 = \cdots = C_{16}^2 = 50;$ $b_{i1} = 0.21, b_{i2} = \cdots = b_{i5} = 0.082, b_{i6} = \cdots = C_{i9} = 0.06,$ $b_{i10} = \cdots = b_{i13} = 0.041, b_{i14} = \cdots = b_{i16} = 0.020$
table 5.5	$\rho_i = 0.2, i = 1, 2, \cdots, 16,$ $C_i^2, b_{ij}, i = 1, 2, \cdots, 16,$ same as in table 5.4

TABLE 5.2
Approximate and simulation results for P(n) and $P_1(n_1)$, switch size: 8×8, buffer size=8.

n	P(n)		n_1	$P_1(n_1)$	
	approx.	simul.		approx.	simul.
0	.1666	.1608±.0054	0	.3329	.3264±.0065
1	.3004	.2999±.0061	1	.1993	.1971±.0037
2	.0469	.0642±.0022	2	.0595	.0572±.0011
3	.0640	.0572±.0013	3	.0509	.0480±.0007
4	.0501	.0452±.0007	4	.0501	.0445±.0010
5	.0419	.0408±.0011	5	.0599	.0524±.0011
6	.0406	.0410±.0009	6	.0830	.0821±.0024
7	.0462	.0476±.0009	7	.1067	.1225±.0051
8	.2432	.2433±.0105	8	.0576	.0699±.0025
cell loss	.16	.20±.01			
m.q.l	3.38	3.51±.0719			

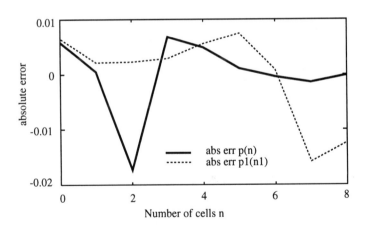

FIG. 5.1. *Absolute errors for the results in Table 5.2*

TABLE 5.3
Approximate and simulation results for P(n) and $P_1(n_1)$, switch size: 8×8, buffer size=16.

n	P(n)		n_1	$P_1(n_1)$	
	approx.	simul.		approx.	simul.
0	.1640	.1591±.0054	0	.3143	.3092±.0066
1	.2741	.2750±.0064	1	.1774	.1754±.0034
2	.0376	.0528±.0016	2	.0473	.0455±.0011
3	.0525	.0460±.0012	3	.0366	.0355±.0010
4	.0401	.0341±.0010	4	.0294	.0280±.0009
5	.0298	.0272±.0007	5	.0246	.0234±.0007
6	.0243	.0228±.0006	6	.0214	.0203±.0006
7	.0210	.0201±.0006	7	.0195	.0188±.0008
8	.0190	.0182±.0007	8	.0185	.0175±.0005
9	.0178	.0172±.0006	9	.0184	.0172±.0007
10	.0173	.0165±.0007	10	.0193	.0176±.0007
11	.0173	.0172±.0007	11	.0220	.0190±.0007
12	.0181	.0182±.0005	12	.0281	.0229±.0008
13	.0197	.0202±.0008	13	.0402	.0331±.0010
14	.0230	.0240±.0009	14	.0605	.0625±.0035
15	.0299	.0320±.0012	15	.0795	.0977±.0057
16	.1945	.1993±.0121	16	.0433	.0564±.0032
cell loss	.139	.148±.008			
m.q.l	5.97	6.21±.16			

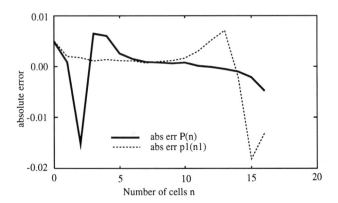

FIG. 5.2. *Absolute errors for the results in Table 5.3*

TABLE 5.4

Approximate and simulation results for P(n) and $P_1(n_1)$, switch size: 16×16, buffer size=16.

n	P(n)		n_1	$P_1(n_1)$	
	approx.	simul.		approx.	simul.
0	.1852	.1978±.0087	0	.6738	.6861±.0083
1	.3264	.3364±.0088	1	.2319	.2265±.0041
2	.2025	.2341±.0053	2	.0541	.0508±.0024
3	.1399	.1143±.0048	3	.0170	.0156±.0014
4	.0673	.0548±.0034	4	.0078	.0071±.0008
5	.0332	.0254±.0022	5	.0044	.0038±.0004
6	.0162	.0128±.0015	6	.0029	.0024±.0003
7	.0089	.0073±.0010	7	.0020	.0017±.0003
8	.0055	.0045±.0005	8	.0015	.0014±.0003
9	.0037	.0029±.0003	9	.0012	.0011±.0002
10	.0026	.0020±.0003	10	.0011	.0010±.0001
11	.0020	.0015±.0002	11	.0009	.0009±.0002
12	.0015	.00012±.0002	12	.0007	.0008±.0002
13	.0012	.0011±.0002	13	.0002	.0005±.0001
14	.0010	.0009±.0002	14	.0	.0002±.0
15	.0009	.0008±.0002	15	.0	.0 ±.0
16	.0021	.0022±.0006	16	.0	.0 ±.0
cell loss	.00181	.00165 ±.00062			
m.q.l	1.89	1.81 ±.06			

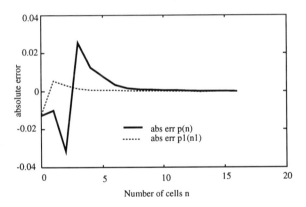

FIG. 5.3. *Absolute errors for the results in Table 5.4*

TABLE 5.5
Approximate and simulation results for $P(n)$ and $P_1(n_1)$, switch size: 16×16, buffer size=16.

n	P(n)		n_1	$P_1(n_1)$	
	approx.	simul.		approx.	simul.
0	.0281	.0293±.0023	0	.3684	.3719±.0132
1	.1083	.1125±.0078	1	.2487	.2474±.0040
2	.1204	.1850±.0083	2	.1110	.1082±.0017
3	.1450	.1649±.0049	3	.0571	.0553±.0015
4	.1219	.1187±.0026	4	.0374	.0351±.0016
5	.0979	.0815±.0017	5	.0286	.0261±.0014
6	.0721	.0546±.0014	6	.0242	.0215±.0012
7	.0534	.0382±.0017	7	.0222	.0193±.0016
8	.0403	.0291±.0015	8	.0215	.0183±.0018
9	.0316	.0234±.0014	9	.0213	.0188±.0020
10	.0258	.0199±.0014	10	.0204	.0196±.0018
11	.0221	.0172±.0014	11	.0174	.0203±.0021
12	.0197	.0156±.0014	12	.0122	.0180±.0018
13	.0183	.0146±.0015	13	.0065	.0127±.0014
14	.0174	.0146±.0013	14	.0023	.0059±.0008
15	.0166	.0158±.0013	15	.0005	.0015±.0002
16	.0602	.0651±.0071	16	.0	.0002±.0
cell loss	.039	.031±.03			
m.q.l	5.	5.15±.20			

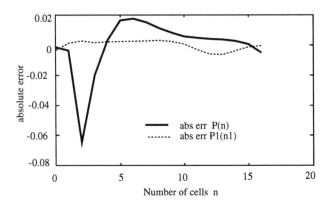

FIG. 5.4. *Absolute errors for the results in Table 5.5*

REFERENCES

[1] A. Baiocchi, N. Melazzi, M. Listanti, A. Roveri, and R. Winkler, *Loss performance analysis of an ATM multiplexer loaded with high-speed on-off sources*, IEEE J. SAC **9** (1991), 388–393.

[2] P. Boyer, M. R. Lehnert, and P. J. Kuehn, *Queueing in an ATM basic switch element*, Tech. Report CNET-123-030-CD-CC, CNET, France, 1988.

[3] H. Bruneel and B. Steyaert, *Tail distribution of shared buffer queue contents*, Tech. report, RUG Internal Research Report, Nr. 4, 900522.

[4] M. Devault, J.-Y. Cochennec, and M. Servel, *The "Prelude" ATD experiment: Assessments and future prospects*, IEEE J. SAC **6** (1988), no. 9, 1528–1537.

[5] A. E. Eckberg and T.-C. Hou, *Effects of output buffer sharing on buffer requirements in an ATDM packet switch*, INFOCOM '88, pp. 459–466.

[6] H. Heffes, *A class of data traffic processes-covariance function characterization and related queueing results*, Bell Sys. Tech. J. **59** (1980), 897–929.

[7] H. Heffes and D. M. Lucantoni, *A Markov modulated characterization of packetized voice and data traffic and related statistical multiplexer performance*, IEEE J. SAC **4** (1986), 856–868.

[8] F.P. Kelly, *Reversibility and stochastic netwroks*, John Wiley & Sons, Inc., 1979.

[9] D.D. Kouvatsos and N.P. Xenios, *MEM for arbitrary queueing networks with multiple general services and repetitive-service blockings*, Perform. Eval. **10** (1989), 169–195.

[10] H. Kuwahara, N. Endo, M. Ogino, and T. Kozaki, *A shared buffer memory switch for an ATM exchange*, Int. Conf. on Communications (Boston, MA), pp. 4.4.1–4.4.5.

[11] H. Lee, K. Kook, C. S. Rim, K. Jun, and S.-K. Lim, *A limited shared output buffer switch for ATM*, Fourth Int. Conf. on Data Communication Systems and Their Performance (Barcelona), pp. 163–179.

[12] R. Nagarajan, J. F. Kurose, and D. Towsley, *Approximation techniques for computing packet loss in finite-buffered voice multiplexers*, IEEE J. SAC **9** (1991), 368–377.

[13] I. Norros, J. W. Roberts, A. Simmonian, and J. T. Virtamo, *Loss performance analysis of an ATM multiplexer loaded with high-speed on-off sources*, IEEE J. SAC **9** (1991), 378–387.

[14] H. G. Perros and R. Onvural, *On the superposition of arrival processes for voice and data*, Fourth Int. Conf. on Data Communication Systems and Their Performance (Barcelona), pp. 341–357.

[15] G.H. Petit, A. Buchheister, A. Guerrero, and P. Parmentier, *Performance evaluation methods applicable to an ATM multi-path self-routing switching network*, Teletraffic and datatraffic in a period of change (Jensen and Iversen, eds.), North Holland, 1991, pp. 917–922.

[16] G.H. Petit and E. M. Desmet, *Performance evaluation of shared buffer multiserver output queue switches used in ATM*, 7th ITC Specialist Seminar (New Jersey), p. paper 7.1.

[17] G. Pujolle and H.G.Perros, *Queueing systems for modelling ATM networks*, Int'l conf. on the Performance of Distributed Systems and Integrated Comm. Networks (Koyto).

[18] B. Sengupta, *A queue with superposition of arrival streams with and application to packet voice technology*, Performance '90 (King, Mitrani, and Pooley, eds.), North Holland, 1990, pp. 53–59.

[19] F.A. Tobagi, *Fast packet switch architectures for broadband integrated services networks*, Proc. IEEE **78** (1990), 1133–1167.

[20] H. Yamashita, H. G. Perros, and S. Hong, *Performance modelling of a shared buffer ATM switch architecture*, Teletraffic and datatraffic in a period of change (Jensen and Iversen, eds.), North-Holland, 1991.

ALGORITHMS FOR INFINITE MARKOV CHAINS WITH REPEATING COLUMNS

GUY LATOUCHE*

Abstract. We consider Markov chains on an infinite state space, with a lower block-Hessenberg transition matrix. If the columns repeat as in the case of the $GI/PH/1$ queue, then the stationary probability vector is matrix-geometric. The major difficulty in computing the stationary vector resides in the determination of the solution of a non-linear matrix equation.

Various algorithms may be used for that purpose. They are discussed, and their computational efficiency is compared. While most of the algorithms examined are well known, some are presented here for the first time.

1. Introduction. We consider a Markov chain $\{X_t, t \in N\}$ on the bidimensional infinite state space $S = \{(i,j), i \geq 0, 1 \leq j \leq m\}$, with transition matrix P given by

$$(1.1) \qquad P = \begin{bmatrix} B_0 & A_0 & 0 & 0 & \cdots \\ B_1 & A_1 & A_0 & 0 & \cdots \\ B_2 & A_2 & A_1 & A_0 & \cdots \\ B_3 & A_3 & A_2 & A_1 & \cdots \\ \vdots & \vdots & \vdots & \vdots & \ddots \end{bmatrix},$$

where B_i, A_i, for $i \geq 0$, are matrices of order m. We assume that m is finite and that the Markov chain is irreducible and aperiodic.

We denote by $\boldsymbol{\pi}$ the stationary probability vector associated to P, when the Markov chain is ergodic, and we write $\boldsymbol{\pi} = (\boldsymbol{\pi}_0, \boldsymbol{\pi}_1, \boldsymbol{\pi}_2, \cdots)$, where each $\boldsymbol{\pi}_i$, for $i \geq 0$, is an m-vector. It is well known (see M.F. Neuts [12] for instance) that the subvectors satisfy the relation $\boldsymbol{\pi}_i = \boldsymbol{\pi}_0 R^i$, for $i \geq 0$, where R is a matrix of order m, solution of the equation $R = \sum_{\nu \geq 0} R^\nu A_\nu$.

In general, the matrix R may only be computed by an iterative algorithm. Thus, the computation of R is the most costly step in the determination of the stationary vector $\boldsymbol{\pi}$. Many different algorithms may be used, in order to evaluate that matrix. It is our purpose here to describe them and to compare their efficiency.

The main index of efficiency which we use for these comparisons is, quite naturally, the execution time of each algorithm. Since this depends on the detailed structure of the matrix P, we have considered a number of different examples. The execution time also depends on the programming style, the programming language and the computer which is used; obviously, these factors have influenced our experimental results. For that reason, we also consider as indices of efficiency, the number of iterations

* Université Libre de Bruxelles, Séminaire de Théorie des Probabilités, CP212, Boulevard du Triomphe, 1050 Bruxelles, Belgium

needed by each algorithm and the mathematical complexity of each iteration. In a number of cases, direct comparisons can be made between the
values of these indices for different algorithms; such comparisons confirm
the conclusions based on the numerical experimentation.

The algorithms are grouped in three families. In each, a basic algorithm
is justified by purely probabilistic arguments. Upon these basic algorithms,
various improvements are then brought, which are justified by analytic
considerations mostly.

The paper is organised as follows. In the next section, we give a brief
presentation of a simple proof for the matrix-geometric result. That proof
has been reported in details in Latouche [9]; it yields an equation which
forms the basis for one of the most efficient algorithm in our collection.
That algorithm is described in Section 3. In Section 4, we describe algorithms which may be termed as 'traditional': the basic algorithm in this
family has been recommended in Neuts [12], and improvements on this
have been proposed in Neuts [13] and Ramaswami [17]. The last family
of algorithms is described in Section 5. Briefly stated, they are improved
variants of the algorithms described in Kao [6].

Although none of the algorithms, that we have investigated, appears
to uniformly dominate the others, we are nevertheless able to draw on our
theoretical and experimental results (described in Section 7) and formulate
general conclusions in Section 8.

We conclude this introduction by precisely defining the limits of our
comparisons which are based on the macro-structure only of the stochastic
matric P, as given by the equation (1.1).

The algorithms can readily be adapted to the case of *continuous-time*
Markov chains for which the infinitesimal generator has the structure (1.1).
We believe that the qualitative comparisons observed here will remain valid,
but this has not been verified yet.

Secondly, if the matrix P is upper-Hessenberg as below,

$$(1.2) \qquad P = \begin{bmatrix} B_0 & B_1 & B_2 & B_3 & \cdots \\ A_0 & A_1 & A_2 & A_3 & \cdots \\ 0 & A_0 & A_1 & A_2 & \cdots \\ 0 & 0 & A_0 & A_1 & \cdots \\ \vdots & \vdots & \vdots & \vdots & \ddots \end{bmatrix},$$

then an essential role is played by the solution of another matrix equation:
$G = \sum_{\nu \geq 0} A_\nu G^\nu$. The algorithms of the sections 2 to 4 still apply, mutatis
mutandis; it is possible that algorithms may be found, similar to those of
Section 5. However, it is far from certain that the same comparisons will
hold. Some unexpected results have been reported in that area by Gün [4].

Finally, it is almost certain that some of our reported comparisons will
be invalidated if the matrix P has a more particular structure yet and if
the algorithms are suitably adapted. For instance, if the Markov chain

is a Quasi-Birth-and-Death process (in which case $A_i = 0$, for all $i \geq 3$, and $B_i = 0$ for $i \geq 2$), then it can be shown that the basic algorithms of Sections 3 and 5 are identical. Another example is the $GI/PH/1$ queue examined at epochs of arrivals, for which further improvements might be obtained, by using the uniformisation approach described in Lucantoni and Ramaswami [11,20].

2. Preliminaries. We begin by recalling a few well established results, which we shall need in the remainder of this paper. Firstly, the following theorem has been proved in Neuts [12], Chapter 1 (Lemma 1.2.3 and Theorem 1.2.1).

THEOREM 2.1. *If the Markov chain P is positive recurrent, then there exists a matrix R of order m such that*

$$(2.1) \qquad \qquad \pi_n = \pi_0 R^n, \qquad \text{for all } n \geq 0,$$

and the vector π_0 is the unique solution of the system

$$(2.2) \qquad \qquad \pi_0 B[R] = \pi_0,$$
$$(2.3) \qquad \qquad \pi_0 (I - R)^{-1} \mathbf{1} = 1,$$

where the stochastic matrix $B[R]$ is defined by

$$(2.4) \qquad \qquad B[R] = \sum_{\nu \geq 0} R^\nu B_k.$$

Furthermore, R_{ij} is the expected number of visits to the state $(n + 1, j)$ before the first return to the level n, given that the chain P starts in the state (n, i). Finally, the matrix R is the minimal nonnegative solution to the matrix equation

$$(2.5) \qquad \qquad R = \sum_{\nu \geq 0} R^\nu A_\nu.$$

It is the unique such solution in the set of nonnegative matrices with spectral radius strictly less than one.   □

We define the level n as the subset of states $\{(n, j), 1 \leq j \leq m\}$ where the first index is fixed.

Secondly, let S be partitioned in two subsets, E and E^c. We partition π as (π_E, π_{E^c}), and the matrix P as

$$P = \begin{bmatrix} T & H \\ D & Q \end{bmatrix}.$$

Then the following theorem results, which is a reformulation of Propositions 5.11, 6.4, Lemmas 6.6, 6.7, and Corollary 6.22 in Kemeny et al. [7].

THEOREM 2.2. *Assume that the Markov chain P is positive recurrent. The matrix $T + HND$ is then the stochastic transition matrix for a positive recurrent Markov chain, where the matrix N is defined by $N = \sum_{k \geq 0} Q^k$ and is the minimum nonnegative right inverse and left inverse of $(I - Q)$. Furthermore, the vector π_E is a solution of the system*

$$(2.6) \qquad\qquad \pi_E(T + HND) = \pi_E,$$

and the vector π_{E^c} is given by

$$(2.7) \qquad\qquad \pi_{E^c} = \pi_E HN.$$

Finally, the following probabilistic interpretations hold.

1. N_{ij}, for $i, j \in E^c$, is the expected number of visits to state j, before reaching any state in E, given that the initial state is i.

2. $(HN)_{ij}$ is the expected number of visits to $j \in E^c$, before returning to any state in E, given that the initial state is $i \in E$.

3. $(ND)_{ij}$ is the probability that, upon return to E, the first state visited is $j \in E$, given that the initial state is $i \in E^c$.

4. $(T + HND)_{ij}$ is the probability that, upon return to E, the first state visited is $j \in E$, given that the initial state is $i \in E$. Thus, $T + HND$ is the transition matrix from states in E to states in E, either directly, or after a sojourn of unspecified duration in E^c.

□

We choose $E = 0 \cup \ldots \cup n$, and observe that Q and N are independent of n. We partition the matrix N into blocks $N_{k,k'}$ which contain the expected number of visits to the states of level $n + k'$ before the first visit to level n or below, given that the initial state is in level $n + k$, for any value of $n \geq 0$.

The following equations have been proved in [9]:

$$
\begin{aligned}
(2.8) \qquad N_{1,k+1} &= N_{11} A_0 N_{1k}, && \text{for all } k \geq 1, \\
(2.9) \qquad A_0 N_{1k} &= R^k, && \text{for all } k \geq 1, \\
(2.10) \qquad N_{11} &= I + \sum_{\nu \geq 1} N_{1\nu} A_\nu.
\end{aligned}
$$

To compute the matrix R, an algorithm which naturally suggests itself is to use the equation (2.5) and to proceed by successive substitutions. Indeed, it has been shown in [12] that the following sequence of matrices monotonically converges to R:

$$
\begin{aligned}
(2.11) \qquad R_0 &= A_0, \\
(2.12) \qquad R_{k+1} &= \sum_{\nu \geq 0} R_k^\nu A_\nu, && \text{for all } k \geq 0.
\end{aligned}
$$

In practice, of course, it will be necessary to truncate the series in the right-hand side of (2.12) at some index M, where M is such that for every j, j', with $1 \leq j, j' \leq m$, the tail $\sum_{\nu \geq M+1}(A_\nu)_{jj'}$ is negligible.

Thus, we obtain a first algorithm which we call the algorithm N (see below). The sum $\sum_{0 \leq \nu \leq M} R_k^\nu A_\nu$ is computed with Horner's algorithm for polynomials, and if we measure the complexity by the total number of multiplications and divisions of two floating point numbers, it is readily seen that the complexity of this algorithm is given by

$$(2.13) \qquad C_N = J_N M m^3,$$

where J_N is the number of required iterations. We may not conclude that $C_N = O(m^3)$, since J_N itself might be a function of m, although this is unlikely.

> **Algorithm N—** The natural algorithm
> $k := 0$;
> $R_0 := A_0$;
> **repeat**
> $k := k + 1$;
> $S := A_M$;
>   **for** $i := M - 1$ **downto** 0 **do** $S := R_{k-1} S + A_i$;
> $R_k := S$;
> **until** $\|R_k - R_{k-1}\| < \epsilon$

In addition to the matrices A_0 to A_M, this algorithm requires two matrices, to store R_{k-1} and R_k at each iteration, and at least one vector, as temporary storage when performing matrix multiplications. Thus, the space complexity is given by

$$(2.14) \qquad S_N = (M + 3)m^2 + m,$$

if we omit the space needed by the program itself, which is $O(1)$.

3. Algorithms Based on Taboo Probabilities. We now introduce the same matrix U as we did in [8]. For any fixed $n \geq 1$, $U_{jj'}$ is the taboo probability that the Markov chain enters the state (n, j') upon its next visit to the level n, without visiting any state at the levels 0 to $n-1$, given that the initial state is (n, j). We emphasize here that the matrix U is independent of n.

THEOREM 3.1. *The matrices U and R are related as follows:*

$$(3.1) \qquad U = \sum_{\nu \geq 0} R^\nu A_{\nu+1},$$

$$(3.2) \qquad R = A_0 (I - U)^{-1}.$$

The matrix U is the minimal nonnegative solution to the matrix equation

$$(3.3) \qquad U = \sum_{\nu \geq 0} [A_0(I - U)^{-1}]^\nu A_{\nu+1}.$$

Proof It clearly results from the definition of U and from the last statement in Theorem 2 that $U_{jj'} = (T + HND)_{(n,j);(n,j')}$, and therefore that

$$U = A_1 + \sum_{\nu \geq 1} A_0 N_{1\nu} A_{\nu+1},$$

from which (3.1) result, by (2.9). Since $U_{jj'}$ is also the probability of moving from $(n+1,j)$ to $(n+1,j')$ under taboo of the levels 0 to n, therefore we have that $N_{11} = (I-U)^{-1}$; note that the inverse exists because the Markov chain P is irreducible. By (2.9) again, we then obtain (3.2).

It is now easily observed that U is a solution of the matrix equation (3.3). To prove that it is the smallest, we consider another solution and call it $\tilde{U}$; we also define $\tilde{R} = A_0(I - \tilde{U})^{-1}$. We may write $\tilde{R} = A_0 + \tilde{R}\tilde{U} = \sum_{\nu \geq 0} \tilde{R} A_\nu$, which proves that $\tilde{R} \geq R$, since R is the minimal nonnegative solution of (2.5). Since $\tilde{U} = \sum_{\nu \geq 0} \tilde{R}^\nu A_{\nu+1}$, we obtain that $\tilde{U} \geq U$. □

REMARK 3.2. This theorem generalises some results which we obtained in [8]. More specifically, the equations (3.1, 3.2, 3.3) respectively correspond to the equations (9, 3, 8) there.

REMARK 3.3. A simple probabilistic argument shows that, for any level n and for all $\nu \geq 1, ((I - U)^{-1} A_{\nu+1})_{jj'}$ is the probability that upon its last departure from level n, the Markov chain moves directly to level $n-\nu$, in the state $(n - \nu, j')$, given that the initial state is (n,j).

REMARK 3.4. This leads us to a probabilistic interpretation of the equation (3.3) which we rewrite as follows:

$$U = A_1 + \sum_{\nu \geq 1} A_0[(I - U)^{-1}A_0]^{\nu-1}(I - U)^{-1}A_{\nu+1}.$$

The νth term in the series corresponds to the event that the Markov chain returns to the level n under taboo of the levels below, *and* that it re-enters the level n from the level $n + \nu$. Indeed, the factor A_0 indicates that the Markov chain moves from n to $n+1$. The first factor $(I-U)^{-1}A_0$ indicates that upon its *last* exit from $n+1$, the Markov chain moves to $n+2$, at which time $n+1$ becomes part of the taboo set of states. Each succeeding factor $(I - U)^{-1}A_0$ indicates that the Markov chain is progressively forced to remain at $n+3$ or above, $n+4$ or above, etc. up to $n + \nu$. Finally, the last factor $(I - U)^{-1}A_{\nu+1}$ indicates that the Markov chain eventually makes a downwards jump from $n + \nu$ to n.

From the equation above, we also obtain the following equation, which we shall use later:

$$(3.4) \qquad (I - U)^{-1} = I + \sum_{\nu \geq 0}[(I - U)^{-1}A_0]^\nu (I - U)^{-1}A_{\nu+1}.$$

To show this, we write

$$I - U = I - A_1 - \sum_{\nu \geq 1} A_0[(I - U)^{-1}A_0]^{\nu-1}(I - U)^{-1}A_{\nu+1},$$

we then pre-multiply both sides by $(I - U)^{-1}$ and re-arrange the terms of the resulting equation.

We shall now define two matrix operators $\mathcal{F}$ and $\mathcal{G}$ as follows:

$$(3.5) \qquad \mathcal{F}Y = \sum_{\nu \geq 0} Y^{\nu} A_{\nu+1}, \qquad \text{for } 0 \leq Y \leq R,$$

$$(3.6) \qquad \mathcal{G}Z = A_0(I - Z)^{-1}, \qquad \text{for } 0 \leq Z \leq U.$$

We observe that the series in (3.5) does converge for $0 \leq Y \leq R$, and that $\mathcal{F}Y_1 \leq \mathcal{F}Y_2$ whenever $0 \leq Y_1 \leq Y_2 \leq R$. Also, since $(I - Z)^{-1} = \sum_{i \geq 0} Z^i$ for $0 \leq Z \leq U$, therefore $\mathcal{G}Z_1 \leq \mathcal{G}Z_2$ whenever $0 \leq Z_1 \leq Z_2 \leq U$. Finally, we have that $\mathcal{F}R = U$ and $\mathcal{G}U = R$.

The equations in Theorem 3.1 suggest another procedure to evaluate the matrix R. We define the matrix sequence $\{(U_k, R_k), k \geq 0\}$, where

$$(3.7) \qquad U_0 = A_1,$$

$$(3.8) \qquad R_0 = A_0(I - U_0)^{-1},$$

$$(3.9) \qquad U_{k+1} = \mathcal{F}R_k = \sum_{\nu \geq 0} R_k^{\nu} A_{\nu+1},$$

$$(3.10) \qquad R_{k+1} = \mathcal{G}U_{k+1} = A_0(I - U_{k+1})^{-1}, \qquad \text{for } k \geq 0.$$

In order to keep the notations reasonably simple, we use the same symbols R_k here as in the recurrence equation (2.12). It is only when comparing algorithms that we shall use different notations.

The following property holds.

THEOREM 3.5. *The sequences $\{U_k, k \geq 0\}$ and $\{R_k, k \geq 0\}$ monotonically converge from below to the matrices U and R, respectively.*

Proof We give the detailed formal proof of this theorem because it is representative of arguments which are repeatedly used in the sequel.

Firstly, we prove that $U_k \leq U$ and $R_k \leq R$, for all $k \geq 0$. Clearly, $U_0 \leq U$, and therefore $(I - U_0)^{-1} = \sum_{i \geq 0} U_0^i \leq \sum_{i \geq 0} U^i = (I - U)^{-1}$, hence $R_0 \leq R$. If we make the induction assumption that $U_k \leq U$ and $R_k \leq R$, then we immediately conclude from the monotonicity of $\mathcal{F}$ and $\mathcal{G}$ that $U_{k+1} = \mathcal{F}R_k \leq \mathcal{F}R = U$, and that $R_{k+1} = \mathcal{G}U_{k+1} \leq \mathcal{G}U = R$.

Secondly, we have that $U_1 \geq U_0$, since $R_0 \geq 0$. Thus, $R_1 = \mathcal{G}U_1 \geq \mathcal{G}U_0 = R_0$. It is easy to prove that if $U_k \geq U_{k-1}$ and $R_k \geq R_{k-1}$, then $U_{k+1} \geq U_k$ and $R_{k+1} \geq R_k$, which proves the monotone behaviour of the two sequences. The sequences being monotone and bounded are converging.

Let us define $\tilde{U} = \lim_{k \to \infty} U_k$ and $\tilde{R} = \lim_{k \to \infty} R_k$. Clearly, we must have $\tilde{U} \leq U$ and $\tilde{R} \leq R$; moreover, $\tilde{R} = A_0(I - \tilde{U})^{-1}$. We may now repeat the argument at the end of the proof of Theorem 3.1 to conclude that $\tilde{R} \geq R$ and $\tilde{U} \geq U$, which concludes the proof of the theorem. $\square$

In terms of the matrix R only, the equations (3.9) and (3.10) give

$$(3.11) \qquad R_{k+1} = A_0(I - \mathcal{F}R_k)^{-1} = A_0(I - \sum_{\nu \geq 0} R_k^\nu A_{\nu+1})^{-1}.$$

This equation has already been noticed (see Neuts [12], pp.22 and 37, or Ramaswami [17] p.257) but it has not been much investigated, apparently.

The resulting algorithm is given below. As we did before, we truncate the series in (3.9) at the index $\nu = M - 1$.

> **Algorithm U**— The algorithm based on the matrix U
> $k := 0$;
> $U_0 := A_1$;
> $R_0 := A_0(I - U_0)^{-1}$;
> **repeat**
> $k := k + 1$;
> $S := A_M$;
>   **for** $i := M - 1$ **downto 1 do** $S := R_{k-1}S + A_i$;
> $U_k := S$;
> $R_k := A_0(I - U_k)^{-1}$
> **until** $\|R_k - R_{k-1}\| < \epsilon$

If we assume that the matrix inversion is performed by some version of Gaussian elimination, then the corresponding complexity is $4/3m^3 - m/3$ (see Isaacson and Keller [5]), thus the complexity of this algorithm based on the relations between R and U is

$$(3.12) \qquad C_U = J_U[(M + 1/3)m^3 - 1/3m] + 4/3m^3 - 1/3m,$$

where J_U is the number of iterations. In order to perform the matrix inversions, a temporary storage matrix is needed, and the space complexity is

$$(3.13) \qquad S_U = (M + 4)m^2 + m.$$

If we compare this to (2.14), we see that Algorithm U requires one additional matrix.

The comparison between the time complexities C_N and C_U is not as straightforward. Here, there is a fixed cost of one Gaussian elimination, and each iteration costs approximately an additional $1/3$ of matrix multiplication. The total complexity C_U however may be smaller than C_N, if the number J_U of iterations is sufficiently smaller than J_N.

Before proceeding with the comparison of C_U and C_N, we make two comments about the algorithm above. Firstly, we might have used the stopping criterion $\|U_k - U_{k-1}\| < \epsilon$, which might result in fewer iterations (for a related discussion, see Gün [4]). However, to study the choice of stopping criterion is beyond the scope of the present study. We therefore assume that the same stopping criterion is used for all the algorithms. Secondly, the choice of inversion algorithm is important, since it may influence the overall numerical stability of the algorithm. The discussion of this issue is postponed till Section 6.

We now turn our attention to the comparison between J_U and J_N. On the basis of the next two lemmas, we may expect J_U to be significantly smaller than J_N. It will be useful here to have different notations for the two algorithms. We shall denote by $\{R_k^{(U)}, k \geq 0\}$ the sequence of matrices defined by (3.8,3.10), or equivalently (3.11), and by $\{R_k^{(N)}, k \geq 0\}$ the natural sequence defined by (2.12).

LEMMA 3.6.   *One has that $R_k^{(N)} \leq A_0(I - \mathcal{F}R_{k-1}^{(N)})^{-1}$ for all $k \geq 1$, from which it results that*

$$R_k^{(N)} \leq R_k^{(U)} \leq R, \qquad \text{for all } k \geq 0.$$

Proof   We may write (2.12) as $R_{k+1}^{(N)} = A_0 + R_k^{(N)}\mathcal{F}R_k^{(N)}$. Since $R_0^{(N)} = A_0$, we conclude that each iterate is the product of A_0 by some matrix, and we write $R_k^{(N)} = A_0 V_k$, where $V_0 = I$ and $V_{k+1} = I + V_k\mathcal{F}R_k^{(N)}$, for $k \geq 0$. Since $V_1 = I + \mathcal{F}R_0^{(N)} \leq (I - \mathcal{F}R_0^{(N)})^{-1}$, the statement is proved for $k = 1$. We now assume that $V_k \leq (I - \mathcal{F}R_{k-1}^{(N)})^{-1}$, and we consider the next iteration.

Since $R_{k-1}^{(N)} \leq R_k^{(N)} \leq R$, therefore $\mathcal{F}R_{k-1}^{(N)} \leq \mathcal{F}R_k^{(N)}$ by the monotonicity property for $\mathcal{F}$, and $(I - \mathcal{F}R_{k-1}^{(N)})^{-1} \leq (I - \mathcal{F}R_k^{(N)})^{-1}$. This implies that $V_k \leq (I - \mathcal{F}R_k^{(N)})^{-1}$, and therefore that $V_{k+1} = I + V_k\mathcal{F}R_k^{(N)} \leq (I - \mathcal{F}R_k^{(N)})^{-1}$, which completes the proof of the first statement.

It is clear that $R_0^{(U)} \geq R_0^{(N)}$, since $A_0(I - U_0)^{-1} \geq A_0$. If $R_k^{(U)} \geq R_k^{(N)}$, since $R_{k+1}^{(U)} = A_0(I - \mathcal{F}R_k^{(U)})^{-1}$ by Equation (3.11), then

$$R_{k+1}^{(U)} \geq A_0(I - \mathcal{F}R_k^{(N)})^{-1} \geq R_{k+1}^{(N)},$$

where the first inequality holds by the induction assumption and by the monotonicity property for $\mathcal{F}$. This completes the proof.

□

Strictly speaking, this does not guarantee that $J_U < J_N$. In fact, it is even conceivable that we might find $J_U > J_N$ for some value of ϵ: if the natural sequence converges very slowly to R, it is possible that algorithm N stops before the sequence has reached the neighbourhood of R. For a short discussion about this possibility, see Gün [4]. However, the first part of the lemma states that algorithm U is more efficient at *each* iteration; even if we would chose $R_0^{(N)} = R_0^{(U)}$, we would still have that $R_k^{(N)} \leq R_k^{(U)}$ for all $k \geq 1$. We have found that $J_U < J_N$ in all examples which we considered.

The convergence is usually slow when the Markov chain nearly fails to be ergodic, or stated otherwise, when the expected first passage time from the level $n+1$ to levels below is very large. Now, for every basic iterative algorithm in this paper, we may interpret R_k as the expected sojourn at the level $n+1$, starting from the level n, under taboo of the levels 0 to n, where the expectation is taken on a subset of the possible sample paths.

In section 5 for instance, R_k is computed by allowing the Markov chain to move freely between the levels $n+1$ to $n+k$, but not beyond. For other methods, however, the restrictions are not based on probabilistic argument, and therefore are difficult to describe in a concise manner. We shall illustrate this on the first two iterations defined by the equations (3.7) to (3.10).

At the first iteration, $U_0 = A_1$, which means that one only takes into consideration the sample paths where the Markov chain continuously remains in level $n+1$ before moving to any other level.

At the next iteration, we have that

$$U_1 = A_1 + \sum_{\nu \geq 1} A_0 [(I - A_1)^{-1} A_0]^{\nu-1} (I - A_1)^{-1} A_{\nu+1}.$$

Thus, we observe that multiple returns to $n+1$ are allowed (see the remark 3.4); the Markov chain may visit any level $n+1+\nu$ with $\nu \geq 1$, with the restriction that such levels may only be entered from below. For instance, the term with $\nu = 2$ in the series above indicates that the Markov chain moves upwards from $n+1$ to $n+2$ (with probabilities given by A_0), then after a while moves upwards to $n+3$ (with probabilities given by $(I - A_1)^{-1} A_0$) but does not return to $n+2$: it moves down to $n+1$ (with probabilities given by $(I - A_1)^{-1} A_3$). Similarly for the other terms in the series, the only downwards jumps which are allowed are from some level $n+1+\nu$ to $n+1$.

At each successive iteration, less restrictions are imposed on the behaviour of the Markov chain in the levels $n+1$ and above, but this becomes very complicated to describe.

Now, let us denote by θ the first passage time from $n+1$ to n or below. If $E[\theta]$ is small, then the probability mass is likely to be concentrated on a small number of sample paths, for which the Markov chain does not venture far above $n+1$. If this is case, then the important sample paths will soon be captured and the sequence $\{R_k, k \geq 0\}$ will rapidly converge to R. On the other hand, if $E[\theta]$ is large, then the probability mass is dispersed over a large number of sample paths, and the algorithm will need a large number of iterations before the restrictions are sufficiently weakened.

We are therefore lead to the conclusion that in order to accelerate the convergence of a given algorithm, it is necessary to use an initial matrix R_0 close to R, and that it seems difficult to find one by probabilistic argument only.

We now turn our attention to a simple modification of the algorithm, which has been suggested by Neuts [13] for the basic algorithm of Section 4, and which often substantially reduces the number of iterations.

Under certain circumstances, one may determine the maximal eigenvalue η of R, and its corresponding left eigenvector $\boldsymbol{u}$, without having to compute R at all. Neuts suggests that one should initialize the recurrence with the matrix $R_0' = \eta \boldsymbol{u} \cdot \boldsymbol{1}$, then every iterate will have the same eigenvalue η and left eigenvector $\boldsymbol{u}$. It is hoped that R_0' is sufficiently close to R, so that the number of iterations is small. We give more details below.

We define the matrix $A(z) = \sum_{\nu \geq 0} z^\nu A_\nu$. We assume that $A(1)$ is irreducible, which implies that $A(z)$ is irreducible for $z > 0$, and we denote by $\chi(z)$ the maximal eigenvalue of $A(z)$. We then have the following theorem, which summarizes results in Neuts [12] Section 1.3.

THEOREM 3.7.   *Assume that the Markov chain is irreducible and ergodic. If the matrix $A(1)$ is irreducible and stochastic, then the equation $z = \chi(z)$ has two solutions in $0 \leq z \leq 1$; one is $z = 1$, the other is $z = \eta$.*

Moreover, the matrix R is the unique nonnegative solution with maximal eigenvalue η of the equation (2.5) .

Finally, if $\boldsymbol{u}$ is a left eigenvector of $A(\eta)$ associated to the maximal eigenvalue $\chi(\eta) = \eta$, then $\boldsymbol{u}$ is a left eigenvector of R corresponding to the eigenvalue η.

$\square$

The stated conditions are not necessary, but they are satisfied in many applications. We now prove the following result.

THEOREM 3.8.   *If η is the maximal eigenvalue of R, and $\boldsymbol{u}$ is the corresponding positive left eigenvector normalized by $\boldsymbol{u}\boldsymbol{1} = 1$; if in addition the matrix R is irreducible, then the sequence $\{R_k', k \geq 1\}$ of matrices defined by the equations (3.9) and (3.10), with $R_0' = \eta \boldsymbol{u} \cdot \boldsymbol{1}$, converges to the matrice R.*

Proof   Firstly, we have to show that all the matrices share the eigenvalue η and eigenvector $\boldsymbol{u}$; since $\boldsymbol{u}$ is positive, η is the maximal eigenvalue. Clearly, this is true for R_0', and we may assume that it holds for R_k'. Now, simple calculations show that

$$
\begin{aligned}
\boldsymbol{u}\mathcal{F}R_k' &= \boldsymbol{u}\left(\sum_{\nu \geq 1} \eta^{\nu-1} A_\nu\right), \\
&= \frac{1}{\eta}\boldsymbol{u}(A(\eta) - A_0), \\
&= \boldsymbol{u} - \frac{1}{\eta}\boldsymbol{u}A_0.
\end{aligned}
$$

Then, we see that $\boldsymbol{u}A_0 = \eta\boldsymbol{u}(I - \mathcal{F}R_k')$, and therefore $\boldsymbol{u}R_{k+1}' = \eta\boldsymbol{u}$.

Thus, the matrices in the sequence remain in the compact set of nonnegative matrices with spectral radius equal to η and corresponding eigenvector $\boldsymbol{u}$. Therefore, the sequence has at least one accumulation point. We denote by $\tilde{R}$ one such point, and we choose a subset of indices $\{k_1, k_2, \cdots\}$ such that the subsequence $\{R_{k_i}', i \geq 1\}$ converges to $\tilde{R}$.

GUY LATOUCHE

We also use (3.9, 3.10) to define the sequence $\{R_k^*, k \geq 0\}$, initialised with $R_0^* = 0$. It is readily seen that $R_1^* = A_0(I - U_0)^{-1}$; thus, except for the starting value, this is the sequence analyzed in Theorem 3.5 and we know that it converges to R. In particular, we have that the subsequence $\{R_{k_i}^*, i \geq 1\}$ converges to R.

Since $R_0^* \leq R_0'$, we may show by induction, using the monotonicity property of $\mathcal{F}$ and $\mathcal{G}$, that $R_k^* \leq R_k'$ for all k. Therefore, we have that $R \leq \tilde{R}$. If R is irreducible, since R and $\tilde{R}$ have the same maximal eigenvalue, then necessarily $R = \tilde{R}$ by Varga [22] Theorem 2.1. Thus we have shown that the sequence $\{R_k', k \geq 0\}$ may have only one accumulation point, equal to R, which proves the Theorem. This argument to prove convergence is due to Ramaswami [18]. $\qquad\qquad\square$

COROLLARY 3.9. *The sequences* $\{R_k, k \geq 0\}$ *of Theorem 3.5 and* $\{R_k', k \geq 0\}$ *of Theorem 3.8 are related as follows:* $R_k \leq R_{k+1}'$. $\qquad\square$

We may accordingly modify the algorithm U and replace the first three statements by the following

$$\text{Evaluate } \eta \text{ and } \boldsymbol{u} \text{ from } A(z);$$
$$R_0 \leftarrow \eta \boldsymbol{u} \cdot \boldsymbol{1}$$

We shall call this the algorithm U'.

We cannot compare a priori the complexities C_U and $C_{U'}$ of the two algorithms. The complexity of each iteration remains unchanged, and we may expect that $J_{U'} \leq J_U$, although the preceding corollary only suggests that $J_{U'} \leq J_U + 1$. However, the overhead is no longer a simple matrix inversion, and will depend on how fast we find the solution η of the equation $z = \chi(z)$. This question is examined in Section 6. Thus, we rely on experimentation only, in concluding that the algorithm U' is one of the fastest known to date.

Since we shall be comparing several algorithms, we shall use a graphical summary of our findings. At this time, we have that

$$U' \cdots\cdots\blacktriangleright U \longrightarrow N$$

where the arrow between the nodes U and N indicates that $J_U \leq J_N$, the dot-arrow between U' and U indicates that $J_{U'} \leq J_U + 1$. The full graph is given at the end of Section 5.

4. Algorithms Based on Changes of Levels. The algorithms which have traditionally been advocated in the literature are based on a simple transformation of (2.5) which may be written as

$$(4.1) \qquad\qquad R = \sum_{\substack{\nu \geq 0 \\ \nu \neq 1}} R^\nu A_\nu (I - A_1)^{-1}.$$

The corresponding recurrence equations are

$$(4.2) \qquad R_0 = A_0(I - A_1)^{-1},$$

$$(4.3) \qquad R_{k+1} = \sum_{\substack{\nu \geq 0 \\ \nu \neq 1}} R_k^{\nu} A_\nu (I - A_1)^{-1},$$

and the algorithm may be written as

Algorithm T— The traditional algorithm
$A_0' := A_0(I - A_1)^{-1}$;
for $i := 2$ **to** M **do** $A_i' := A_i(I - A_1)^{-1}$;
$k := 0$;
$R_0 := A_0'$;
repeat
 $k := k + 1$;
 $S := 0$;
 for $i := M$ **downto** 2 **do** $S := R_{k-1}(S + A_i')$;
 $R_k := R_{k-1}S + A_0'$
until $\|R_k - R_{k-1}\| < \epsilon$

The computational complexity of this algorithm is given by

$$(4.4) \qquad C_T = J_T M m^3 + (M + 4/3)m^3 - 1/3m.$$

If we compare C_T to C_N and C_U, we see that the traditional algorithm is at least as efficient as Algorithm N if the number J_T of iterations is less than $J_N - 1$. Also, each iteration of Algorithm T is slightly more efficient than the iterations of Algorithm U, but it is likely that $J_U < J_T$ (see the next lemma). Thus, we expect to have $C_U < C_T$, except (possibly) when M is small or J_T is close to J_U.

The space complexity is given by

$$(4.5) \qquad S_T = (M + 4)m^2 + m$$

if we only keep the set of matrices $\{A_i' = A_i(I - A_1)^{-1}\}$ and drop the set $\{A_i\}$. Thus, we see that $S_T = S_U = S_N + m^2$.

We shall now denote by $\{R_k^{(T)}, k \geq 0\}$ the sequence of matrices defined by (4.2, 4.3), and we shall denote, as we did previously, by $\{R_k^{(N)}, k \geq 0\}$ and $\{R_k^{(U)}, k \geq 0\}$ the sequences defined by the equations (2.12) and (3.10) respectively.

LEMMA 4.1. *One has that $R_k^{(N)} \leq R_k^{(T)} \leq R_k^{(U)}$, for all $k \geq 0$. In other words, the natural sequence converges more slowly to R than the traditional sequence, which in turn converges more slowly than the sequence obtained by Algorithm U.*

Proof We define the matrix operator $\mathcal{H}$ as follows:

$$\mathcal{H}Y = \sum_{\nu \geq 1} Y^\nu A_{\nu+1} = \mathcal{F}Y - A_1.$$

Then, the equations (4.2) and (4.3) may be written as

$$R_0^{(T)} = A_0(I - A_1)^{-1},$$
$$R_{k+1}^{(T)} = A_0(I - A_1)^{-1} + R_k^{(T)}\{\mathcal{H}R_k^{(T)}\}(I - A_1)^{-1}.$$

Since $\mathcal{H}Y_1 \leq \mathcal{H}Y_2$ whenever $0 \leq Y_1 \leq Y_2 \leq R$, one may repeat the argument in Theorem 3.5 to prove that the sequence $\{R_k^{(T)}, k \geq 0\}$ monotonically converges from below to the matrix R.

Clearly, we have that $R_0^{(N)} \leq R_0^{(T)} = R_0^{(U)}$. We shall now make the induction assumption that $R_k^{(N)} \leq R_k^{(T)} \leq R_k^{(U)}$. Starting from (3.11), we successively obtain that

$$R_{k+1}^{(U)} = A_0(I - \mathcal{F}R_k^{(U)})^{-1} = A_0(I + (I - \mathcal{F}R_k^{(U)})^{-1}\mathcal{F}R_k^{(U)})$$
$$= A_0 + R_{k+1}^{(U)}\mathcal{F}R_k^{(U)} = A_0 + R_{k+1}^{(U)}A_1 + R_{k+1}^{(U)}\mathcal{H}R_k^{(U)}$$

Therefore,

$$R_{k+1}^{(U)} = A_0(I - A_1)^{-1} + R_{k+1}^{(U)}\{\mathcal{H}R_k^{(U)}\}(I - A_1)^{-1}$$
$$\geq A_0(I - A_1)^{-1} + R_k^{(U)}\{\mathcal{H}R_k^{(U)}\}(I - A_1)^{-1}$$
$$\geq A_0(I - A_1)^{-1} + R_k^{(T)}\{\mathcal{H}R_k^{(T)}\}(I - A_1)^{-1} = R_{k+1}^{(T)},$$

where the first inequality results from Theorem 3.5, the second inequality results from the induction assumption and the monotonicity of $\mathcal{H}$.

To prove that $R_{k+1}^{(N)} \leq R_{k+1}^{(T)}$, we successively write that

$$R_{k+1}^{(T)} = A_0 + R_k^{(T)}\mathcal{H}R_k^{(T)} + R_{k+1}^{(T)}A_1$$
$$\geq A_0 + R_k^{(T)}\mathcal{H}R_k^{(T)} + R_k^{(T)}A_1 = A_0 + R_k^{(T)}\mathcal{F}R_k^{(T)}$$
$$\geq A_0 + R_k^{(N)}\mathcal{F}R_k^{(N)} = R_{k+1}^{(N)}.$$

□

As a consequence of this lemma, we may expect to have the following inequalities: $J_U \leq J_T \leq J_N$.

REMARK 4.2. It is interesting to note that the equation (4.1) may also be obtained by a purely probabilistic argument. We define a sojourn at level $n+1$ as an interval $[t_1, t_1 + 1, \cdots, t_2]$ such that $X_t \in n+1$ for all $t, t_1 \leq t \leq t_2, X_{t_1-1} \notin n+1$ and $X_{t_2+1} \notin n+1$. For every k and $j, 1 \leq k, j \leq m$, $(I - A_1)^{-1}_{kj}$ is the expected number of visits to the state $(n+1, j)$ during a sojourn at level $n+1$ which begins in the state $(n+1, k)$.

We also define the matrix E of order m, where E_{ik} is the expected number of times that a sojourn at level $n+1$ occurs, beginning in the state $(n + 1, k)$, under taboo of the levels 0 to n, given that the initial state is $(n + 1, i)$. We obtain by a simple argument that $R = A_0 E(I - A_1)^{-1}$, by combining the expected number of sojourns and the expected number of visits during a sojourn.

In order to determine the matrix E, we consider the censored Markov chain observed at the epochs of transition from one level to another. This is a new Markov chain with transition matrix P^* given by

$$P^* = \begin{bmatrix} 0 & C_0^* & 0 & 0 & \cdots \\ B_1^* & 0 & A_0^* & 0 & \cdots \\ B_2^* & A_2^* & 0 & A_0^* & \cdots \\ B_3^* & A_3^* & A_2^* & 0 & \cdots \\ \vdots & \vdots & \vdots & \vdots & \ddots \end{bmatrix}$$

where $B_i^* = (I - A_1)^{-1}B_i$, for $i \geq 1$, $A_i^* = (I - A_1)^{-1}A_i$, for $i \neq 1$, and $C_0^* = (I - B_0)^{-1}A_0$.

If we denote by U^* the taboo transition matrix for this new Markov chain, it is clear that $E = (I - U^*)^{-1}$, hence we have that $R = A_0(I - U^*)^{-1}(I - A_1)^{-1}$ or, by adapting (3.4) to the new Markov chain,

$$\begin{aligned}
R &= A_0\{I + \sum_{\nu \geq 1}[(I - U^*)^{-1}A_0^*]^\nu(I - U^*)^{-1}A_{\nu+1}^*\}(I - A_1)^{-1} \\
&= A_0(I - A_1)^{-1} + \\
&\quad \sum_{\nu \geq 1} A_0[(I - U^*)^{-1}(I - A_1)^{-1}A_0]^\nu(I - U^*)^{-1}(I - A_1)^{-1} \\
&\qquad \times A_{\nu+1}(I - A_1)^{-1} \\
&= A_0(I - A_1)^{-1} + \sum_{\nu \geq 1}[A_0(I - U^*)^{-1}(I - A_1)^{-1}]^{\nu+1}A_{\nu+1}(I - A_1)^{-1} \\
&= A_0(I - A_1)^{-1} + \sum_{\nu \geq 2} R^\nu A_\nu(I - A_1)^{-1},
\end{aligned}$$

which is identical to (4.1).

The modification described at the end of Section 3 also applies, of course, and the sequence of matrices defined by (4.3) with $R_0 = \eta u \cdot 1$ also converges to the matrix R. The proof is similar to that of Theorem 3.8. We shall denote by T' the corresponding algorithm, which has the same complexity per iteration as Algorithm T and the same overhead as Algorithm U'.

The same comparison can be made between the algorithms T and T' as between U and U' (see Corollary 3.9), and therefore we conclude without giving the detailed proof, that $J_{T'} \leq J_T + 1$.

REMARK 4.3. Ramaswami [17] has proposed an approach based on the Newton-Kantorovich method. The various algorithms that we describe solve an equation of the type $\mathcal{N}R = 0$ by using the iteration $R_{k+1} = R_k - \mathcal{N}R_k$. The Newton iteration is defined by

$$R_{k+1} = R_k - \{\mathcal{N}'(R_k)\}^{-1}\mathcal{N}R_k,$$

where $\mathcal{N}'(X)$ denotes the derivative of the operator $\mathcal{N}$, evaluated at X.

Under suitable conditions, this scheme sharply reduces the number of iterations. Unfortunately, this reduction is balanced by a much higher cost per iteration, since it is necessary to evaluate the expression $\mathcal{N}'(R_k)\}^{-1}\mathcal{N}R_k$.

It is proposed in [17] to use sub-inverse operators, in the hope that the cost of each iteration is not too high, and yet the number of iterations is significantly reduced. We have experimented with this idea, and have observed some limited improvements over the basic algorithms U and T. Details may be found in [10].

5. Algorithms Based on Censored Markov Chains. The problem of finding the stationary probability vector for the matrix P in (1.1) is also addressed in Grassmann and Heyman [3] and Kao [6]. The common approach there has three characteristics. Firstly, it is considered that since only a finite number M of matrices A_i are used in actual computations, then one may assume right at the beginning that $A_i = 0$ for $i \geq M+1$. Secondly, the algorithm directly determines the matrix $T + HND$ (see Theorem 2); the stationary vector is then to be determined by Gaussian elimination; the matrix R is obtained as a by-product. Finally, it is assumed that $B_i = 0$ for $i \geq M$.

From the same starting point, we shall proceed by purely probabilistic arguments. We shall eventually obtain an algorithm which is more efficient and more general than the algorithm in [6].

We shall thus restrict our attention to the special case where we assume that $A_i = 0$ for $i \geq M + 1$, but we do not make any assumption about the matrices $B_i, i \geq M$, thereby allowing the Markov chain to directly drop to level $\mathbf{0}$ in one transition, from any other level.

Since the matrix H has nonzero elements in its last block of rows only, therefore only the last block of rows in $T + HND$ will differ from T. If we partition the state space as in Section 2, at any level $\boldsymbol{n}$ where $n \geq M$ (recall that the results in that section are independent of n), then the last block of rows of T is given by

$$T_{n,.} = [B_n \; 0 \; \cdots \; 0 \; A_M \; \cdots \; A_1]$$

Also, for every state in any of the levels $\mathbf{1}$ to $\boldsymbol{n} - \boldsymbol{M}+\mathbf{1}$, the corresponding column of the matrix D is identically zero, and the corresponding columns of T and $T + HND$ are equal. Thus, if we write the last block of rows of $T + HND$ as follows,

$$(T + HND)_{n,.} = [C_n \; 0 \; \cdots \; 0 \; U^{(M-1)} \; \cdots \; U^{(0)}],$$

we have that

$$\begin{aligned}
[C_n \; 0 \; \cdots \; 0 \; U^{(M-1)} \; \cdots \; U^{(0)}] = \\
[B_n \; 0 \; \cdots \; 0 \; A_M \; \cdots \; A_1] + [A_0 \; 0 \; 0 \; \cdots]ND.
\end{aligned}$$

Since $A_0 N_{1k} = R^k$ (by the equation (2.9)), we find that

$$[A_0\ 0\ 0\ \cdots]N = [R\ R^2\ R^3\ \cdots],$$

and we eventually obtain the following equations

$$C_n = B_n + \sum_{\nu \geq 1} R^\nu B_{n+\nu},$$

$$U^{(i)} = A_{i+1} + \sum_{1 \leq \nu \leq M-i-1} R^\nu A_{i+1+\nu}, \qquad \text{for } 0 \leq i \leq M-1,$$

which may also be written as

$$(5.1) \qquad C_n = \sum_{\nu \geq 0} R^\nu B_{n+\nu},$$

$$U^{(M-1)} = A_M,$$

$$U^{(i)} = A_{i+1} + RU^{(i+1)}, \qquad \text{for } 0 \leq i \leq M-2.$$

Since the matrix $U^{(0)}$ is the matrix U defined in the section 3 (we used this in the proof of Theorem 3.1), then we may write the last equation as $U^{(i)} = A_{i+1} + A_0(I - U^{(0)})^{-1} U^{(i+1)}$. If we are only interested in determining the matrix R or equivalently the matrix $U = U^{(0)}$, then we may write globally the last two equations as follows:

$$[U^{(M-1)}\ U^{(M-2)}\ \cdots\ U^{(0)}] =$$
$$(5.2) \qquad [A_M\ A_{M-1}\ \cdots\ A_1] + A_0(I - U^{(0)})^{-1}[0\ U^{(M-1)}\ \cdots\ U^{(1)}].$$

From this equation, we obtain an iterative scheme, which is stated and proved in the next theorem. The proof follows along the same line as that of Theorem 2.1 in Benseba [1].

It will sometimes be convenient to use a more compact notation: we define the matrices U and A with m rows and mM columns, respectively by $U = [U^{(M-1)}U^{(M-2)}\ldots U^{(0)}]$, and $A = [A_M A_{M-1} \cdots A_1]$.

THEOREM 5.1. *If the Markov chain is irreducible and ergodic, then the matrix U is the limit of the non-decreasing sequence $\{U_k, k \geq 0\}$ defined by*

$$(5.3) \qquad [U_0^{(M-1)}\ U_0^{(M-2)}\ \cdots\ U_0^{(0)}] = [A_M\ A_{M-1}\ \cdots\ A_1],$$

$$[U_{k+1}^{(M-1)}\ U_{k+1}^{(M-2)}\ \cdots\ U_{k+1}^{(0)}] = [A_M\ A_{M-1}\ \cdots\ A_1]$$
$$(5.4) \qquad + A_0(I - U_k^{(0)})^{-1}[0\ U_k^{(M-1)}\ \cdots\ U_k^{(1)}], \qquad \text{for } k \geq 0.$$

Furthermore, the matrix U is the minimal nonnegative solution of the equation (5.2). Finally, the sequence $\{R_k, k \geq 0\}$ where

$$(5.5) \qquad R_k = A_0(I - U_k^{(0)})^{-1}, \qquad \text{for } k \geq 0$$

monotonically converges to R.

Proof We firstly define the subsets of states l_n and L_n, where $l_n = 0 \cup 1 \cup \cdots \cup n$, and $L_n = n+1 \cup n+2 \cup \cdots$, for $n \geq 0$. We also define the first passage times T_n to l_n: $T_n = \inf\{t \geq 1 : X_t \in l_n\}$; these are a.s. finite since the Markov chain is ergodic. We define the events $E_n(i,j) = [X_{T_n} = (i,j)]$, for $n \geq 0, 0 \leq i \leq n, 1 \leq j \leq m$. Finally, we define the events $E_n^{n+k}(i,j) = [X_{T_n} = (i,j), X_t \notin L_{n+k}, \text{ for } 1 \leq t \leq T_n - 1]$ where the Markov chain is not allowed to move beyond the level $n+k$ before it enters l_n for the first time.

It is obvious that for any fixed n, i and j, the events $E_n^{(n+k)}(i,j)$ monotonically converge to the event $E_n(i,j)$ as k tends to infinity:

$$E_n^n(i,j) \subset E_n^{n+1}(i,j) \subset \cdots \subset E_n^{n+k}(i,j) \subset \cdots \subset E_n(i,j),$$

We clearly have that $U_{jj'}^{(i)} = P[E_n(n-i,j')|X_0 = (n,j)]$ for any $n \geq M$. If we define $U_{k;jj'}^{(i)} = P[E_n^{n+k}(n-i,j')|X_0 = (n,j)]$, then it is obvious that the sequence $\{U_k, k \geq 0\}$ is non-decreasing and converges to U. It is also obvious that the matrix U_0 is given by (5.3).

For $k \geq 1$, we obtain by conditioning on X_1 that

$$(U_k^{(i)})_{jj'} = (A_{i+1})_{jj'} + \sum_{1 \leq s \leq m} (A_0)_{js} P[E_n^{n+k}(n-i,j')|X_0 = (n+1,s)],$$

or in matrix notations, $U_k = A + A_0 G_k$, where $G_k = [G_k^{(M-1)} G_k^{(M-2)} \cdots G_k^{(0)}]$, with $(G_k^{(i)})_{jj'} = P[E_n^{n+k}(n-i,j')|X_0 = (n+1,s)]$.

Starting from X_0 in $n+1$, and conditioning on the state at time T_{n+1}, we obtain by the strong Markov property that

$$P[E_n^{n+k}(n-i,j')|X_0 = (n+1,j)] =$$
$$P[E_{n+1}^{n+k}(n-i,j')|X_0 = (n+1,j)] +$$
$$\sum_{1 \leq s \leq m} P[E_{n+1}^{n+k}(n+1,s)|X_0 = (n+1,j)]P[E_n^{n+k}(n-i,j')|X_0$$
$$= (n+1,s)]$$

In matrix form, this is written as $G_k^{(i)} = U_{k-1}^{(i+1)} + U_{k-1}^{(0)} G_k^{(i)}$, and we obtain that

(5.6) $\qquad G_k^{(i)} = (I - U_{k-1}^{(0)})^{-1} U_{k-1}^{(i+1)}, \qquad \text{for } 0 \leq i \leq M - 2.$

Observe that the inverse exists because the Markov chain is irreducible.

It is impossible to move from $n+1$ to $n+1-M$ without visiting one of the intermediary levels, therefore $G_k^{(M-1)} = 0$. This, combined with (5.6) and $U_k = A + A_0 G_k$ gives (5.4).

To prove the second statement of the theorem, we proceed as usual: if $V = [V^{(M-1)} V^{(M-2)} \cdots V^{(0)}]$ is another solution, then necessarily $V \geq A = U_0$. We then prove by induction that $V \geq U_k$ for all k.

Since $\{U_k^{(0)}, k \geq 0\}$ is an increasing sequence converging to $U^{(0)} = U$, then $\{R_k, k \geq 0\}$ is also an increasing sequence, converging to $A_0(I - U)^{-1} = R$, which concludes the proof.

$\square$

REMARK 5.2. It is clear that the matrices $\{G_k, k \geq 0\}$ also converge to a matrix G such that $G^{(i)} = (I - U)^{-1}U^{(i+1)}$, for $0 \leq i \leq M - 2$, $G^{(M-1)} = 0$. That matrix is of independent interest and has the following interpretation: for any $n \geq M + 1$, the probability that, starting from the state (n, j), the first state visited in the levels 0 to $n-1$ is (n', j'), is equal to $G_{jj'}^{(n-n'-1)}$, for $n+1-M \leq n' \leq n-1$. For $1 \leq n' \leq n-M$, that probability is 0, and it is simple to verify that for $n' = 0$ it is equal to $((I-U)^{-1}C_n)_{jj'}$, where C_n is given by (5.1).

The matrix $G^{(0)}$ plays an important role for Markov chains with the structure (1.2), see Neuts [14] .

In order to determine the matrix R, we have a new set of iterative equations. In expanded form:

$$(5.7) \qquad U_0^{(i)} = A_{i+1}, \qquad \text{for } i = 0, \cdots, M - 1,$$

$$(5.8) \qquad R_0 = A_0(I - U_0^{(0)})^{-1},$$

$$(5.9) \qquad U_{k+1}^{(M-1)} = A_M,$$

$$(5.10) \qquad U_{k+1}^{(i)} = A_{i+1} + R_k U_k^{(i+1)}, \qquad \text{for } i = M - 2, \cdots, 0,$$

$$(5.11) \qquad R_{k+1} = A_0(I - U_{k+1}^{(0)}), \qquad \text{for } k \geq 0.$$

The corresponding algorithm is given below.

Algorithm C — The algorithm based on censored Markov chains
$k := 0$;
for $i := 0$ **to** $M - 1$ **do** $U_0^{(i)} := A_{i+1}$;
$R_0 := A_0(I - U_0^{(0)})^{-1}$;
repeat
  $k := k + 1$;
  for $i := 0$ **to** $M - 2$ **do** $U_k^{(i)} := A_{i+1} + R_{k-1}U_{k-1}^{(i+1)}$;
  $U_k^{(M-1)} := A_M$;
  $R_k := A_0(I - U_k^{(0)})^{-1}$
until $\|R_k - R_{k-1}\| < \epsilon$

The computational and space complexities are respectively given by

$$(5.12) \qquad C_C = J_C[(M + 1/3)m^3 - 1/3m] + 4/3m^3 - 1/3m,$$

and

$$(5.13) \qquad S_C = (2M + 4)m^2 + m.$$

If we compare this to the equations (3.12, 3.13) for Algorithm U, we observe that both algorithms have the same complexity per iteration, but that S_C

is approximately equal to $2S_U$. We show below that J_C is likely to be larger than J_U and our numerical experience indicates that J_C is often larger than J_N even, thus we conclude that this algorithm is definitely not the best one.

We denote by $\{R_k^{(C)}, k \geq 0\}$ the sequence defined by (5.8, 5.11), and (as before) by $\{R_k^{(U)}, k \geq 0\}$, the sequence defined by (3.10).

LEMMA 5.3. *One has that* $R_k^{(C)} \leq R_k^{(U)}$, *for all* $k \geq 0$.

Proof The proof is indirect. The equations (5.9) to (5.11) suggest a Gauss-Seidel type of improvement, where each iterate for $U^{(i)}$ is computed from the latest iterate for $U^{(i+1)}$:

$$
\begin{aligned}
\tilde{R}_0^{(C)} &= A_0(I - A_1)^{-1}, \\
\tilde{U}_{k+1}^{(M-1)} &= A_M, \\
(5.14) \quad \tilde{U}_{k+1}^{(i)} &= A_{i+1} + \tilde{R}_k^{(C)}\tilde{U}_{k+1}^{(i+1)}, \qquad \text{for } i = M-2, \cdots, 0, \\
\tilde{R}_{k+1}^{(C)} &= A_0(I - \tilde{U}_{k+1}^{(0)})^{-1}, \qquad \text{for } k \geq 0.
\end{aligned}
$$

If we expand (5.14), we eventually find that $\tilde{U}_{k+1}^{(0)} = \sum_{0 \leq \nu \leq M-1}(\tilde{R}_k^{(C)})^\nu A_{\nu+1}$ so that, given our assumption that $A_i = 0$ for $i \geq M+1$, we actually have that $\tilde{R}_k^{(C)} = R_k^{(U)}$, for all $k \geq 0$. Thus, we need to prove that $R_k^{(C)} \leq \tilde{R}_k^{(C)}$, for all $k \geq 0$.

Since $\tilde{R}_0^{(C)} = R_0^{(C)}$, the claimed inequalities hold for $k = 0$, and we make the first induction hypothesis that $R_k^{(C)} \leq \tilde{R}_k^{(C)}$ for some k.

At the beginning of the next iteration, we have that $U_{k+1}^{(M-1)} = \tilde{U}_{k+1}^{(M-1)} = A_M$. Thus, we make the second induction hypothesis that $U_{k+1}^{(i)} \leq \tilde{U}_{k+1}^{(i)}$ for some i. Then we have that

$$
\begin{aligned}
U_{k+1}^{(i-1)} &= A_i + R_k^{(C)}U_k^{(i)} \\
&\leq A_i + R_k^{(C)}U_{k+1}^{(i)}, \qquad \text{by Theorem 5.1,} \\
&\leq A_i + R_k^{(C)}\tilde{U}_{k+1}^{(i)}, \qquad \text{by the second induction assumption,} \\
&\leq A_i + \tilde{R}_k^{(C)}\tilde{U}_{k+1}^{(i)} = \tilde{U}_{k+1}^{(i-1)},
\end{aligned}
$$

by the first induction assumption. Then, $U_{k+1}^{(0)} \leq \tilde{U}_{k+1}^{(0)}$, and $R_{k+1}^{(C)} \leq \tilde{R}_{k+1}^{(C)}$, which completes the proof.

□

In the discussion after Lemma 3.6, we have indicated that the probabilistic significance of the iterated $\{R_k, k \geq 0\}$, for a given iterative sequence, is usually difficult to describe. One clear exception is the sequence $\{R_k^{(C)}, k \geq 0\}$ (see the proof of Theorem 5.1). Given the argument in the proof of the lemma above, we believe (but have not attempted to prove) that for a given k, $R_k^{(U)}$ captures trajectories such that the Markov chain moves freely in the levels $n+1$ to $n+k$ and is allowed to travel *upwards* through the levels in L_{n+k}.

It is quite natural to try and bring to the algorithm C the same improvements as to the algorithms U and T. We have not found a way to force the matrices $\{R_k, k \geq 0\}$ from (5.5) to remain within the set of nonnegative matrices with maximal eigenvalue η and left eigenvector u. However, there is an alternative, suggested by Kao [6], in the particular case when $B_n = 0$ for all $n \geq M$. Then $C_n = 0$ for all $n \geq M$, and since both P and $T+HND$ are stochastic matrices, we have that $A1 = U1 = 1$. Thus, it seems that some improvement may be expected if U_0 is properly initialized. This is shown below.

THEOREM 5.4. *Assume that* $B_n = 0$ *for all* $n \geq M$. *Let the matrix* U_0 *be chosen such that* $U_0 \geq A$, *and* $U_0 1 = 1$. *For this choice of initial matrix, the sequence* $\{R'_k, k \geq 0\}$ *of matrices defined by the equations (5.4, 5.5) converges to the matrix* R. *Moreover, this sequence is related as follows to the sequence* $\{R_k, k \geq 0\}$ *of Theorem 5.1: we have that* $R_k \leq R'_k$ *for all* k.

Proof We shall denote by $U'_k, k \geq 0$, the matrices which are obtained by the present choice of initial matrices, and keep the notation U_k for the matrices of Theorem 5.1. By assumption, $U_0 \leq U'_0$ and it is a simple matter to prove by induction that $U_k \leq U'_k$, for all k, from which it results that $R_k \leq R'_k$ for all k.

We now make the induction assumption that $U'_k 1 = 1$ for some k. This is true for $k = 0$ by hypothesis. We then obtain that

$$U'_{k+1}1 = \sum_{1 \leq i \leq M} A_i 1 + A_0(I - U'^{(0)}_k)^{-1} \sum_{1 \leq i \leq M-1} U'^{(i)}_k 1$$

$$= 1 - A_0 1 + A_0(I - U'^{(0)}_k)^{-1}(1 - U'^{(0)}_k 1)$$

$$= 1 - A_0 1 + A_0(I - U'^{(0)}_k)^{-1}(I - U'^{(0)}_k)1 = 1.$$

Thus, the matrices $\{U'_k, k \geq 0\}$ remain in the compact set of nonnegative matrices with rowsums equal to one. Therefore, the sequence has at least one accumulation point. We denote by U^* one such point and we readily conclude that $U^* \geq U$. Since $U \geq 0$, and since $U^*1 = U1 = 1$, implies that $U^* = U$.

The proof is now complete: the sequence $\{U'_k, k \geq 0\}$ necessarily converges to U, and the sequence $\{R_k, k \geq 0\}$ converges to R.

□

We may accordingly modify the algorithm C and replace the second statement by some initialization which satisfies the condition of the theorem above. We shall call this the algorithm C'. The complexity per iteration is the same as that of Algorithm C (see (5.12)), and we may expect that $J_{C'} \leq J_C$.

REMARK 5.5. The theorem 5.4 leaves open the choice of initial matrices. One such choice is $U^{(i)}_0 = A_{i+1}, 1 \leq i \leq M - 1, U^{(0)}_0 = A_1 + A_0$.

For that choice, the algorithm C' produces the same sequence $\{U_k, k \geq 0\}$ as the algorithm in Kao [6]. However, the complexity of the algorithm in [6] is $(3/2M - 2/3)m^3 - (M/2 - 1)m^2 - 1/3m$ at each iteration, which is larger than the complexity $(M + 1/3)m^3 - 1/3m$ for each iteration of the algorithm C'. Other choices are available, however, and we have found that the following one yields sequences which converge slightly faster: $U^{(M-1)} = A_M, U^{(i)} = \Delta A_{i+1}, 0 \leq i \leq M - 2$, where Δ is a normalising diagonal matrix. This is the initialization which we have chosen in our numerical comparisons.

The inequalities that we have proved, between the number of iterations required by the various algorithms, are summarized in the precedence graph below.

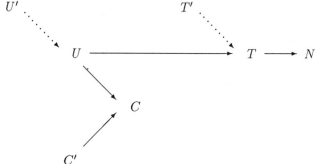

Clearly, this is a partial order only.

Our computational experience indicates that the algorithm C' is about as fast as the algorithm U'. However, it requires a much larger workspace, since $S_{C'} = (2M + 4)m^2 + m = S_{U'} + Mm^2$.

6. Programming Details. The algorithms which we have described are all very simple and may be easily translated in any programming language. For instance, we have used PASCAL in our exploratory investigations and FORTRAN 77 for the main study to be described in the next section. Most of the operations such as matrix additions or products are entirely routine and do not call for any comment.

The various matrix inversions are performed by Gaussian elimination. Typically, we have to evaluate expressions of the form $A(I - V)^{-1}$, where V is a substochastic matrix such that $(I - V)^{-1}$ exists. A standard recommendation (see Isaacson and Keller [5] for instance) is to solve the system $X(I - V) = A$, at a cost of approximately $4/3$ the cost of a matrix multiplication. In addition, if the factors used during the elimination phase are stored, then a second expression $B(I - V)^{-1}$ with the same matrix V can be obtained at the same cost as a matrix multiplication.

During the elimination phase, we have to do row-elimination and there is no need to perform any pivoting, as mentioned in Grassmann et al. [2]. Furthermore, we may use the special structure of the coefficient matrix

$I - V$, so that the program executes arithmetic operations on nonnegative numbers only. This is a desirable property for the following reason.

The algorithms appear to be robust against numerical instabilities due to rounding errors, since they only involve nonnegative quantities. Indeed, one interprets either as probabilities or as expected numbers of visits the entries in the various matrices which are computed. Thus, it is desirable that no numerical instability is caused by the matrix inversion.

The last algorithmic step which we have to describe is the evaluation of the maximal eigenvalue η and eigenvector u which are needed by the algorithms U' and T'. This should be done as efficiently as possible, since it constitutes an added overhead to the execution time of U' and T'.

Recall that η is the unique solution of $z = \chi(z)$ in $(0,1)$. The solution is found iteratively by successive bisections of the interval $(0,1)$. At each iteration we compute the maximal eigenvalue $\chi(z)$ of $A(z)$ for some z, and examine the sign of $z - \chi(z)$, in order to determine which subinterval is to be retained.

In order to evaluate $\chi(z)$, Neuts ([12], page 39) recommends an algorithm due to Elsner, but we prefer the following approach, which we have found to converge slightly faster. We firstly use the power method (see Ralston and Rabinowitz [16], Chapter 7) in order to find a rough approximation of $\chi(z)$ and its associated eigenvector $u(z)$. We secondly use the inverse power method, with Rayleigh quotient iteration ([16], Section 7.2) which then converges very rapidly. It is possible that more efficient procedures exist, but we believe that our experimental results are clear enough, so that we need not pursue this subject here.

In actual fact, the algorithm does not calculate the exact eigenvalue η, of course. Therefore, we may rely on the theorem 3.8 to ensure the convergence of the algorithms U' and T' only if the computed value for η is very close to being exact. For that reason, we have chosen 10^{-12} as the final required precision on the eigenvalues, even though we require a precision of 10^{-8} on the matrix R. Indeed, we have found in a few cases that the convergence of the algorithms U' and T' is slow if η is calculated with a precision of 10^{-8} only.

We finally mention that we have used a relative norm when comparing two successive matrix iterates:

$$\|A - B\| = m^2 \max_{1 \leq i,j \leq m} |A_{ij} - B_{ij}| / \sum_{1 \leq i,j \leq m} (|A_{ij}| + |B_{ij}|).$$

7. Numerical Experiment.

In order to compare the algorithms, we have applied them to a collection of Markov chains obtained by considering $GI/PH/1$ queues at epochs of arrivals; these constitute the paradigm for Markov chains with the structure indicated by (1.1).

Our selection of test problems is a subset of those designed in Ramaswami and Latouche [19]. We have considered the same 3 values for the traffic coefficient, the same 11 distributions for service times and 3 of the distributions for the interarrival times (E_5, M and H_2), see [19] for details. These 99 test problems cover some of the worst (in terms of speed of convergence) as well as some of the best cases reported there. The values of m and M respectively range from 1 to 26 and 4 to 76: we have in each case truncated the series $\sum_{\nu \geq 0} A_\nu$ so that the remainder is less than 10^{-4}, then re-normalized the matrices so that the sum $\sum_{0 \leq \nu \leq M} A_\nu$ becomes stochastic.

For the main experiment, a program was written in FORTRAN 77 and executed on a CRAY X-MP14. By turning off the optimization options of the compiler, we forced the program to execute in a purely sequential manner; furthermore, we did not use any library subroutine. In a final experiment, we did perform some code optimization; this will be described in the last part of this section.

The execution times which we report do not include the overhead due to I/O operations and the computation of the matrices A_i; they do include the overhead due to all the other computations required by each algorithm. For instance, the execution time for Algorithm U' includes the evaluation of η; the execution time for Algorithm T' includes both the evaluation of η and the computation of the matrices A_i', etc.

In this way, we feel confident that the comparisons between the execution times are not dependent on the computing environment, and that they are tightly related to the algorithms.

As a final verification, we wanted to verify that the programmed algorithms produce similar results, so that it is meaningful to compare their execution times. Ideally, we should examine the differences between the various approximations and the exact matrix R. Since we do not know the latter, we have used the following procedure. We have computed the approximation given by a Newton-Kantorovich algorithm with $\epsilon = 10^{-12}$ and have used it as the solution of reference. Our choice was dictated by the fact that the iterations so produced are closer to R than those of the other monotonically convergent algorithms. We then solved the same problem by every algorithm, with ϵ ranging from 10^{-4} to 10^{-12}. This was repeated for a small subset of our test problems. The differences between the various approximations and the solution of reference were always between ϵ and $10^2 \epsilon$.

Thus, it appears that the algorithms do not really have the same precision, for a given value of ϵ. However, we also observed that the fastest and the slowest algorithms respectively were the most and the least precise. Therefore, we feel that we need not investigate this further here. We choose $\epsilon = 10^{-8}$ and assume that all the algorithms will give solutions which are exact to 6 significant digits.

7.1. Comparisons within each family. The graphs displayed below are all log-log scatter plots, where we have drawn the line $y = x$ for visual reference.

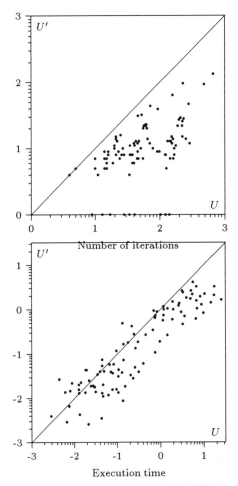

Fig. 7.1. *Scatter plots for the algorithms U' against U.*

We firstly compare on Figure 7.1 the two algorithms based on the matrix U. The plot on the left confirms the theoretical results: we clearly see that $J_{U'}$ is smaller than J_U, often by a considerable factor. The plot on the right indicates that the comparisons for the execution times are not as sharp. Let us denote by T_A the execution time for any algorithm A. As expected, $T_{U'}$ is generally smaller than T_U, by factors greater than 10 sometimes; however, we see a number of cases where $T_{U'}$ is larger than T_U, despite the facts that $J_{U'}$ is smaller than J_U. This indicates that the overhead of Algorithm U' is not negligible, and also that it is important to evaluate the maximal eigenvalue η as efficiently as possible.

The same remarks apply to the two traditional algorithms , as can be seen on Figure 7.2.

We compare on Figure 7.3 the algorithms C and C', and we observe

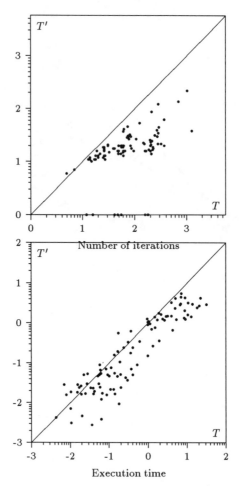

FIG. 7.2. *Scatter plots for the algorithms T' against T.*

very large differences both in the number of iterations and in the execution times: $T_{C'}$ is smaller than T_C by factors up to 100. In fact, these two algorithms are at opposite extremes of our set, where C is the slowest and C' one of the fastest. We also observe that a reduction in the number of iterations directly translates in a reduction for the execution time, since the overhead for Algorithm C' is negligible: we only need to re-normalize the matrix $\boldsymbol{U}_0$.

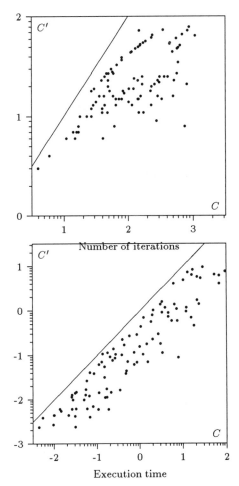

FIG. 7.3. *Scatter plots for the algorithms C' against C.*

7.2. Comparison of the basic algorithms. We next compare the four basic algorithms described in the sections 2 to 5.

Judging from Figure 7.4, there is no sharp difference between the algorithms C and N, although the latter seems to be faster on the whole, especially for cases with a long execution time.

The figures 7.5 and 7.6 confirm our theoretical results and show that there is a slight reduction in the number of iterations when mowing from the algorithm N to T and to U. We observe (see the plot on the right of Figure 7.6) that T_U is generally smaller than T_T, despite the slightly greater complexity of each iteration of the algorithm U.

We also know that $J_U \leq J_C$. This is confirmed on Figure 7.7 where we see that the reduction factor may be as large as 4, and similarly for the execution times.

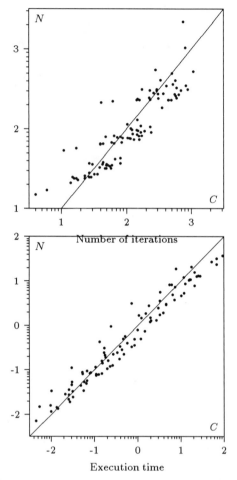

FIG. 7.4. *Scatter plots for the algorithms N against C.*

7.3. Comparison of the improved algorithms. Here, we compare the improved algorithms from different families.

The comparison between the algorithms U' and T' (Figure 7.8) reveals the same relationships as between the algorithms U and T (Figure 7.6). Our final comparison between the algorithms C' and U' does not allow us to draw a clear conclusion. On the one hand, it appears that $J_{U'}$ is usually smaller than $J_{C'}$, with some noticeable exceptions to this rule (see the left plot on Figure 7.9). On the other hand, there is a small majority of cases where $T_{C'} < T_{U'}$.

7.4. Synthesis. We shall not draw scatter plots for all 21 pairs of algorithms, but shall summarize our experimental results as precedence graphs. We shall draw an arrow from A to B to indicate that Algorithm

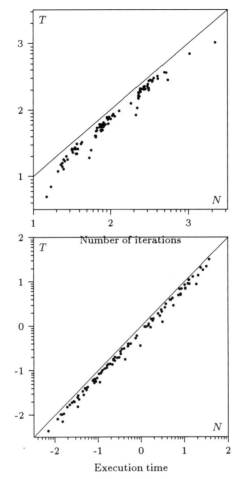

FIG. 7.5. *Scatter plots for the algorithms T against N.*

A is always (or nearly always) faster than B; a broken arrow indicates that A is very often faster than B, while a dot-arrow indicates that the relationship holds more often than not. Clearly, this synthesis is slightly subjective.

Our findings about the number of iterations can be summarized as follows; we must mention that the difference is very small in some cases.

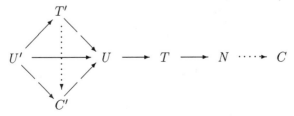

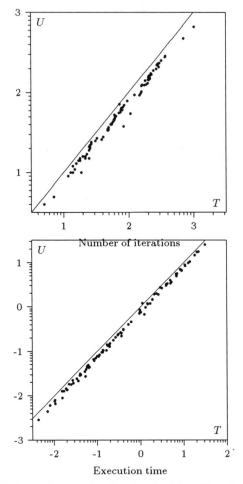

FIG. 7.6. *Scatter plots for the algorithms U against T.*

Observe that some of the relations which we draw above have not been shown on the preceding figures; they are substantiated by plots which are collected in [10].

Our experiment indicates the following hierarchy for the execution times:

$$C'$$

$$U' \longrightarrow T' \longrightarrow U \longrightarrow T \longrightarrow N \cdots\!\blacktriangleright C$$

Again, some of the differences are very small.

7.5. Vectorization. As a final experiment, we wanted to measure the improvements which could result from using the vectorization features of the CRAY computer. This seemed to be promising, since matrix operations

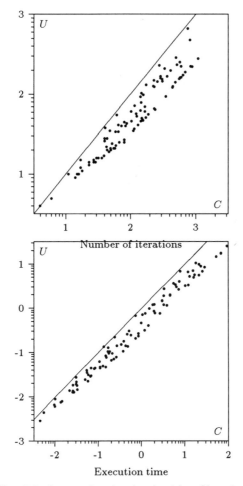

FIG. 7.7. *Scatter plots for the algorithms U against C.*

are a prime target for such optimization.

We firstly used the same program, turning on the optimization options of the compiler. We did observe some slight improvement, since the average speed went up from 8.92 Mflops for the version without optimization to 10.42 Mflops for the version with compiler optimization.

Next, we noticed that 90 percent of the computational time was spent in our subroutine for matrix multiplications. We replaced it by the *sgemm* subroutine from the BLAS package; we also replaced our matrix inversion subroutines used in the inverse power iteration algorithm by the subroutines *sgefa* and *sgesl* from the LINPACK package. The average execution speed increased ten-fold to 103.80 Mflops.

Our procedure for computing the maximal eigenvalue η then became totally inefficient, since the overhead for the algorithm U' became as much

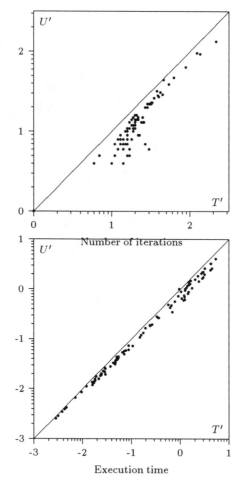

Fig. 7.8. *Scatter plot for the algorithms U' against T'.*

as 10 times greater than the remainder of the execution time. In this final experiment, we therefore only used the power method. Nevertheless, the overhead for the algorithms U' and T' remains somewhat large, and this has a negative effect on the relative performance of these two algorithms.

Overall, the hierarchy which we drew above between the seven algorithms is not much perturbed. We mostly observe a strengthening of the dominant position of Algorithm C' in terms of execution time.

8. Conclusions. We have described and analysed seven different algorithms for computing the matrix R. Our findings about their mathematical complexity is summarised in the following table, where we indicate the number of matrices required by each algorithm and the number of matrix multiplications (this dominates by far the execution time, except in cer-

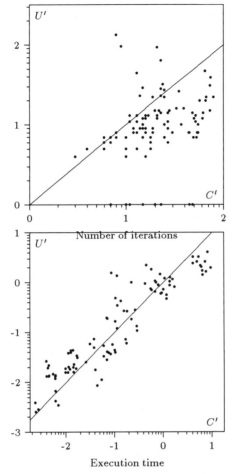

FIG. 7.9. *Scatter plots for the algorithms U' against C'.*

tain cases for the algorithms U' and T', when the evaluation of η takes a significant time).

Algorithms	Number of matrix products	Number of matrices
N	$J_N M$	$M + 3$
U and U'	$J_U(M + 1/3) + 4/3$	$M + 4$
T and T'	$J_T M + M + 4/3$	$M + 4$
C and C'	$J_C(M + 1/3) + 4/3$	$2M + 4$

The algorithm U' usually requires the smallest number of iterations; the algorithm C' is frequently the fastest.

None of the algorithms uniformly dominates all the others. For instance, the three fastest algorithms (U', C' and T') all require some additional assumption on the matrix P in order to be applicable. Moreover, the algorithm C' has a very high space complexity. Finally, we consider on Figure 8.1 the comparison between the algorithms U' and N (which is one of the slowest) as a last illustration. While $T_{U'}$ is usually smaller than T_N, with factors up to 70, yet there are a few cases where $T_N < T_{U'}$.

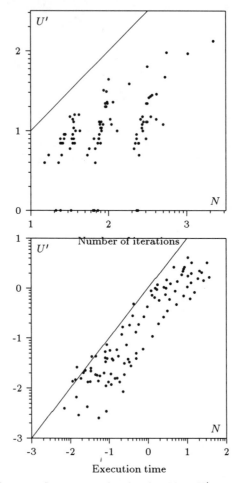

FIG. 8.1. *Scatter plots for the algorithms U' against N.*

The Newton-Kantorovich approach yields a reasonably consistent, but very modest improvement. This is a more pessimistic statement than those expressed in Gün [4] and Ramaswami [17]. Perhaps the difference in conclusions is due to the difference in examples which were considered.

Finally, we observed that the algorithms take advantage of vectorization features in a very natural way.

Clearly, it would be interesting to identify factors which would indicate when a particular algorithm can be expected to perform best. This is beyond the scope of the present paper.

REFERENCES

[1] Benseba, N. Durées de séjour dans un système de deux files en série. LITh, Université Libre de Bruxelles, TR.225, 1991.

[2] Grassmann, W.K., Taksar, M.J., and Heyman, D.P. Regenerative analysis and steady state distributions for Markov-chains. *Oper. Res.*, **33**, 1985, 1107–1116.

[3] Grassmann, W.K., and Heyman, D.P. Computation of steady-state probabilities for infinite-state Markov chains with repeating rows. Submitted for publication.

[4] Gün, L. Experimental results on matrix-analytical solution techniques — Extensions and comparisons. *Commun. Statist. — Stochastic Models*, **5**, 1989, 669–682.

[5] Isaacson, E. and Keller, H. B. *Analysis of Numerical Methods.* John Wiley & Sons, Inc., New York, 1966.

[6] Kao, E.P. Using state reduction for computing steady state probabilities of queues of $GI/PH/1$ types. *ORSA Journal on Computing*, **3**, 1991, 231–240.

[7] Kemeny, J.G., Snell, J.L., and Knapp, A.W. *Denumerable Markov Chains.* Van Nostrand, N.J., 1966.

[8] Latouche, G. A note on two matrices occuring in the solution of Quasi-Birth-and-Death processes. *Commun. Statist. — Stochastic Models*, **3**, 1987, 251–257.

[9] Latouche, G. A simple proof for the matrix-geometric theorem. *Applied Stochastic Models and Data Analysis*, to appear.

[10] Latouche, G. Comparison of algorithms for infinite Markov chains with repeating columns. Séminaire de Théorie des Probabilités, Université Libre de Bruxelles, TR.91/1, 1991.

[11] Lucantoni, D.M. and Ramaswami, V. Efficient algorithms for solving the non-linear matrix equations arising in phase type queues. *Commun. Statist. — Stochastic Models*, **1**, 1985, 29–52.

[12] Neuts, M.F. *Matrix-Geometric Solutions in Stochastic Models. An Algorithmic Approach.* The Johns Hopkins University Press, Md, 1981.

[13] Neuts, M.F. The caudal characteristic curve of queues. *Adv. Appl. Prob.*, **18**, 1986, 221–254.

[14] Neuts, M.F. *Structured Stochastic Matrices of M/G/1 Type and Their Applications.* Marcel Dekker Inc., New York, 1989.

[15] Ortega, J.M. and Rheinboldt, W.C. *Iterative Solution of Nonlinear Equations in Several Variables*, Academic Press, New York, 1970.

[16] Ralston, A. and Rabinowitz, P. *A First Course in Numerical Analysis.* McGraw-Hill, New York, 1978.

[17] Ramaswami, V. Nonlinear matrix equations in applied probability — Solution techniques and open problems. *SIAM Review*, **30**, 1988, 256–263.

[18] Ramaswami, V. Private communication, 1991.

[19] Ramaswami, V. and Latouche, G. An experimental evaluation of the matrix-geometric method for the $GI/PH/1$ queue. *Commun. Statist.— Stochastic Models*, **5**, 1989, 629–667.

[20] Ramaswami, V. and Lucantoni, D.M. Stationary waiting time distribution in queues with phase type service and in Quasi-Birth-and-Death processes. *Commun. Statist. — Stochastic Models*, **1**, 1985, 125–136.

[21] Stewart, G.W. *Introduction to Matrix Computations.* Academic Press, New York, 1973.

[22] Varga, R.S. *Matrix Iterative Analysis.* Prentice-Hall, N.J., 1962.

CRAY-2 MEMORY ORGANIZATION AND INTERPROCESSOR MEMORY CONTENTION

ROBERT W. NUMRICH*

Abstract. This paper describes a simulation study of interprocessor memory contention for a shared memory, vector multiprocessor like the CRAY-2. When programs execute together on such a system, each program's performance, relative to its performance on a single dedicated processor, degrades because of contention among processors for shared memory. From the results of the simulation study, the paper proposes analytic forms for the asymptotic steady state behavior of throughput, time delay, and efficiency as functions of hardware parameters. The results suggest criteria for evaluating hardware designs and an index of quality for comparing different designs.

1. Introduction. Explanation of interprocessor memory contention for shared memory, vector multiprocessors remains an important open problem in the design and evaluation of computer architectures. This paper presents an empirical model, based on simulation, to help explain the behavior of numerical programs executing on shared memory, multiprocessor, vector computers like the CRAY-2. Although specific to the CRAY-2, the analysis suggests that the results ought to apply, at least qualitatively, to any shared memory vector computer.

To illustrate the nature of the problem, consider the following experiment [10,15]. Measure the performance of a given program on a single processor by its time of execution or by its computational rate. Next run two copies of the same program on two processors and compare the performance of each copy with its performance on one processor. Contention for shared memory between the two processors usually results in lower performance in the second case. The performance degrades further as three, four or more copies of the program execute at the same time.

Define an efficiency as the ratio of performance for multiple copies to the performance for a single copy. Figure 1.1 shows the result of such an experiment for three different programs. The solid curves show the drop in efficiency for a machine with slow memory; the dotted curves show the drop for the same programs for a machine with fast memory. The efficiency curves change from essentially concave functions of the number of processors p for slow memory to essentially convex functions of p for fast memory. Moreover, the relative positions of the three programs change: mxm is least efficient for slow memory but most efficient for fast memory. The dramatic differences in the positions and shapes of these curves result from a complicated mixture of hardware and software parameters. The list of parameters includes the kind of computation being done, the method of programming, and the quality of code generated by the compiler. It also includes the number of active processors, the number of memory ports

---

* Software Division, Cray Research, Inc., Eagan, MN

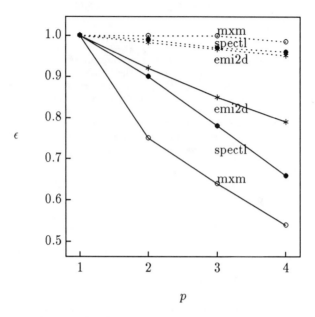

FIG. 1.1. *An example of performance degradation for three programs on the CRAY-2. Program mxm(o) is matrix multiplication; program spectl(•) is a spectral weather code; and program emi2d(∗) is an electromagnetic plasma code. The lower solid lines represent efficiency for a slow memory CRAY-2. The upper dotted lines represent efficiency for the same programs on a fast memory CRAY-2*

per processor, the number of memory banks, the memory bank reservation time, the specific design of the interconnection network between processors and memory, and the priority scheme used to resolve conflicts among processors.

No theoretical model adequately explains the observed behavior. This paper suggests an empirical model that explains some of the observed behavior. As a brief outline of the paper's organization, Section 2 describes the organization of the CRAY-2 memory system and Section 3 describes a simulator that models its behavior. Section 4 describes a dimensional analysis of the input and output parameters of the simulator. Sections 5 and 6 define dimensionless time delay and throughput functions measured by the simulator. Section 7 discusses the consequences of Little's Law for the simulation and Section 8 defines efficiency. Section 9 describes the results of the simulation and proposes asymptotic forms for the behavior of system efficiency under either light traffic or heavy traffic conditions. Section 10 proposes a method for using the results of the simulation to define hardware design criteria. Section 11 compares the results from this paper with previous work. Section 12 concludes with a short summary of results and a long list of open questions.

2. CRAY-2 Memory Organization. The CRAY-2 system contains four processors interconnected with shared memory divided into four quadrants. Each quadrant contains one quarter of the total number of memory banks in the system. Consecutive memory addresses interleave first across the four quadrants and then across the banks within each quadrant. In other words, starting from address zero in bank zero of quadrant zero, the first four consecutive addresses lie in bank zero in consecutive quadrants, the next four addresses lie in bank one of consecutive quadrants, and so forth. Traversal of the banks through consecutive addresses repeats the starting bank address only after all other banks have been visited.

A single stage perfect shuffle network [40] connects the four processors with the four quadrants. Let $i = 0, 1, 2, 3$ indicate one of the four processors and let $j = 0, 1, 2, 3$ indicate one of the four quadrants. Each processor i has four buffers p_{ij}, one for each quadrant, and each quadrant j has four buffers q_{ji}, one for each processor. The perfect shuffle network connects processor buffer p_{ij} to quadrant buffer q_{ji}.

A simple phased scheme [27] establishes a priority protocol to prevent more than one processor from accessing the same memory bank at the same time. Let

$$(2.1) \qquad \Phi_{ij}(t) = (t - i - j)(mod 4)$$

be the phase function for processor i and quadrant j at time t. When $\Phi_{ij}(t) = 0$, processor i may pass a request from buffer p_{ij} across the network to quadrant buffer q_{ji}. When $\Phi_{ij}(t) = 1$, a request in buffer q_{ji} may pass forward to a memory bank in quadrant j. Since the phase function assumes unique values for each (i, j) pair at each time t, only one processor at a time may request data from any particular bank in the memory system.

When a request arrives at a memory bank at time t_0, the bank honors it if free and rejects it if busy. A rejection is called a bank conflict. If no conflict occurs, the bank sets a reservation flag for τ clock periods. The bank rejects all subsequent requests until time $t = t_0 + \tau$. By setting $\tau = 1(mod 4)$, the processor making the original request gets lowest priority when the bank becomes free.

When a bank conflict occurs, the request in buffer q_{ji} remains there for four clock periods and attempts another bank request at its next phase time. This request cycle repeats every four clock periods until the bank is free. During this time, subsequent requests to quadrant j from processor i cannot cross the network. The issue control mechanism for processor i shuts down when it encounters the next request to quadrant j. Such a conflict is called a quadrant conflict.

By the time the issue control mechanism detects a quadrant conflict, as many as ten references may have already entered the issue control pipeline. The issue control mechanism contains a backup loop of length ten to hold delayed requests. Requests destined for quadrants other than the one causing the conflict are also delayed because all requests are serviced in their correct order. Delayed requests may traverse the backup loop more than once depending on how long it takes to resolve the underlying bank conflict.

The phased priority scheme for the CRAY-2 system exhibits an elegant simplicity. Its weakness occurs for strides [1] not equal to one. With stride two, for example, each processor returns to the same memory quadrant every other clock period. Since each processor is on phase to the same quadrant only once every four clock periods, quadrant conflicts occur.

3. Simulation Model. Analysis of the performance of real programs in terms of hardware parameters remains too difficult for any theoretical framework. This paper resorts to simulation of a very simple program. It does nothing but make vector requests to memory as often as possible. Bailey [1] used a similar simulation technique to analyze the performance of the CRAY-2 memory system. At the time, Bailey had limited information about the exact details of the memory port operation, the details of the priority scheme, and the details of the interconnection network.

The simulator used in this paper represents an exact clock-by-clock emulation [27] of the CRAY-2 memory port, the interconnection network and the memory system including the phased priority scheme and the resolution of quadrant and bank conflicts. The simulator generates stride one vector requests of fixed vector length with random starting addresses. Each processor issues independent requests and contends for resources through the interconnection network and at the memory banks. When each request is satisfied, the simulator records the time required for that request and maintains a frequency distribution for these completion times. It then issues another request, independently for each processor, with a new random starting address. After a fixed number of requests for each processor, picked large enough to reach steady state equilibrium, the simulator measures the total time required to satisfy them and computes system throughput as the ratio of total words moved to total time taken.. From the minimum time observed to satisfy one request, it also computes an average time delay per word. The simulator repeats the process for many combinations of hardware parameters as described in Section 4.

4. Dimensional Analysis. Scaling relationships [21,1,8,9] support intuitive ideas of the dependence of memory contention on hardware parameters. For example, memory contention is more severe with more processors. For a fixed number of processors, it is more severe with slow memory than with fast memory. On the other hand, memory contention is less severe with more memory banks or with reduced activity from each processor.

---

[1] Stride is the fixed spacing between consecutive addresses, or banks, during a vector request. If a_i is the address for request i, then $a_{i+1} = a_i + s$ is the address for request $i + 1$.

The methods of dimensional analysis [11,6,22,20,5,4] make precise statements about these intuitive ideas. Table 4.1 lists a set of important quantities when considering memory contention in vector multiprocessors. Length $[L]$ and time $[T]$ represent the two primary dimensional units relevant to this problem. For the CRAY-2 system, the normal unit of length is one data word, equal to 64 bits, and the normal unit of time is the clock period, the discrete interval of time that synchronizes the system, equal to 4.1×10^{-9} sec.

The first important parameter is the number of active processors p. It is a dimensionless quantity equal to 1,2,3 or 4 on the CRAY-2 system. On systems with more than one memory port per processor, p is the product of the number of active processors times the number of active memory ports per processor.

The next important parameter is the number of memory banks m with dimension $[L]$. Each bank delivers one word [2] at each request. Each processor sees a resource of m words that it must share with the other processors.

The next important parameter is the bank reservation time τ with dimension $[T]$. When any processor makes a successful request to a memory bank at clock period t, say, then the earliest clock period that any processor may make a successful request to the same memory bank is at $t + \tau$.

The next important parameter is the vector length N with dimension $[L]$. Each vector processor makes requests to memory in blocks of N words. The interval between requests is T_0 with dimension $[T]$. The request interval is bounded below by the port reservation time, the minimum time required by the instruction issue mechanism to service a request of length N. On the CRAY-2 system, as on most vector processors, the port reservation time consists of a fixed startup time T_S plus a time proportional to the vector length so that

$$(4.1) \qquad\qquad T_0 \geq T_S + N D_0.$$

The constant D_0 with dimension $[L^{-1}T]$ is the time required to service each word of the vector request. On the CRAY-2 system, $T_S = 8$ and $D_0 = 1$. On multiple pipeline vector processors with n_p pipelines, $D_0 = 1/n_p$.

Two important derived quantities are the individual processor request rate

$$(4.2) \qquad\qquad P_0 = N/T_0$$

---

[2] The parameter m might more accurately be considered the product of two parameters, a quantity m equal to the number of banks with dimension [0] and a quantity W equal to the bank width with dimension $[L]$. Bank width is the amount of data each bank can deliver in a single request. For the CRAY-2 memory system, the bank width is one word, $W = 1$, so the product mW has been replaced by m with dimension $[L]$ for simplicity. For memory systems with wider or narrower bank widths, the factor W must be considered explicitly.

and the system request rate

$$(4.3) \qquad\qquad R_0 = pN/T_0$$

both with dimension $[LT^{-1}]$. Each processor p tries to make a request for N words every T_0 clock periods.

With these input parameters, the simulator produces derived output quantities such as throughput and time delay as defined in Sections 5 and 6. Let

$$(4.4) \qquad\qquad F = f(R_0, N, \tau, m)$$

be some measured property of the system. Experience shows [3] that these derived quantities depend on the request rate R_0, the vector length N, the bank reservation time τ, and the number of memory banks m. The assumption that f depends on the number of processors p and the request interval T_0 only implicitly through its dependence on the request rate R_0 is verified by the results of the simulation.

The objective of the simulation study is to discover an empirical form for the function f. Suppose the measured quantity F has dimension $[L^{-1}T]$, for example, an average time delay per word. Then, by the Pi Theorem of dimensional analysis and by the principle of dimensional homogeneity, [4, Chapter 4], pick any two of the four quantities (R_0, N, τ, m) as primary units and scale quantities in such a way that all functions and variables become dimensionless. In this case, with m as the unit of length and τ as the unit of time, the Pi Theorem implies

$$(4.5) \qquad\qquad F = (\tau/m)f((\tau/m)R_0, N/m, 1, 1).$$

Let

$$(4.6) \qquad\qquad u = (\tau/m)R_0$$

be the scaled request rate, let

$$(4.7) \qquad\qquad v = N/m$$

be the ratio of vector length to the number of memory banks, and let

$$(4.8) \qquad\qquad \phi(u, v) = f(u, v, 1, 1)$$

be a dimensionless function of u and v. Then Equation (4.5) becomes

$$(4.9) \qquad\qquad \phi(u, v) = (m/\tau)F.$$

---

[3] The function f also depends on the stride s with dimension $[L]$. All results in this paper are for $s = 1$ so the dependence on stride has been ignored.

Scaling the measured quantity F by the factor m/τ yields a dimensionless function $\phi(u, v)$ of two dimensionless variables u and v. Similarly, a measured rate G with dimension $[LT^{-1}]$, when scaled by the factor τ/m such that

$$(4.10) \qquad \gamma(u, v) = (\tau/m)G,$$

yields a dimensionless function $\gamma(u, v)$ of two dimensionless variables u and v. The problem now lies in a two parameter space rather than a four parameter space.

The choice of fundamental units m for length and τ for time is somewhat arbitrary. Another obvious choice for the length unit is the vector length N and for the time unit the request interval T_0. The two choices are equally useful within their own context. The first choice emphasizes the properties of the hardware and the second choice emphasizes the properties of the programs. One choice is better than another only if it adds simplicity or insight to the analysis of the problem. Furthermore, the equations describing memory performance look slightly different depending on the point of view adopted. Section 5 adopts the point of view of variable programs executing on fixed hardware. From that point of view, N and T_0 seem to be the natural set of primary units. But this choice leads to an explicit dependence on p and leaves the analysis in a three parameter space. Section 6, on the other hand, adopts the point of view of variable hardware executing fixed programs. In this case, m and τ seem to be the natural set of primary units and reduce the problem to a two parameter space. Section 7 demonstrates the relationship between the two points of view through Little's Law under the assumption of steady state equilibrium.

5. Program Throughput and Program Time Delay. This section considers memory performance from the program point of view. With no interference, a program by itself requests vectors from memory at the rate defined by Equation (4.2). When interference occurs, the satisfaction of each request may take longer than the minimum time T_0. Let $\bar{T}$ be the average time to complete a vector request of length N and let

$$(5.1) \qquad P = N/\bar{T}$$

be the program's actual request rate under the influence of memory conflicts.

Time delays [37,14] are a convenient device for analyzing the average time $\bar{T}$. Let T_i be the time required to service request i and let D_i be the time delay such that

$$(5.2) \qquad T_i = T_0 + D_i.$$

The simulator determines T_0 as the minimum time encountered during the simulation and computes time delay D_i as the difference between the measured time T_i and T_0.

TABLE 4.1
Dimensional Analysis

Quantitiy	Symbol	Dimension
Number of Processors	p	$[0]$
Memory Banks	m	$[L]$
Bank Reservation Time	τ	$[T]$
Vector Length	N	$[L]$
Interval Between Vector Requests	T_0	$[T]$
Program Request Rate	$P_0 = N/T_0$	$[LT^{-1}]$
System Request Rate	$R_0 = pN/T_0$	$[LT^{-1}]$
Time Delay for Request i	D_i	$[T]$
Time to Complete Request i	$T_i = T_0 + D_i$	$[T]$
Average Time to Complete Request	$\bar{T}$	$[T]$
Average Program Time Delay per Word	D	$[L^{-1}T]$
Program Throughput	$P = N/\bar{T}$	$[LT^{-1}]$
Vector Requests per Processor	M	$[0]$
Time to Complete M Requests	$T_M = MT_0 + pNM\Delta$	$[T]$
Average System Time Delay per Word	Δ	$[L^{-1}T]$
System Throughput	$R = pNM/T_M$	$[LT^{-1}]$
Average Banks Busy	$\bar{m}$	$[L]$
Fraction of Banks Busy	$\bar{m}/m$	$[0]$
Ratio Vector Length to Banks	N/m	$[0]$
Dimensionless Program Throughput	$\sigma = (\tau/m)P$	$[0]$
Dimensionless Program Time Delay	$d = (m/\tau)D$	$[0]$
Program Efficiency	$\epsilon = P/P_0 = T_0/\bar{T}$	$[0]$
Dimensionless System Request Rate	$u = (\tau/m)R_0$	$[0]$
Dimensionless System Throughtput	$\rho = (\tau/m)R$	$[0]$
Dimensionless System Time Delay	$\delta = (m/\tau)\Delta$	$[0]$
System Efficiency	$\epsilon = R/R_0 = \rho/u$	$[0]$

Suppose p active processors each make M vector requests of length N. Then [4]

$$(5.3) \qquad \bar{T} = \frac{1}{pM} \sum_{i=1}^{pM} T_i$$

is the average time to service a vector request. Substitution of Equation (5.2) for T_i into Equation (5.3) yields

$$(5.4) \qquad \bar{T} = T_0 + \frac{1}{pM} \sum_{i=1}^{pM} D_i.$$

Let

$$(5.5) \qquad D = \frac{1}{pNM} \sum_{i=1}^{pM} D_i$$

---

[4] More accurately, all definitions involving the parameter M assume a limiting process for which the system approaches a steady state equilibrium. In Section 5, the average time to service a vector request defined by Equation (5.3) is the limit

$$\bar{T} = \lim_{M \to \infty} \frac{1}{pM} \sum_{i=1}^{pM} T_i.$$

For this definition to make sense, the limit

$$D = \lim_{M \to \infty} \frac{1}{pNM} \sum_{i=1}^{pM} D_i.$$

must exist for average time delay per word defined by Equation (5.5). Similarly, in Section 6, the limit

$$\lim_{M \to \infty} \frac{T_M}{M}$$

must exist in order to define system throughput by Equation (6.1) as the limit

$$R = \lim_{M \to \infty} \frac{pN}{T_M/M},$$

and the time delay per word by Equation (6.3) as the limit

$$\Delta = \lim_{M \to \infty} \frac{T_M/M - T_0}{pN}.$$

The steady state assumption of Equation (7.6)

$$\lim_{M \to \infty} \frac{1}{pM} \sum_{i=1}^{pM} T_i = \lim_{M \to \infty} \frac{T_M}{M}$$

is a statement of the ergodic hypothesis that the ensemble average over many trials T_i is equal to the statistical average for a long simulation with a very large value for M. In practice, the simulator picks a large value for M, usually $M = 1000$, and verifies the steady state condition from Little's Law as described in Sections 7 and 9.

be the average time delay per word [4] so that

(5.6) $$\bar{T} = T_0 + ND.$$

Then from Equation (4.1),

(5.7) $$\bar{T} \geq T_S + N(D_0 + D).$$

Since D has dimension $[L^{-1}T]$, the Pi Theorem implies that the scaled quantity

(5.8) $$d(u, v) = (m/\tau)D$$

is a dimensionless function of the dimensionless variables u and v. Similarly, since P has dimension $[LT^{-1}]$, the scaled program throughput

(5.9) $$\sigma(u, v) = (\tau/m)P$$

is a dimensionless function of the dimensionless variables u and v.

Equation (5.6) establishes a relationship between time delay and throughput. Divide by the vector length N and use definitions (4.2) and (5.1) for P_0 and P to obtain

(5.10) $$D = \frac{1}{P} - \frac{1}{P_0}.$$

Multiply Equation (5.10) by the scale factor m/τ to obtain

(5.11) $$d = \frac{1}{\sigma} - \frac{p}{u}.$$

Rearrangement of Equation (5.11) yields

(5.12) $$\frac{1}{\sigma} - d = \frac{p}{u}.$$

The difference between two functions that depend on the two variables u and v is in fact independent of v.

6. System Throughput and System Time Delay. This section considers memory performance from the hardware point of view. Suppose p processors each make M vector requests of length N. With no interference, the system requires time MT_0 to satisfy M requests. With interference, it requires a longer time T_M. Define system throughput [4]

(6.1) $$R = \frac{pNM}{T_M}$$

as the ratio of the total number of words moved divided by the total time to move them.

Split the time T_M into two pieces such that

(6.2) $$T_M = MT_0 + pNM\Delta.$$

The first term corresponds to the time without interference and the second term defines a system time delay per word [4]

(6.3) $$\Delta = \frac{T_M - MT_0}{pNM}.$$

Rearrangement of Equation (6.3) yields

(6.4) $$\Delta = \frac{1}{pNM/T_M} - \frac{1}{pN/T_0}.$$

From definition (6.1) for R and definition (4.3) for R_0, Equation (6.4) becomes

(6.5) $$\Delta = \frac{1}{R} - \frac{1}{R_0}.$$

Since the system throughput R has dimension $[LT^{-1}]$, the Pi Theorem implies that the scaled function

(6.6) $$\rho(u, v) = (\tau/m)R$$

is a dimensionless function of the dimensionless variables u and v. In queueing theory terminology [25, page 18], the dimensionless variable ρ is the utilization factor. Similarly, the system time delay per word Δ has dimension $[L^{-1}T]$. The scaled function

(6.7) $$\delta(u, v) = (m/\tau)\Delta$$

is a dimensionless function of the dimensionless variables u and v.
    Scaling Equation (6.5) by m/τ yields

(6.8) $$\delta = \frac{1}{\rho} - \frac{1}{u}.$$

Rearrangement of Equation (6.8) yields

(6.9) $$\frac{1}{\rho} - \delta = \frac{1}{u}.$$

The difference between two functions that depend on the variables u and v is in fact independent of v.

7. Little's Law. The results of the simulation contain simple, but important, examples of the application of Little's Law [25, page 17]. In steady state equilibrium, average service rate equals the average number of customers divided by the average service time per customer. This law applies to the system as a whole or to any one of its parts separately. Consider first the system of memory banks. The average number of customers in the system equals the average number $\bar{m}$ of memory banks busy at any clock period. The service time per customer equals the constant bank reservation time τ. Then Little's Law becomes

(7.1) $R = \bar{m}/\tau.$

Divide Equation (7.1) by the number of banks m and multiply by the bank reservation time τ to obtain

(7.2) $(\tau/m)R = \bar{m}/m.$

Comparison of Equation (7.2) with Definition (6.6) for the dimensionless throughput ρ yields the result

(7.3) $\rho = \bar{m}/m.$

Scaled system throughput equals the average fraction of banks busy. The simulator measures the two quantities $\bar{m}$ and ρ independently. Relationship (7.3) provides a check that the simulation has reached equilibrium.

For a second application of Little's Law, consider the behavior of a single program. The average number of customers equals N and the average service time equals $\bar{T}$. Then Definition (5.1), $P = N/\bar{T}$, is itself a statement of Little's Law for each program.

For a third application, consider the system as a whole. The average number of customers equals pN and the average service time equals $\bar{T}$. Then

(7.4) $R = pN/\bar{T}$

is another statement of Little's Law. Furthermore, substitution of Definition (6.1) for R into Equation (7.4) yields

(7.5) $pNM/T_M = pN/\bar{T}$

and rearrangement yields [4]

(7.6) $\bar{T} = T_M/M.$

At equilibrium, the total time T_M is the sum of M average times $\bar{T}$.

Equation (7.6) establishes a relationship between the program point of view from Section 5 and the system point of view from Section 6. For throughput, compare Equation (4.3) with Equation (4.2) to obtain

(7.7) $R_0 = pP_0$

and compare Equation (5.1) with Equation (7.4) to obtain

(7.8) $$R = pP.$$

At equilibrium, system throughput is the sum of p individual program throughputs.

For time delay, substitute $\bar{T}$ from Equation (5.6) into Equation (7.6) to obtain

(7.9) $$T_M = M(T_0 + ND).$$

Multiply the second term in Equation (7.9) by unity in the form p/p to obtain

(7.10) $$T_M = MT_0 + pNM(D/p).$$

Compare Equation (7.10) with Equation (6.2) to conclude

(7.11) $$\Delta = D/p.$$

At equilibrium, individual program time delays overlap so that the system time delay is $1/p$ times the program time delay.

Finally, multiply Equation (7.7) times Equation (7.11) to obtain

(7.12) $$R_0\Delta = P_0D$$

and multiply Equation (7.8) times Equation (7.11) to obtain

(7.13) $$R\Delta = PD.$$

In summary, at equilibrium, the two points of view differ essentially by a factor of p. In terms of dimensionless quantities, multiply Equation (7.8) by τ/m to obtain

(7.14) $$\rho = p\sigma$$

and multiply Equation (7.11) by m/τ to obtain

(7.15) $$\delta = d/p.$$

The product of Equations (7.14) and (7.15) yields

(7.16) $$\rho\delta = \sigma d.$$

8. System Efficiency. System efficiency ϵ has several interpretations depending on point of view. From the program point of view,

(8.1) $$\epsilon = P/P_0,$$

the ratio of program throughput to program request rate, called the performance degradation factor in [42]. Or equivalently, dividing Equation (5.1) by Equation (4.2),

$$(8.2) \qquad \epsilon = T_0/\bar{T},$$

the ratio of its minimum request interval to its actual request interval, called the stretching factor in [37]. From the system point of view,

$$(8.3) \qquad \epsilon = R/R_0,$$

the ratio of system throughput R to the system request rate R_0. At equilibrium, the program and system definitions are equivalent. In fact, divide Equation (7.8) by Equation (7.7) to obtain

$$(8.4) \qquad R/R_0 = P/P_0.$$

In terms of scaled quantities, Equation (8.3) becomes

$$(8.5) \qquad \epsilon = \rho/u.$$

Equation (6.8) yields two equivalent expressions for efficiency. Multiplication by ρ followed by rearrangement yields

$$(8.6) \qquad \epsilon = 1 - \rho\delta.$$

Or multiplication by u followed by rearrangement yields

$$(8.7) \qquad \epsilon = \frac{1}{1 + u\delta}.$$

Equivalently, in terms of unscaled quantities, Equation (6.5) yields

$$(8.8) \qquad \epsilon = 1 - R\Delta = (1 + R_0\Delta)^{-1} = (1 + pN\Delta/T_0)^{-1}.$$

Or, using Equations (7.11), (7.12), and (7.13),

$$(8.9) \qquad \epsilon = 1 - PD = (1 + P_0D)^{-1} = (1 + ND/T_0)^{-1}.$$

9. Simulation Results. This section describes the results of the simulation. Two criteria exist to check for equilibrium. First, according to Equation (7.3), normalized system throughput ρ equals the average fraction of banks busy $\bar{m}/m$. Secondly, substitution of Equation (7.15) into Equation (6.9) yields

$$(9.1) \qquad u(1/\rho - d/p) - 1 = 0.$$

The simulator measures ρ, $\bar{m}$, δ and d independently. Table 9.1 shows some sample results that verify the steady state assumption. The mea-

TABLE 9.1
Sample Simulation Results

p	N	m	τ	T_0	N/m	u	ρ	$\bar{m}/m$	δ	d	ε	$u(1/\rho - d/p) - 1$
1	32	64	13	40	0.5	0.1625	0.1590	0.1590	0.1349	0.1337	0.9785	0.0002
2	32	64	13	40	0.5	0.3250	0.2908	0.2903	0.3621	0.7031	0.8947	0.0034
3	32	64	13	40	0.5	0.4875	0.3915	0.3899	0.5030	1.4888	0.8031	0.0033
4	32	64	13	40	0.5	0.6500	0.4617	0.4587	0.6275	2.4512	0.7103	0.0096
1	64	128	17	72	0.5	0.1181	0.1167	0.1167	0.0955	0.0945	0.9888	0.0001
2	64	128	17	72	0.5	0.2361	0.2234	0.2233	0.2418	0.4521	0.9460	0.0037
3	64	128	17	72	0.5	0.3542	0.3184	0.3182	0.3171	0.9089	0.8990	0.0050
4	64	128	17	72	0.5	0.4722	0.3931	0.3926	0.4262	1.6404	0.8324	0.0076
1	64	64	21	72	1.0	0.2917	0.2814	0.2814	0.1245	0.1241	0.9650	0.0001
2	64	64	21	72	1.0	0.5833	0.4943	0.4940	0.3088	0.6145	0.8473	0.0009
3	64	64	21	72	1.0	0.8750	0.6062	0.6049	0.5068	1.4734	0.6928	0.0137
4	64	64	21	72	1.0	1.1667	0.6627	0.6601	0.6519	2.5760	0.5680	0.0092
1	32	32	25	40	1.0	0.6250	0.5157	0.5155	0.3391	0.3388	0.8251	0.0002
2	32	32	25	40	1.0	1.2500	0.6849	0.6794	0.6600	1.3114	0.5480	0.0054
3	32	32	25	40	1.0	1.8750	0.7296	0.7209	0.8374	2.4939	0.3891	0.0114
4	32	32	25	46	1.0	2.1739	0.7707	0.7620	0.8376	3.3379	0.3545	0.0067
1	64	32	29	72	2.0	0.8056	0.6956	0.6955	0.1963	0.1961	0.8635	0.0001
2	64	32	29	72	2.0	1.6111	0.7836	0.7795	0.6554	1.3075	0.4864	0.0027
3	64	32	29	116	2.0	1.5000	0.8396	0.8345	0.5244	1.5674	0.5597	0.0029
4	64	32	29	83	2.0	2.7952	0.8716	0.8663	0.7896	3.1453	0.3118	0.0092
1	32	16	33	57	2.0	1.1579	0.7921	0.7913	0.3988	0.3986	0.6841	0.0002
2	32	16	33	79	2.0	1.6709	0.8663	0.8622	0.5558	1.1089	0.5185	0.0023
3	32	16	33	68	2.0	2.9118	0.9097	0.9066	0.7558	2.2605	0.3124	0.0068
4	32	16	33	79	2.0	3.3418	0.9415	0.9385	0.7629	3.0469	0.2817	0.0039

sured error represented by Equation (9.1) is typically less than 1 per cent. One simulation showed a 4 per cent error. The equilibrium condition is almost exactly verified for small values of u with worsening agreement as u increases. Running the simulator with larger values of M gives better agreement for large u.

Consider the distribution of individual program time delays D_i about the average value. Let $pdf(D_i)$ be the probability distribution function such that

$$(9.2) \qquad D = \frac{1}{N} \sum_{i=1}^{pM} pdf(D_i)D_i.$$

Figure 9.1 shows measured probability distribution functions from a typical simulation. Figure 9.1a shows the distribution for a single processor. Delays occur in discrete increments of eleven clock periods corresponding to the backup mechanism within the CRAY-2 memory port. As described in Section 2, bank conflicts may induce quadrant conflicts forcing the issue control mechanism into a ten clock period backup loop. A delayed request exits the loop at the eleventh clock period giving the observed distribution. The vertical line at $D_i = 0$, approximately equal to $1 - u = 0.6$, represents the probability of no delay. The other vertical lines represent the probabilities for multiple trips through the backup loop. They are approximately equal. When a bank conflict occurs, the bank flag is equally likely to be set anywhere between 1 and τ clock periods. For $D_i < \tau$,

$$(9.3) \qquad pdf(D_i) = \delta_{0D_i}(1 - u) + (1 - \delta_{0D_i})u/\lfloor \tau/11 \rfloor$$

is a reasonable distribution function where δ_{0D_i} is the Kronecker delta and $\lfloor x \rfloor$ is the floor of x.

The probability distributions for two, three and four processors shown in Figures 9.1b,c, and d are more complicated than the single processor distribution. The main distribution corresponds to the same backup mechanism described for the single processor case. But Equation (9.3) is no longer a good representation for the distribution function. In addition, minor distributions under the main distribution correspond to other details of the issue control mechanism [27]. With multiple processors active, each processor may miss its phase time by 1,2, or 3 clock periods or may go through a delayed entry loop corresponding to a partial backup causing a 6 clock period delay. Predicting these probability distributions is a major challenge for a Markov chain model or for a queueing model.

The quantities of most interest from the simulation are the throughput ρ, the time delay δ, and the efficiency ϵ. Figure 9.2 shows the results of the simulation for ρ and δ. Despite scatter in the data, these functions clearly separate into distinct functions of u for fixed values of v verifying assumptions (6.6) and (6.7) from dimensional analysis.

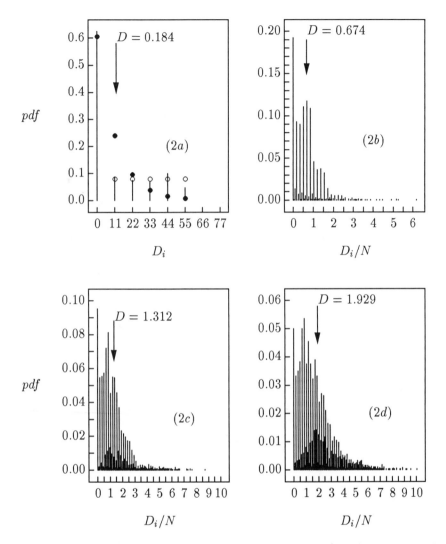

FIG. 9.1. *Probability density functions (pdf) for program delay times D_i. The vertical lines represent the simulated results for a sample set of $M = 1000$ requests per processor with vector length $N = 64$ and request interval $T_0 = 72$. For the results shown here, $\tau = 57$ for the memory bank reservation time and $m = 128$ for the number of memory banks so that $u/p = N\tau/T_0 m = 0.396$. Figure (2a) shows results for one active processor, (2b) for two active processors, (2c) for three active processors, and (2d) for four active processors. The arrow in each figure marks the value of the average time delay per word D. For the single processor results in (2a), the bullets represent the geometric density function $(1-u)u^{D_i/11}$ and the circles represent the density function $(1-u)\delta_{0D_i} + (1-\delta_{0D_i})u/\lfloor \tau/11 \rfloor$ where δ_{0D_i} is the Kronecker delta and $\lfloor x \rfloor$ is the floor of x.*

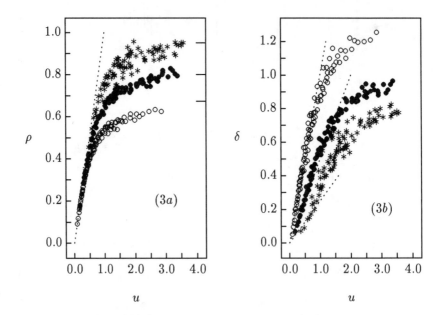

FIG. 9.2. *System throughput ρ (3a) and system time delay δ (3b) as functions of scaled request rate u for fixed values of v = N/m = 2.0(*), 1.0(●), and 0.5(○). The dotted line in (3a) has slope one and the tick marks on the right hand scale are the asymptotic values $\rho_\infty(v) = 0.80v^{1/4}$. The dotted lines in (3b) have slope 1/2v.*

By definition, both the throughput ρ and the time delay δ go to zero as the request rate u goes to zero. More precisely, as δ approaches zero, the relationship between ρ, δ, and u expressed by Equation (6.8) implies

(9.4)
$$\frac{1}{\rho} - \frac{1}{u} \to 0, \qquad u \to 0.$$

Consequently,

(9.5)
$$\rho \to u, \qquad u \to 0$$

as verified by the dooted line with slope one in Figure 9.2a. The system efficiency, from Equation (8.5), approaches unity

(9.6)
$$\epsilon = \rho/u \to 1, \qquad u \to 0.$$

The more important observation is the rate at which the time delay δ approaches zero. Figure 9.2b shows dotted lines near the origin with slope equal to $1/2v$. A reasonable approximation for the time delay δ near the origin is

(9.7)
$$\delta \to u/2v, \qquad u \to 0.$$

The time delay and the throughput both increase linearly with u near the origin. Depending on the value of v, the time delay grows more or less rapidly than the throughput.

The behavior of ρ and δ near the origin determines the behavior of the efficiency near the origin. For light traffic, $u < 1$, efficiency behaves like

(9.8)
$$\epsilon \to 1 - u^2/2v, \qquad u \to 0$$

from Equation (8.6) or like

(9.9)
$$\epsilon \to \frac{1}{1 + u^2/2v}, \qquad u \to 0$$

from Equation (8.7). The two asymptotic forms are equivalent for $u \ll \sqrt{2v}$. Figure 9.3a shows that approximation (9.9) represents the efficiency in the light traffic region reasonably well.

Under heavy traffic conditions, $u > 1$, system throughput saturates to some value

(9.10)
$$\rho_\infty(v) = \lim_{u \to \infty} \rho(u, v)$$

dependent on the parameter v. From Equation (6.8), the system time delay saturates to the reciprocal value

(9.11)
$$\delta_\infty(v) = \lim_{u \to \infty} (1/\rho - 1/u) = 1/\rho_\infty(v).$$

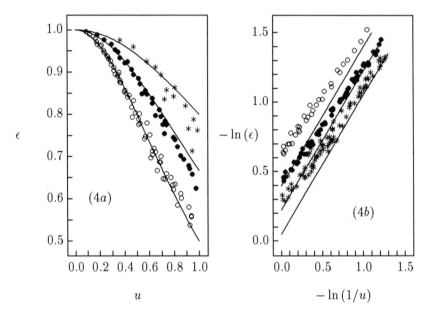

FIG. 9.3. *System efficiency ϵ as a function of scaled request rate u for light traffic ($u < 1$) in (4a) and a double logarithmic plot of efficiency versus $1/u$ for heavy traffic ($u > 1$) in (4b) for fixed $v = N/m = 2.0(*), 1.0(\bullet), and 0.5(\circ)$. The solid curves in (4a) are the functions $\epsilon = (1 + u^2/2v)^{-1}$. Extrapolation of the lines in (4b) from infinity to $u = 1$ yields an estimate for the saturation bandwidth $\rho_\infty(v)$. The lines in (4b) have slopes equal to one and intercepts corresponding to $\rho_\infty(v) = 0.80v^{1/4}$.*

Since efficiency depends on the product of ρ and δ as described by Equation (8.6), it approaches zero in the heavy traffic limit

$$(9.12) \qquad \lim_{u \to \infty} \epsilon = \lim_{u \to \infty} (1 - \rho\delta) = 1 - \rho_\infty \delta_\infty = 0.$$

To estimate the saturation limit $\rho_\infty(v)$, note that for large u Equation (8.7) approaches the form

$$(9.13) \qquad \epsilon \to \rho_\infty(v)(1/u), \qquad u \to \infty$$

because of the reciprocal relationship given by Equation (9.11). For fixed v, a plot of $-\ln(\epsilon)$ versus $-\ln(1/u)$ in the limit $u \to \infty$ yields straight lines with slope equal to one and intercept equal to $-\ln(\rho_\infty(v))$. Figure 9.3b shows such a plot. Extrapolation to infinity is risky but

$$(9.14) \qquad \rho_\infty(v) = 0.80v^{1/4}.$$

represents the saturation limit reasonably well. Tick marks on the right hand scale of Figure 9.2a show values for $\rho_\infty(v)$. Throughput has essentially reached its saturation value at $u = 3$ but time delay has not.

10. Hardware Design Criteria. The results of the simulation suggest methods for establishing hardware design criteria. Efficiency defined by Equation (8.5) is a function of the two independent variables

$$(10.1) \qquad u = p(N/m)(\tau/T_0)$$

and

$$(10.2) \qquad v = N/m.$$

The first obvious criterion restricts hardware parameters such that the machine operates in the light traffic region, $u < 1$, for all values of p including $p = p_{max}$, the maximum number of processors. From Equation (10.1), the light traffic condition is equivalent to the condition

$$(10.3) \qquad m/\tau > p_{max} N/T_0.$$

The maximum memory delivery rate, m/τ, must be greater than the maximum system request rate, $p_{max}N/T_0$. Or in terms of memory speed,

$$(10.4) \qquad \tau < (\frac{m}{p_{max}N})T_0.$$

For the CRAY-2 with $m = 128$, $N = 64$, $T_0 = 72$ and $p_{max} = 4$, $\tau < 36.0$.

Inequality (10.3) keeps the machine in the light traffic region for all p. But the simulation results contain more information about the behavior of efficiency in this region. In the two parameter space (u, v), the total differential of the efficiency

(10.5)
$$d\epsilon = \frac{\partial \epsilon}{\partial u} du + \frac{\partial \epsilon}{\partial v} dv$$

measures its sensitivity to changes in hardware parameters. Of particular interest is the change in efficiency as the number of active processors changes. Since v is independent of p,

(10.6)
$$\frac{\partial \epsilon}{\partial p} = \frac{\partial \epsilon}{\partial u} \frac{\partial u}{\partial p}$$

and

(10.7)
$$\frac{\partial u}{\partial p} = (N/m)(\tau/T_0).$$

In the light traffic region, the derivative with respect to p is easy to evaluate. In terms of unscaled variables, from Equation (9.8), system efficiency behaves like

(10.8)
$$\epsilon = 1 - \frac{1}{2} p^2 (\frac{N}{m})(\frac{\tau}{T_0})^2.$$

In particular, with all other parameters fixed, system efficiency decreases like the square of the number of active processors. The rate of decrease

(10.9)
$$\frac{d\epsilon}{dp} = -p(\frac{N}{m})(\frac{\tau}{T_0})^2$$

is equal to the first derivative with respect to p and the curvature

(10.10)
$$\frac{d^2\epsilon}{dp^2} = -(\frac{N}{m})(\frac{\tau}{T_0})^2$$

is given by the second derivative. In the light traffic region, the system efficiency degrades gracefully as the number of processors p increases as a convex function of p. Clearly, a good design requires $m > N$ and $T_0 > \tau$.

The fact that system efficiency in the light traffic region decreases like the square of the bank reservation time has been the subject of some discussion in the literature [1,15,41,8,9]. As Equation (10.8) shows, however, the major contribution to the decrease in efficiency comes from the quadratic dependence on the number of processors p. For most vector computers, $m > N$ and $T_0 > \tau$ so that the other factors have only a small effect on the efficiency in the light traffic region. The big effect comes from the collision of two processors at the same memory bank resulting in a quadratic term in p. Higher order collisions among processors only occur in the heavy traffic region.

The behavior of the first derivative suggests another reasonable design criterion. For each value of p, require efficiency to decrease by less than $1/p_{max}$,

$$(10.11) \qquad\qquad \epsilon(p) - \epsilon(p-1) > -1/p_{max},$$

when p increases by one. Approximate the difference on the left side of Equation (10.11) by the first derivative given by Equation (10.9) to obtain

$$(10.12) \qquad\qquad p(\frac{N}{m})(\frac{\tau}{T_0})^2 < \frac{1}{p_{max}}.$$

Since this inequality holds for all p including $p = p_{max}$, the design criterion becomes

$$(10.13) \qquad\qquad \tau < \frac{1}{p_{max}}\sqrt{\frac{m}{N}}T_0.$$

For the CRAY-2, this second criterion requires $\tau < 25.5$, a more demanding requirement than the first.

In fact, an even more demanding criterion is appropriate. At $p = p_{max}$, require efficiency

$$(10.14) \qquad\qquad \epsilon(p_{max}) \geq 1 - \alpha$$

to remain larger than some acceptable value determined by a constant $0 \leq \alpha \leq 1$. Then memory speed is determined by

$$(10.15) \qquad\qquad \tau < \frac{1}{p_{max}}\sqrt{\frac{2\alpha m}{N}}T_0.$$

Efficiency remains above one half for $\alpha = 1/2$, equivalent to the previous criterion (10.13). It remains above three quarters for $\alpha = 1/4$ with a more demanding requirement on memory speed, $\tau < 18.0$.

11. Comparison with Other Work. Several combinatorial models, probabilistic models and queueing models have been published [21,42,37,12,16,34,3,26,2,39,32,36,33,7,13,19,24] describing synchronous processors accessing multiple memory modules. For a review of this early work through about 1980 see [42]. These models do not apply to shared memory, vector multiprocessors which appeared for the first time in the mid 1980's. Since then, a few attempts [1,23,8,41,10,9,17,18,28,29,30,14,15,35] have been made to explain the memory interference observed on these machines. For a review of this work through about 1990 see [9].

Bucher and Calahan [8,9] adopt a point of view and interpretation similar to the approach presented in this paper. In modified form corresponding to the notation of the present paper, they write efficiency [5]

$$(11.1) \qquad \epsilon = \frac{1}{1 + ND/T_0}$$

corresponding to the last expression in Equation (8.9). They suggest the following form [5] for the time delay

$$(11.2) \qquad \frac{ND}{T_0} = \frac{u-1}{2} + \sqrt{\frac{(u-1)^2}{4} + (\frac{\tau}{T_0})(\frac{u}{2})}.$$

In the light traffic region,

$$(11.3) \qquad \frac{ND}{T_0} \to \frac{1}{2}(\frac{\tau}{T_0})u = \frac{1}{2}p(\frac{N}{m})(\frac{\tau}{T_0})^2, \qquad u \to 0.$$

And in the heavy traffic region,

$$(11.4) \qquad \frac{ND}{T_0} \to u = p(\frac{N}{m})(\frac{\tau}{T_0}), \qquad u \to \infty.$$

To compare with the results reported in this paper, note that

$$(11.5) \qquad \frac{ND}{T_0} = \frac{pN}{T_0}(\frac{\tau}{m})(\frac{m}{\tau})\frac{D}{p} = u\delta.$$

In the light traffic region, from Equation (9.7),

$$(11.6) \qquad u\delta \to \frac{u^2}{2v} = \frac{1}{2}p^2(\frac{N}{m})(\frac{\tau}{T_0})^2, \qquad u \to 0.$$

And in the heavy traffic region, from Equations (9.14) and (9.11),

$$(11.7) \qquad u\delta \to \frac{1}{0.80}\frac{u}{v^{1/4}} = 1.25p(\frac{N}{m})^{3/4}(\frac{\tau}{T_0}), \qquad u \to \infty.$$

Compare the observed results from this paper represented by Equation (11.6) for the light traffic region with the theoretical result of Bucher and Calahan represented by Equation (11.3). They differ by a factor of p. One way to fix the difference is to postulate that Equation (11.3) actually

---

[5] More accurately, Bucher and Calahan write $\epsilon = (1+D)^{-1}$ with

$$D = \frac{u_0 - 1}{2} + \sqrt{\frac{(u_0-1)^2}{4} + \tau(\frac{u_0}{2})}$$

and $u_0 = p\tau/m$. Surprisingly, vector length N does not appear in their result. They assume $N/T_0 = 1$ for a vector machine. From Inequality (4.1), however, this ratio is less than one. This slight difference seems to be important in the analysis.

represents the system time delay Δ rather than the program time delay D. From Equation (7.11)

$$(11.8) \qquad \frac{N\Delta}{T_0} = \frac{ND}{pT_0} = \frac{u-1}{2} + \sqrt{\frac{(u-1)^2}{4} + (\frac{\tau}{T_0})(\frac{u}{2})}$$

giving the extra factor of p required. Bucher and Calahan do not make the distinction between the two time delays so it is difficult to tell which one their formula represents.

In the heavy traffic region, represented by Equations (11.4) and (11.7), both forms predict linear behavior in the factor $p\tau/T_0$ but the similarity ends there. No obvious reconciliation of the two asymptotic forms in the heavy traffic region makes itself apparent.

12. Summary of Results and Open Questions. This paper has presented an analysis of interprocessor memory contention based on the results of an exact simulation of the CRAY-2 memory system. It carried through the analysis based on a dimensional analysis argument that reduced the problem to a two parameter space, one being a dimensionless request rate and the other being the ratio of vector length to the number of memory banks. It proposed asymptotic forms for throughput, time delay and efficiency for both the light traffic and the heavy traffic regions. It also suggested criteria for evaluating hardware design.

But a number of open questions remain to be answered. Development of a Markov chain model or a queueing model to explain the behavior described in this paper remains an open challenge. What is the explanation for the probability distribution functions like those shown in Figure 9.1? What is the condition for a steady state equilibrium to exist? The condition is not $u < 1$ since the data in Table 9.1 indicate that the equilibrium conditions, Equations (7.3) and (9.1), are well satisfied even for $u > 1$. Perhaps $\rho < 1$ is the correct steady state requirement since in that case the memory delivery rate exceeds the system request rate. If so, at what rate does the system approach equilibrium?

None of the previous attempts to explain memory contention on vector processors have predicted any dependence on v. What is the explanation for the dependence on v? What determines the slope of the time delay δ at the origin and what determines the saturation limit for ρ? In particular, why does it depend on the fourth root of v? For fixed v, what are the functions $\rho(u,v)$ and $\delta(u,v)$ for all values of u?

Scatter in the data is a disturbing feature of the simulation results and its cause is unknown. It persists even if averages are computed for multiple simulations at fixed M. It also persists for very long simulations with large values of M [31]. As M increases, the criteria for equilibrium are more closely satisfied, but the scatter persists.

What is the behavior for vector requests with stride not equal to one? Preliminary simulation results indicate quite different behavior from the stride one case.

How do the results presented in this paper relate to other shared memory vector multiprocessors? Do machines like the CRAY X-MP [17,18,29,28,30] or CRAY Y-MP [38,35] show the same qualitative behavior or quite different behavior? How important are particular design details of the memory port and the interconnection network? Work is under way to investigate these questions.

Suppose, however, that other machines behave qualitatively the same and that the only difference appears in the values for the saturation bandwidth $\rho_\infty(v)$ and the slope of the time delay at the origin $\delta_0(v) = \lim_{u \to 0} d\delta/du$. A good design has high throughput and low time delay. An index to measure the design then might be the ratio of these two limiting values

$$(12.1) \qquad\qquad I(v) = \frac{\rho_\infty(v)}{\delta_0(v)}.$$

Big values for $I(v)$ indicate a good design. On the CRAY-2, with $v = N/m = 64/128 = 0.5$, $\rho_\infty(v)$ given by Equation (9.14) and $\delta_0(v) = 1/2v$ given by Equation (9.7), this index has the value $I = 0.67$. What is its value for other machines?

Finally, how do the simulation results relate to the behavior of real programs? Recall the experiment described in Figure 1.1. The simulation results indicate that efficiency curves are convex functions of p in the light traffic region. For the slow memory machine, two of the solid curves in Figure 1.1 are concave functions of p and must correspond to heavy traffic conditions. With faster memory, all three programs seem to be operating in the light traffic region. Unfortunately, not much more can be said. Relating the performance of real programs to hardware parameters remains one of the most important and challenging open questions.

REFERENCES

[1] D. H. BAILEY, *Vector computer memory bank contention*, IEEE Trans. Computers, C-36 (1987), pp. 293–298.

[2] F. BASKETT AND A. J. SMITH, *Interference in multiprocessor computer systems with interleaved memory*, Commun. Assoc. Comput. Mach., 19 (1976), pp. 327–334.

[3] D. P. BHANDARKAR, *Analysis of memory interference in multiprocessors*, IEEE Trans. Computers, C-24 (1975), pp. 897–908.

[4] G. BIRKHOFF, *Hydrodynamics*, Princeton University Press, Princeton, New Jersey, 2nd ed., 1960.

[5] L. BRAND, *The pi theorem of dimensional analysis*, Arch. Rat. Mech. Anal., 1 (1957), pp. 35–45.

[6] P. W. BRIDGMAN, *Dimensional Analysis*, Yale University Press, New Haven, 2nd ed., 1931.

[7] F. A. BRIGGS AND E. S. DAVIDSON, *Organization of semiconductor memories for parallel-pipelined processors*, IEEE Trans. Computers, C-26 (1977), pp. 162–169.

[8] I. Y. BUCHER AND D. A. CALAHAN, *Access conflicts in multiprocessor memories: Queuing models and simulation studies*, Proc. 1990 International Conf. on Supercomputing, (1990), pp. 428–438.

[9] ———, *Models of access delays in multiprocessor memories*. IEEE Trans. Par. Distr. Computing, to appear, 1992.

[10] I. Y. BUCHER AND M. L. SIMMONS, *Measurement of memory access contentions in multiple vector processor systems*, Proc. Supercomputing '91, (1991), pp. 806–817.

[11] E. BUCKINGHAM, *On physically similar systems: Illustrations of the use of dimensional equations*, Physical Review, 4 (1914), pp. 345–376.

[12] P. BUDNIK AND D. J. KUCK, *The organization and use of parallel memories*, IEEE Trans. Computers, C-20 (1971), pp. 1566–1569.

[13] G. J. BURNETT AND E. G. COFFMAN JR., *Analysis of interleaved memory systems using blockage buffers*, Commun. Assoc. Comput. Mach., 18 (1975), pp. 91–95.

[14] D. A. CALAHAN, *Characterization of memory conflict loading on the CRAY-2*, Proc. 1988 Int. Conf. Parallel Processing, I (1988), pp. 299–302.

[15] D. A. CALAHAN AND D. H. BAILEY, *Measurement and analysis of memory conflicts in vector multiprocessors*, in Performance Evaluation of Supercomputers, J. L. Martin, ed., Elsevier, 1988, pp. 83–106.

[16] D. Y. CHANG, D. J. KUCK, AND D. H. LAWRIE, *On the effective bandwidth of parallel memories*, IEEE Trans. Computers, C-26 (1977), pp. 480–489.

[17] T. CHEUNG AND J. E. SMITH, *An analysis of the CRAY X-MP memory system*, Proc. 1984 Int. Conf. Parallel Processing, (1984), pp. 613–622.

[18] ———, *A simulation study of the CRAY X-MP memory system*, IEEE Trans. Computers, C-35 (1986), pp. 613–622.

[19] E. G. COFFMAN, JR., G. J. BURNETT, AND R. A. SNOWDON, *On the performance of interleaved memories with multiple-word bandwidths*, IEEE Trans. Computers, C-20 (1971), pp. 1570–1573.

[20] S. DROBOT, *On the foundations of dimensional analysis*, Studia Mathematica, 14 (1954), pp. 84–99.

[21] I. FLORES, *Derivation of a waiting-time factor for a multiple-bank memory*, J. Assoc. Comp. Mach., 11 (1964), pp. 265–282.

[22] C. M. FOCKEN, *Dimensional Methods and Their Applications*, Edward Arnold and Co., London, 1953.

[23] C. FRICKER, *On memory contention problems in vector multiprocessors*, Tech. Rep. 1034, INRIA, 1989.

[24] C. H. HOOGENDOORN, *A general model for memory interference in multiprocessors*, IEEE Trans. Computers, C-26 (1977), pp. 998–1005.

[25] L. KLEINROCK, *Queueing Systems*, vol. 1, John Wiley, New York, 1975.

[26] D. E. KNUTH AND G. S. RAO, *Activity in an interleaved memory*, IEEE Trans. Computers, C-24 (1975), pp. 943–944.

[27] R. W. NUMRICH, *CRAY-2 common memory*, Tech. Rep. HN-2043, Cray Research, Inc., 1988.

[28] W. OED AND O. LANGE, *On the effective bandwidth of interleaved memories in vector processor systems*, IEEE Trans. Computers, C-34 (1985), pp. 949–957.

[29] ———, *On the effective bandwidth of interleaved memories in vector processor systems*, Proc. 1985 Int. Conf. Parallel Processing, (1985), pp. 33–40.

[30] ———, *Modelling, measurement, and simulation of memory interference in the CRAY X-MP*, Parallel Computing, 3 (1986), pp. 343–358.

[31] K. PAWLIKOWSKI, *Steady-state simulation of queueing processes: A survey of problems and solutions*, ACM Computing Surveys, 22 (1990), pp. 123–170.

[32] C. V. RAMAMOORTHY AND B. W. WAH, *An optimal algorithm for scheduling requests on interleaved memories for a pipelined processor*, IEEE Trans. Computers, C-30 (1981), pp. 787–800.

[33] B. R. RAU, *Program behavior and the performance of interleaved memories*, IEEE Trans. Computers, C-28 (1979), pp. 191–199.

[34] C. V. RAVI, *On the bandwidth and interference in interleaved memory systems*, IEEE Trans. Computers, C-21 (1972), pp. 899–901.

[35] K. A. ROBBINS AND S. ROBBINS, *Bus conflicts for logical memory banks on a CRAY Y-MP type processor system*, Proc. 1991 Int. Conf. Parallel Processing, I (1991), pp. 21–24.

[36] K. V. SASTRY AND R. Y. KAIN, *On the performance of certain multiprocessor computer organizations*, IEEE Trans. Computers, C-24 (1975), pp. 1066–1074.

[37] C. E. SKINNER AND J. R. ASHER, *Effects of storage contention on system performance*, IBM Syst. J., 8 (1969), pp. 319–333.

[38] J. E. SMITH, W. C. HSU, AND C. HSIUNG, *Future general purpose supercomputer architectures*, Proc. Supercomputing '90, (1990), pp. 796–804.

[39] B. SPEELPENNING AND J. NIEVERGELT, *A simple model of processor-resource utilization in networks of communicating modules*, IEEE Trans. Computers, C-28 (1979), pp. 927–929.

[40] H. S. STONE, *Parallel processing with the perfect shuffle*, IEEE Trans. Computers, C-20 (1971), pp. 153–161.

[41] P. TANG AND R. H. MENDEZ, *Memory conflicts and machine performance*, Proc. Supercomputing '89, (1989), pp. 826–831.

[42] D. W. L. YEN, J. H. PATEL, AND E. S. DAVIDSON, *Memory interference in synchronous multiprocessor systems*, IEEE Trans. Computers, C-31 (1982), pp. 1116–1121.